Springer Tracts in Mechanical Engineering

Springer Tracts in Mechanical Engineering (STME) publishes the latest developments in Mechanical Engineering - quickly, informally and with high quality. The intent is to cover all the main branches of mechanical engineering, both theoretical and applied, including:

- Engineering Design
- Machinery and Machine Elements
- Mechanical Structures and Stress Analysis
- Automotive Engineering
- Engine Technology
- Aerospace Technology and Astronautics
- Nanotechnology and Microengineering
- Control, Robotics, Mechatronics
- MEMS
- Theoretical and Applied Mechanics
- Dynamical Systems, Control
- Fluids Mechanics
- Engineering Thermodynamics, Heat and Mass Transfer
- Manufacturing
- Precision Engineering, Instrumentation, Measurement
- Materials Engineering
- Tribology and Surface Technology

Within the scope of the series are monographs, professional books or graduate textbooks, edited volumes as well as outstanding PhD theses and books purposely devoted to support education in mechanical engineering at graduate and post-graduate levels.

Indexed by SCOPUS. The books of the series are submitted for indexing to Web of Science.

Please check our Lecture Notes in Mechanical Engineering at http://www.springer.com/series/11236 if you are interested in conference proceedings.
To submit a proposal or for further inquiries, please contact the Springer Editor **in your country**:

Dr. Mengchu Huang (China)
Email: mengchu.Huang@springer.com
Priya Vyas (India)
Email: priya.vyas@springer.com
Dr. Leontina Di Cecco (All other countries)
Email: leontina.dicecco@springer.com

More information about this series at http://www.springer.com/series/11693

Baoyan Duan · Yiqun Zhang · Jingli Du

Large Deployable Satellite Antennas

Design Theory, Methods and Applications

Springer

Baoyan Duan
School of Electro-Mechanical Engineering
Xidian University
Xi'an, Shaanxi, China

Yiqun Zhang
School of Electro-Mechanical Engineering
Xidian University
Xi'an, Shaanxi, China

Jingli Du
School of Electro-Mechanical Engineering
Xidian University
Xi'an, Shaanxi, China

ISSN 2195-9862 ISSN 2195-9870 (electronic)
Springer Tracts in Mechanical Engineering
ISBN 978-981-15-6032-3 ISBN 978-981-15-6033-0 (eBook)
https://doi.org/10.1007/978-981-15-6033-0

This Springer imprint is published by the registered company Springer Nature Singapore Pte Ltd.
The registered company address is: 152 Beach Road, #21-01/04 Gateway East, Singapore 189721, Singapore

Preface

With the advent of satellites, developed countries have continued to explore the field of spaceborne deployable antennas for over six decades, since the 1960s, and great progress has been achieved. Spaceborne deployable antennas with various structures are continuously emerging, and new forms of deployable antennas are being explored.

In China, since the launch of the first satellite, "Red Eastern", in the 1970s, 40 years have passed. Since then, China has made remarkable achievements in this field. Specifically, since the beginning of the twenty-first century, development in the aerospace industry has experienced several milestones. The successful return of manned spacecraft and the remarkable achievements of the project, the official launch of the Mars exploration mission, the successful docking of Tiangong 2, the successful launch of the Long March 5 heavy-duty rocket, and other achievements demonstrate the steady pace of China's advancement toward becoming a space power.

Whether satellites can work normally and reliably in space depends very much upon their energy sources and related technologies. Moreover, the large spaceborne deployable antennas that act as the satellite's "eyes" and "ears" are vital equipment. With the continuous advancement of deep-space exploration activities, the requirements of spaceborne deployable antennas, such as having a high frequency and large bandwidth, high precision, a large aperture, a lightweight, and a large furled ratio, have increased. They are significantly different from terrestrial antennas. This poses unprecedented challenges to the analysis, design, manufacture, and testing of spaceborne antennas.

Notably, the real start of China's spaceborne deployable antenna research began in the 1980s and has flourished ever since. As the antenna is required to be as large and light as possible, the antenna is very flexible, which is beyond the scope of traditional elastic deformation. Regarding the flexibility problem of a large spaceborne deployable antenna structure, taking the beam element as an example, the early-stage method was based on the rigid-body displacement of the beam element and the elastic deformation of the superimposed beam such that there was an apparent error in the treatment. For the sake of addressing this, research on multi-flexible body dynamics analysis has become necessary.

The smooth and reliable deployment of spaceborne deployable antennas is a prerequisite for the normal operation of satellite systems. There have been tragic examples of the scrapping of satellite systems for being unable to deploy antennas. Addressing this issue requires an in-depth study of the deployment process control and deployment reliability issues based on multi-flexible body dynamics analysis.

Another requirement of spaceborne deployable antennas is that they possess the required shape precision after the antenna is deployed and they must be furled and placed within the rocket fairing during launch. To guarantee that the antenna has sufficient gain, the antenna aperture is generally required to be as large as possible. This leads to a contradiction, i.e., a large deployed aperture and a small furled volume. Achieving this goal calls for knot (machine) construction innovation as well as the study of static and dynamic analysis along with the optimization of design problems of the two states (deployed and furled).

As a special type of antenna, the spaceborne deployable antenna has strict mechanical and structural performance requirements; close attention to the realization of its electromagnetic performance is needed as well as maintenance and performance stability during its service. These issues should be investigated from the perspective of system and multidisciplinary integration. Thus, the first step is to establish an Electromechanical Coupling (EMC) theory model relating the deformation field, electromagnetic field, and temperature field of the antenna. The second step is to understand the Influence Mechanism (IM) of nonlinear mechanical factors on the electromagnetic performance of the antenna. The third step is to construct an integrated design based on the EMC theory model and IM.

To meet the above requirements, this book discusses nonlinear mechanical analysis, dynamic optimization design, multi-flexible body dynamics analysis, deployment process control, shape precision measurement and adjustment, electromagnetic performance analysis, electromechanical integration design, electromagnetic-thermal-control integrated research, and the development of spaceborne antenna integrated design software platform.

This book is based on the application of spaceborne mesh deployable antennas, including membrane antennas and ultra-large space-assembled antennas. Additionally, the future development trends of spaceborne deployable antennas are described. The Chaps. 1–3, 5, 7 and 9 of the book were compiled by Baoyan Duan, the Chapts. 4 and 8 were compiled by Yiqun Zhang, the Chap. 10 was compiled by Jingli Du, and the Chap. 6 was compiled by Jingli Du and Yiqun Zhang.

This book presents a summary of the authors' long-term research and experience with spaceborne deployable antennas, which includes insights and experience in design, model manufacture, and practical engineering. The authors would like to give special thanks to Ze Zhao, Guangda Chen, Shuxin Zhang, Dongwu Yang, Shen Li, Yajing Ma, Jia Hao, Jun Li, Bo Ding, and Liang Xie for previous work and assistance in the preparation of this book. This book is intended to be a valuable reference for scientists, engineers, and technicians engaged in work on spaceborne antennas. The book is also suitable for teachers, senior undergraduates, and graduate students in colleges and universities.

Owing to the limitations of the authors' skills, there may be errors in the book, and readers' criticisms and suggestions are very much appreciated.

Xi'an, China

Baoyan Duan
Yiqun Zhang
Jingli Du

Brief Introduction

This book is mainly concerned with large spaceborne deployable antennas, especially mesh deployable antennas. It also includes membrane antennas and ultra-large space-assembled antennas. This book focuses on space service environment analysis, cable-truss combined structure form-finding analysis and mesh design, antenna deployment dynamics analysis and control, electronic performance analysis and electrical parameter equivalence, shape precision measurement and adjustment, deployment reliability analysis, scaled model development and testing, integrated design platforms, and electrostatic forming technologies of membrane.

The second focus of the book is the development of a software tool called Spaceborne Deployable Antenna Design Software (SDADS). The architecture and application are described in detail.

The third part of the book discusses the construction of the physical prototype, tests and measurements and analysis of experimental results. Mesh antenna prototype, electrostatic forming membrane antenna prototype, and manufacturing and testing equipment are introduced.

This book may be used as a reference by senior undergraduates and postgraduates, research institutes, and engineering and technical personnel engaged in the design of deployable antenna structures.

Contents

About the Authors

Baoyan Duan Academician of Chinese Academy of Engineering, President of Xidian University, China (2002–2012), Full Professor of Electromechanical Engineering, Xidian University, China.

He received B.S., M.S., and Ph.D. degrees in Mechanical Engineering from Xidian University, Xi'an, China, in 1981, 1984, and 1989, respectively. From 1991 to 1994, he studied as a postdoctoral fellow at Liverpool University, U.K., and worked as a visiting scientist at Cornell University, Ithaca, NY, in 2000. He is currently a full professor in the School of Electromechanical Engineering at Xidian University, where he founded the Research Institute on Mechatronics in Electronic Equipment Design.

He is the chairman of the National Antenna Industry Alliance of China and chairman of the Electromechanical Engineering Society of China. He is a fellow of International Engineering and Technology (IET) and the Chinese Institute of Electronics (CIE), and a Member of the International Society for Structural and Multidisciplinary Optimization (ISSMO). He serves as chief editor of Electromechanical Engineering of China, deputy chief editor of the Chinese Journal of Electronics, subject chief editor of Engineering, and editor of more than 10 other international or Chinese academic journals.

He has been dedicating himself to research on electromechanical engineering and initiated a new area of Electromechanical Coupling (EMC) theory for Microwave Electronic Equipment (MEE) in China. He has established a Multifield Coupling Theory Model

(MFCTM) for the electromagnetic, structural deformation and temperature fields of MEE, clarified the Influence Mechanism (IM) of nonlinear mechanical parameters on the electronic performance of MEE, and proposed an integrated design methodology for MEE based on the MFCTM and the IM. The above results have been successfully applied in national major engineering projects such as lunar exploration, the Shenzhou spacecraft, the "Tiantong No. 1" satellite, and the FAST500 m telescope, the largest radio telescope.

He has published, authored or coauthored 200 papers and six books, and authorized 40 patents of inventions. He has received, as the first author, the second prize of the National Award for Science and Technology Progress of China three times (in 2004, 2008, and 2013). In 2009, he was selected as science Chinese person. In 2012, he received the Hong Kong HLHL prize for science and technology progress. In 2017, he received an award for outstanding scientific and technological achievement from the Chinese Academy of Science and the Golden Prize of China for Good Design. In 2018, he received a lifetime achievement award from the Asian Society of Structural and Multidisciplinary Optimization.

He has also received awards from the central government of China, the "May one" prize medal (2003), the working model award (2005), the award for ethics in the teaching profession (2004), the award for outstanding achievements of being home after studying abroad (2003), and the award for distinguished scientific researchers (2011).

CCTV (the Chinese Central TeleVision station) made up and broadcasted a special video program titled **DUAN Baoyan: Minor Discipline and Great Vision** in 2016 (https://v.qq.com/x/page/i08412h2chd.html).

Prof. Yiqun Zhang He was born in 1984 and obtained bachelor's and doctoral degrees in engineering from Xidian University in 2006 and 2013, respectively. He is currently a professor at the School of Electromechanical Engineering of Xidian University, China.

He has been engaged in the analysis and design of large spaceborne deployable structures and mechanisms, which involves the analysis of various nonlinear factors in the development of large spaceborne deployable structures, flexible mechanisms and structure control, and space solar power station-related system structure design.

He has hosted or participated in more than 10 national key projects, such as National Natural Science Foundation and Defense projects. He has received the second prize of the National Science and Technology Progress Award (2013), the first prize of the Science and Technology of China Electronics Society (2012), and the first prize of Science and Technology Progress of Shaanxi Province (2019).

Prof. Jingli Du He was born in 1977 and received bachelor's, master's, and doctoral degrees in engineering from Xidian University in 2000, 2003, and 2006, respectively. He is currently a professor at the School of Electromechanical Engineering of Xidian University, China. From 2012 to 2013, he was a visiting scholar at the University of Delaware and Columbia University in the USA.

He has been engaged in research and design related to spatial flexible structure analysis for two decades, which involves multi-flexible body dynamics analysis and deployment process control for large spatial structures, structural statics and dynamics analysis after unlocking, and optimization design for large spatial structures.

He presided over or participated in more than 10 scientific research projects such as National Natural Science Foundation and Defense projects. The results won the second prize of the National Science and Technology Progress Award (2013), the first prize of Shaanxi Province Science and Technology Progress two times (2014, 2019), and the second prize of the Science and Technology Progress of China Electronics Society (2016).

Chapter 1
Introduction

1.1 Overview

In August 1972, Orbiting Astronomical Observatory No 3 (OAO-3) of the USA was launched. To commemorate the 500-year anniversary of the birth of the great scientist Nicolaus Copernicus (Fig. 1.1), it was named the Copernicus Satellite. Copernicus's heliocentrism gave rise to a revolution in human understanding of the universe. To explore space, launching satellites with antennas is the first step. As a result, spaceborne antennas have been widely used in communication, reconnaissance, navigation, remote sensing, deep-space exploration, radio astronomy and so on because they are the "eyes" and "ears" of satellite systems and play a key role in the implementation of satellite functions. The four typical requirements of spaceborne antennas are: high precision (high frequency), large diameter (high gain), light weight, and large furled ratio. The reasons are given below.

The first two requirements, for large diameter and high precision, are based on the following four application goals: communication, navigation, remote sensing and defense.

In the field of communication, the multidimensional expansion of information space is an emerging development trend. The realization of integrated space- and ground-based information networks depends, to a large extent, on spatial communication capabilities. To achieve faster and better communication links and network services, future communication satellites need to continuously increase signal strength and communication quality, and there is an urgent need for large-diameter spaceborne antennas. For example, the US broadband global area network needs to achieve a transmission rate of up to 492 Kbps, and the diameter of antennas should be greater than 9 m. The antenna diameter of the Boeing satellite systems' maritime communication satellites is more than 12 m. For military satellites, to enhance anti-interference, the gain and directionality must be improved significantly. To achieve this, frequency-hopping and spread-spectrum techniques are necessary. The antenna has to have a higher operating frequency and a wider frequency bandwidth. These

B. Duan et al., *Large Deployable Satellite Antennas*,
Springer Tracts in Mechanical Engineering,
https://doi.org/10.1007/978-981-15-6033-0_1

Fig. 1.1 Nicolaus Copernicus (1473–1543) was an astronomer and mathematician in Poland. His "Astronomical Operation Theory" advocated the heliocentric theory, which was the starting point of modern astronomy and marked the beginning of modern science

factors all naturally pose new challenges for the diameter and profile precision of antennas.

In the field of remote sensing, high-precision spaceborne microwave radiometers are needed to monitor the atmosphere, ocean and land for all-weather and all-day conditions. For example, the SSMIS microwave radiometer (frequency 23.8–183 GHz) installed on the US DMSP meteorological satellite [1] and MTVZA (frequency 17–183 GHz) installed on the Russian METEOR-3 M satellite [2] have played an important role in military meteorological security, weather forecasting, flood detection and other tasks. However, to improve the resolution of observations in the future, it is necessary to increase the diameter of the antenna. For instance, the Doppler weather radar NIS of the United States has an antenna with a diameter of 35 m and an operating frequency of 35 GHz [3].

In the field of navigation satellites and resource satellites, the demand for large-diameter antennas is also increasing. For example, the GPS Block III navigation satellite of the United States and BeiDou Navigation Satellite Systems (BDS) of China all require large-diameter spaceborne antennas. In addition, a space solar power station (SSPS) requires ultra-large antennas with diameters up to 100 m to transmit high-power continuous microwaves from geosynchronous orbit to the ground; for instance, the diameter of a microwave transmitting antenna with a power capacity of 5 GW may be up to 1 km [4].

Regarding defense, since the battlefield environment of modern warfare may change dramatically, such as the combat style changing rapidly, the disguise and deceive abilities have to continuously improve. Therefore, the acquisition of intelligence is much more difficult than before. Meanwhile, the timeliness, accuracy and continuity of intelligence acquisition are increasingly significant. As a result, to quickly intercept and identify a radiation source in a dense and complex electromagnetic signal environment, an electronic reconnaissance system is required to have a wide frequency bandwidth, high sensitivity, large dynamic range, high frequency and direction test accuracy, fast response speed, strong resolution and high positioning accuracy. These requirements naturally place high demands on the diameter and frequency (band) of spaceborne antennas. With the development of electronic reconnaissance satellites, from geostationary orbit to near-polar large elliptical orbits, including multi-orbital integrated reconnaissance satellites with orbital capability, the detection frequency bandwidth continues to widen, and the sensitivity and signal-to-noise ratio requirements of electronic receivers also continue to increase. For this reason, the demand for large-diameter spaceborne antennas is also increased significantly.

The other two requirements of light weight and large furled ratio are derived from the size of the rocket fairing (for example, for the largest Long March 5 rocket so far in China, the available fairing size of the high rail is just 4.5 m in diameter and 4.5 m in height, and the low rail is slightly larger, 4.5 m in diameter and 12 m in height) and the limited emission cost. Since the size of the rocket fairing is limited and the launch cost should be as low as possible, the antenna must be deployable, that is, it is stowed within the rocket fairing during launching and deployed completely and automatically once in orbit. During the unfolding process, especially at the beginning and ending steps, the acceleration cannot be too high, to avoid being scrapped due to excessive impact. To meet the needs of satellites for different purposes, developed countries have studied or are studying a wide variety of deployable antennas, including the National Aeronautics and Space Administration (NASA), Russian Federal Space Agency (RFSA), European Space Agency (ESA), Institute of Space and Astronautical Science (ISAS) and National Space Development Agency (NASDA) of Japan [5–13]. According to the application and research status of deployable antennas in the aerospace field, deployable antennas can be divided into three kinds: reflector antennas, array antennas and microelectromechanical antennas [14, 15]. This chapter will elaborate on the research and application status and future development of these three types of antennas and provide useful references for research on and applications of large deployable antennas.

1.2 Status of Spaceborne Deployable Reflector Antennas

The reflector antenna is the most widely used type of antenna among various application satellites. It can be used as a communication satellite antenna in ultra-high frequency, microwave or even millimeter-wave frequency bandwidths to form shaped beams, spot beams, multibeams and even scanning beams. According to the structural form of the reflector, these satellites can be divided further into four categories: rigid reflector antennas, inflatable reflector antennas, mesh reflector antennas and membrane reflector antennas.

1.2.1 *Rigid Reflector Antenna*

A typical feature of this type of antenna is that the reflector surface of the antenna consists of a central hub and a number of rigid curved plates [16, 17]. Figure 1.2 shows a physical model of a rigid reflector deployable antenna with a diameter of 4.7 m. Its precision is 0.2 mm (30 GHz). It was made by a company named Dornier and ESA in 1987. The rigid reflector deployable antenna is made of a metal plate or carbon fiber reinforced plastic (CFRP) coated with a metal reflective surface. Since the plate can be processed into a more ideal paraboloid, the advantage of this antenna is the high precision of the reflector surface. According to open reports, the surface precision of a rigid reflector deployable antenna with a diameter of 10 m can reach up to 0.13 mm [18]. Since the accuracy of the profile after unfolding can

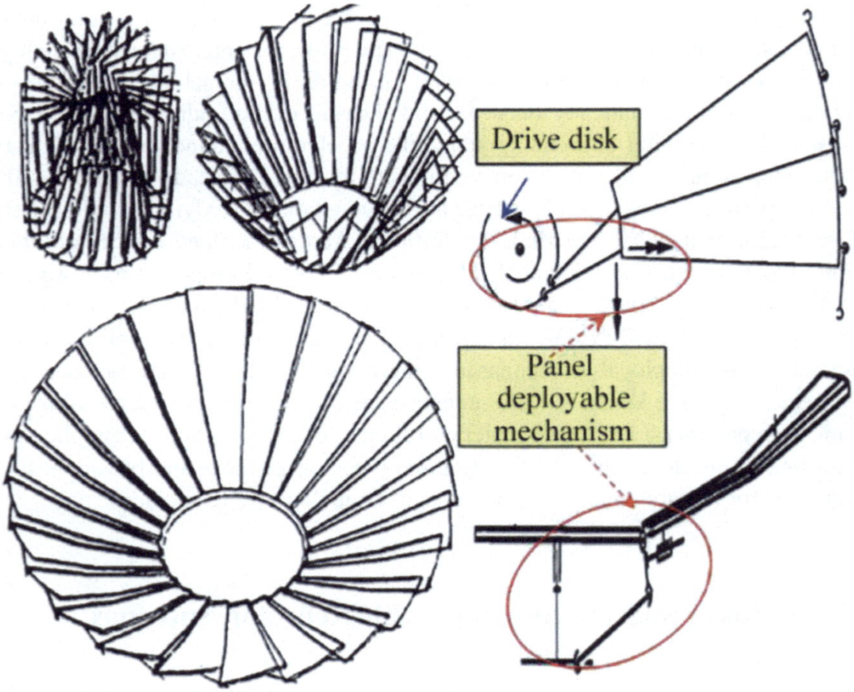

Fig. 1.2 Petal type reflector antenna

be very high, it is a good choice in the field of microwave remote sensing, such as for microwave radiometers and scatterometers, and has been applied in several spacecraft [19]. However, it also has the drawbacks of a heavy structure, high cost, and large folding volume. Currently, it is rarely applied in large-diameter satellite antennas.

1.2.2 *Inflatable Reflector Antenna*

The advantages of inflatable deployable antennas are a high furled rate, large diameter, etc. [3, 20–26]. The disadvantage is that the profile accuracy is limited because the compressor must be sent into space at the same time, which leads to high areal density. This type of antenna is mainly made of flexible materials (a Kevlar membrane or Mylar membrane treated with chemical resin). The unfolding principle is similar to that of inflatable membranes in civil buildings. The structure is inflated to the desired position and shape by internal inflation. Unlike the civilian membrane structure, when the antenna is inflated to a specified position and shape, under illumination of sunlight, the membrane can be hardened by sunlight to maintain the open configuration. Afterwards, even if the internal gas leaks, it will not cause a loss of accuracy for the antenna reflector surface. Research on inflatable reflector antennas began in the early 1960s, but due to the constraints of materials and processes, the development was slow. The new materials, new processes and new demands that emerged in the 1990s have greatly contributed to the rapid development of inflatable reflector antennas. NASA and ESA have made useful explorations and progress in technology. However, commercial applications have not yet been realized, and the materials and hardening processes will remain a popular research topic in the future. A milestone in the development of inflatable reflector deployable antenna was the 14 m-diameter inflatable antenna experiment (IAE) by NASA on an aircraft, on May 29, 1996, as shown in Fig. 1.3.

The upper part of the figure shows the process of transitioning from the furled to deployed state. This signifies the beginning of this kind of antenna structure in possible space applications. Currently, JPL and L'Garde are developing a 25 m-diameter inflatable antenna called advanced radio interferometry between space and earth (ARISE).

1.2.3 *Mesh Reflector Antenna*

At present, the in-orbit large-scale deployable antennas are mainly mesh deployable antennas because they are one of the research focuses of the international aerospace industry in terms of theory, method and experiment. The character of this type of antenna is that the reflector surface is composed of flexible wire mesh so that it is a flexible-force system. Both the tension-acted point and tension level of the supporting

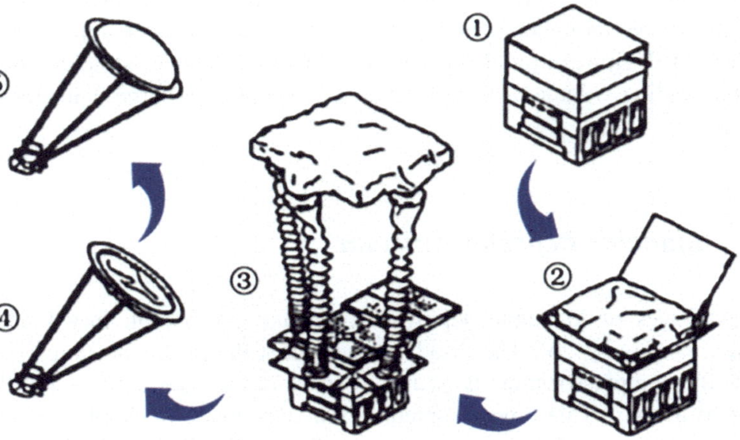

(a) Deployment process of inflatable antenna

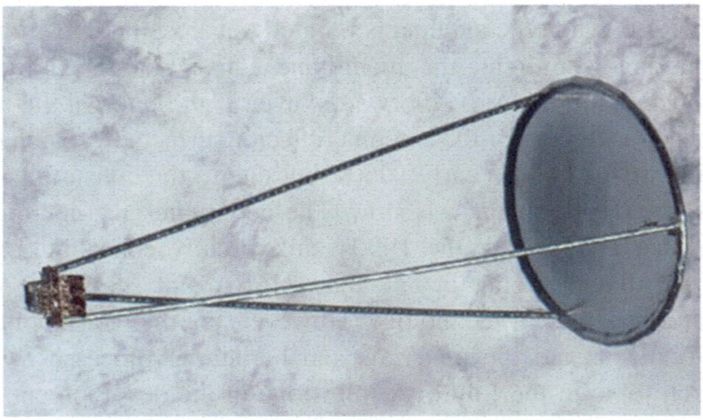

(b) Inflatable antenna in deployed state

Fig. 1.3 The inflatable antenna

net have significant influences on the conformal capability, profile accuracy, smoothness and reliability of deployment. The mesh reflector is ultra-light and easy to fold, has a large furled rate and an easily realized large diameter, and makes it convenient to combine various deployable support structures. However, its shortcoming is that its structure is complicated, and the precision, reliability and repeatability of its profile are low. With the current technological development, the diameter of such antennas can theoretically be up to 50 m, the profile accuracy can reach 200–500 μm, and the radio frequency can be 1.6–40 GHz, satisfying low and high frequency requirements simultaneously [27]. Depending on the form of the support structure for the flexible wire mesh and the way it is driven, the mesh deployable antenna can take a variety of structural forms [28–33].

(a) Thuraya mesh antenna (b) SkyTerra Harris Loop antenna (c) EGS Loop antenna

Fig. 1.4 Loop mesh reflector antenna

1. **Loop antenna**

As shown in Fig. 1.4, the loop antenna is an antenna composed of supporting unfolded truss and cable-net structures. The truss can be at the periphery or center or in a radial direction. This type of antenna can be divided into peripheral truss types, Harris rings, EGS loop antennas, etc. The perimeter truss antenna consists of multiple parallelogram units forming a closed-loop structure. This structure is used in Thuraya, MBSat and SMAP satellites in the United States. The Harris loop antenna consists of an inner ring of a regular polygon shape and two upper and lower mesh diagonal support rods. The 22 m aperture antenna of the US SkyTerra and MSV satellites adopts this form. Each deployable unit of the EGS loop antenna consists of two identical scissor-type expansion mechanisms. Russia and ESA jointly developed this type of antenna with a diameter of 12 m.

2. **Radial rib antenna**

As shown in Fig. 1.5, the radial rib antenna has an umbrella-like shape and is composed of several parabolic tubular carbon fiber ribs fixed to a central hub with a reflective mesh surface connected between the ribs. The 12 m-aperture antenna developed by the Harris Corporation of the United States has folded ribs. The 18 m-

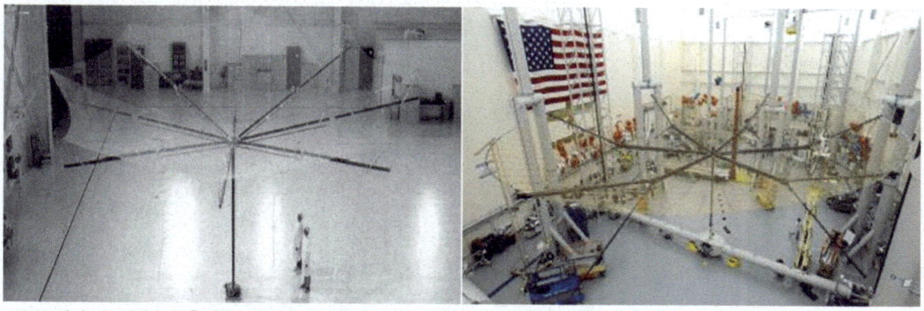

(a) MUOSfolded rib (b) Terre Star Y-shape rib

Fig. 1.5 Radial rib antenna

aperture antenna of the TerreStar satellite developed by the same company has a foldable rib divided into upper and lower sections (in a Y shape) to improve surface accuracy while ensuring the antenna furled ratio.

3. **Frame antenna**

As shown in Fig. 1.6, this type of antenna adopts a modular design, usually consisting of a number of tetrahedral or hexagonal-prism modules. The size and number of modules can be changed to meet the requirements of different diameters. Japan's engineering test satellite ETS-VIII, launched in 2006, carried two 19 m × 17 m frame antennas, which are composed of 14 modules with a diameter of 4.8 m.

4. **Wrapping rib antenna**

This is also a type of umbrella antenna consisting mainly of a central hub, parabolic radial ribs with a lenticular section and a reflective mesh surface, as shown in Fig. 1.7.

Fig. 1.6 ETS-VIII frame antenna

Fig. 1.7 ATS-6 wrapping rib antenna

The US ATS-6 satellite adopts this type of antenna, with a diameter of 9.1 m. Its furled rate advantage is obvious, but the overall rigidity is weak.

1.2.4 Electrostatic Forming Membrane Reflector Antenna

The membrane reflector antenna is a new type of reflector antenna proposed recently. Research institutions in the United States and Russia have taken the lead in the technology of making reflector surfaces with aluminized polyimide membranes. The metal reflective medium is plated on a polyimide membrane to reflect electromagnetic waves [34]. The idea is to use the skirt cable force, gas pressure and electrostatic force to maintain the required precision of the reflector. The forming methods of membrane reflector primarily include the inflation method, cable tensioning method [35, 36], and electrostatic forming method [37–39], according to the difference in external force [40]. The membrane reflector antenna has advantages that traditional reflector antennas do not have, such as high precision, ultra-light weight, small furled volume, and easy folding and unfolding. It can be applied to X, Ku or even higher frequencies to meet the requirements of large diameter, high precision and ultra-light weight for space reflector antennas. It has drawn attention because of its lower areal density and the ability to achieve active profile control by adjusting the electrostatic force. In 2004, Northrop Grumman Company cooperated with Scientific Research Society of America (SRS) to construct a prototype of a 5 m-diameter electrostatic forming deployable membrane antenna. As shown in Fig. 1.8, the feasibility of the electrostatic forming deployable membrane antenna concept was verified.

To obtain a comprehensive understanding of the above methods and approaches, the application ranges of four types of antenna reflector antennas, namely, rigid antennas, membrane antennas, mesh antennas and inflatable antennas, are specifically described in Fig. 1.9. It can be seen from Fig. 1.9 that the rigid reflector antenna has the highest operating frequency, but its diameter is limited to less than 5 m. The mesh antenna can work in the (mobile communication) S/L frequency band, and its diameter can be on the order of tens of meters. The inflatable antenna has a high furled

Fig. 1.8 Electrostatic forming deployable membrane antenna

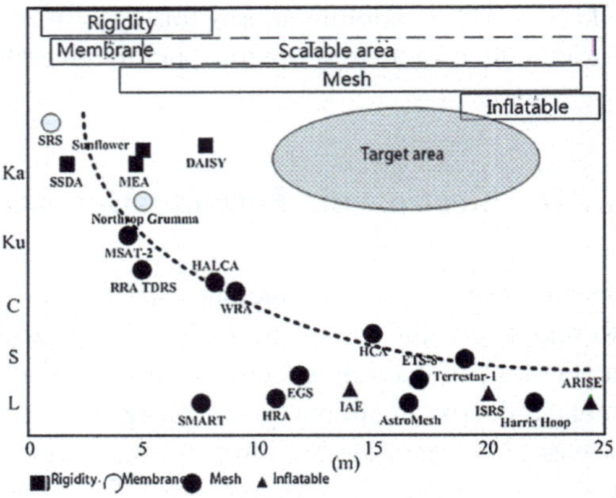

Fig. 1.9 Application range of four types of antenna reflectors

ratio, but its working frequency is the lowest. The membrane reflector antenna is a new idea that can work at very high frequencies. Unfortunately, the current diameter is limited and still in the laboratory stage, although its prospects are promising. It is expected to break through the key techniques of high precision and large diameters.

It can be seen from Fig. 1.9 that the existing reflector antennas are mainly developed toward two goals: a larger scale and a higher precision. The development of spaceborne antennas that meet the requirements of a large diameter, high precision and a light weight (the target area in Fig. 1.9) is a huge challenge in the future.

1.3 Research Status of Spaceborne Array Antennas

Compared with reflector antennas, array antennas have more design freedom—they may be line arrays, plane arrays, conformal arrays, etc.—so they can achieve a high gain, narrow beam, multiple targets, allow for space division multiple access or autonomous control, etc. Among these antennas, the phased array antenna has become the fastest-growing and highest-potential antenna form. It can be used not only for multi-target tracking and anti-missile warning systems but also in shipborne, airborne, and spaceborne weapon systems and electronic countermeasure systems. It can also be applied in space vehicles, satellite communications and air traffic control [41]. In the applications of aerospace high-capacity communication and microwave remote-sensing imaging synthetic aperture radar (SAR), a wider frequency band, multiple beams, and fine-beam ground resolution are required. Therefore, large-scale planar active phase control array antennas have naturally emerged. Due to the size of the carrier, a folded deployable structure is required. At the same time, considering the diversity of user needs, the development cycle, the on-orbit reconstruction capability (i.e., the beam shape, beam position and power & frequency assignment for the beam) and project risks, modular active phased deployable array antennas are often applied.

Spaceborne SAR usually uses a planar phased array antenna with a size of 10–30 m^2. For example, the size of America's SAR antenna in the orbiting marine satellite SEASAT is 10.74 m × 2.16 m [42]. Grumman-Raytheon designed a phased array lens structure with an L-band and a size of 60 m × 25 m. It has two types of deployment: single-axis and two-axis [24, 43]. The prototype of the microstrip SAR array antenna by JPL works in the L-band with a size of 3.3 m × 1.0 m [44]. The Ball company has made a prototype of 13.8 m × 63.6 m [45] that is a single-axis deployable low-orbit SBR active phased array antenna. Owing to the size limitations of launch vehicle fairing, spaceborne SAR antennas are, in most cases, divided into a number of sub-array panels that are folded around the satellite. When the satellite enters the predetermined orbit, the antenna array is deployed outside the satellite cabin with a deployment mechanism.

To meet the requirements of global tactical operations and to track ground targets, the United States proposed a medium-orbit satellite SBR active lens-array antenna that is 3 m × 300 m and operates at the X-band. The reinforced inflatable lens antenna includes a lens antenna, a solar array and a heat radiator, which are mounted on the three faces of the triangular prism, as shown in Fig. 1.10.

The University of Cambridge, UK, has developed a prototype of an offset parabolic cylindrical SAR at the L-band with a line feed array [7]; the physical model is shown in Fig. 1.11. The desired shape of the reflector surface is offset parabolic and is 7.9 m × 3.2 m. The back is formed by a curved carbon fiber composite material sheet connected with a flexible hinge distributed along the edge to form a hollow body, and the front end forms a continuous reflector surface with the required precision. This antenna structure has a very high stiffness-to-mass ratio.

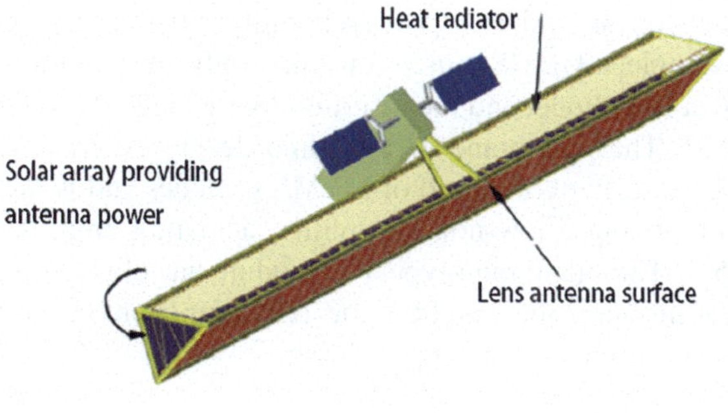

Fig. 1.10 Active lens array antenna

(a) Furled state diagram (b) Deployed state diagram

Fig. 1.11 Parabolic SAR

1.4 Research Status of Spaceborne Microelectromechanical Antennas

The rapid development of integrated information systems with large capacity, multiple functions and ultra-wide frequency bandwidths has given rise to the dramatic increase of antennas and subsystems within one platform. This brings major challenges to the task of designing an integrated information system with reductions in cost, weight and radar cross-section (RCS) and good electromagnetic compatibility (EMC). To overcome this difficulty, integrated technology with radio frequency (RF) and microelectromechanical systems (MEMSs) is an effective method. As a result, RF MEMS devices [46–51] and microelectromechanical antennas [14, 15] have emerged.

RF MEMS devices are miniaturized and integrated devices implemented by MEMS technology for the generation and processing of signals within the frequency range from low frequency to infrared. They are mainly used in RF transceiver front-end circuits. RF MEMS devices include MEMS resonators and filters, mixers, phase shifters, T/R components, and microstrip transmission lines. Their application enables antennas to be unmatched by many conventional devices in terms of geometry, function, physical performance, cost, and other characteristics. It can greatly reduce system size, improve reliability and life, and decrease power consumption. For example, a MEMS switch consumes only one one-thousandth of the power consumed by a PIN diode, and the insertion loss is only about one-third of that of a PIN diode [52]. The membrane T/R module developed by JPL in the United States uses a phase shifter composed of MEMS switches and is integrated with an MMIC circuit comprising a low-noise amplifier, actuating amplifier and power amplifier circuit [53]. The membrane-type T/R module has obvious weight and volume advantages. For instance, the weight of the T/R component of the S-band is just 20 g, but that of the traditional type is up to 200 g.

The application of MEMS in antennas integrates performance with structural design and manufacturing technology. High-performance and low-cost antennas can thereby be produced. New types of antennas can be developed, such as low-cost antennas with a large size, miniature and agile antennas, reconfigurable antennas (phased array antennas, patch and microstrip grid antennas), and multi frequency band antennas.

1.4.1 Microelectromechanical Phased Array Antenna

Due to their low cost, light weight and low loss, microelectromechanical antennas can become large-scale phased array antennas with ultra-light weight; these have broad application prospects in various phased array radars, such as spaceborne radar and shipborne radar, ground-based mobile air defense weapons, conformal multi-band MEMS antenna arrays, MEMS phased array antennas with a box structure, and MEMS grid antennas. In April 2006, with the support of the US Air Force Research Laboratory (AFRL), Radant Technologies produced the world's first X-band MEMS electronic scanning demonstration array [35]. As shown in Fig. 1.12, the array uses more than 25,000 RF MEMS switches to form a lens structure with an aperture of only 0.4 m^2. Compared with traditional active electronic scanning arrays, it can not only improve antenna performance (reducing loss and increasing bandwidth) but also greatly improve the indexes of weight, power consumption and cost.

Fig. 1.12 MEMS electronic scanning demonstration array

1.4.2 Microelectromechanical Reconfigurable Antenna

A reconfigurable antenna utilizes a single antenna or array to achieve multiple functions by simply changing its physical structure or size. It is similar to a multi-antenna sharing a single physical aperture (Fig. 1.13). Since it has a similar radiation mode and gain for all frequency bands, the EMC difficulty can also be overcome. It can be applied in high-frequency satellite communication systems, low-frequency communication systems, electronic intelligence radars, SAR, ground moving-target recognition etc. [54]. Using this technology, the problems of system miniaturization, low power consumption, low cost, high performance and miniaturization of RF modules can also be solved. Moreover, miniaturized, built-in, multi-band and intelligent mobile terminal antennas can be easily implemented.

The applications of MEMSs in reconfigurable antennas include the following four types:

(1) Using MEMS actuators to change the resonant length or reactance value of the antenna to achieve frequency reconfigurability [14];
(2) Using MEMS switches to change the current phase of adjacent elements of antenna array, control the direction beam can realize the reconfigurable pattern [55];
(3) Utilizing a microprocessor to control MEMS switches between components on the antenna template according to the feedback signal so that the electrical structure of the antenna can be changed accordingly. By means of this, the simultaneous reconstruction of frequency and pattern can be achieved [56];

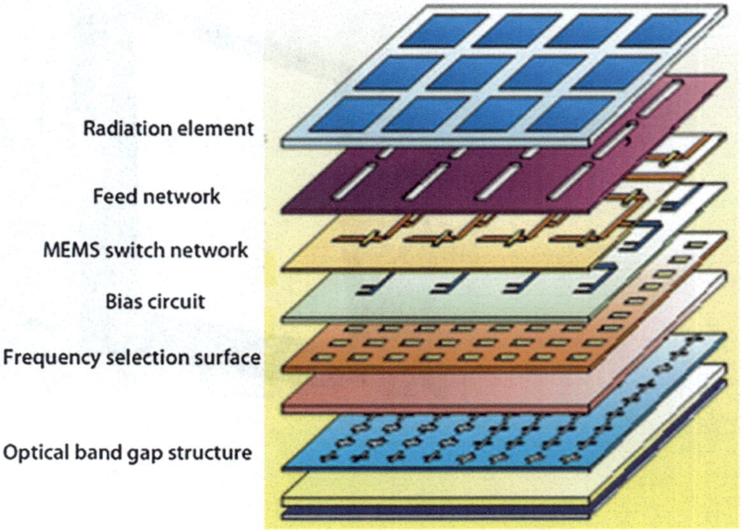

Fig. 1.13 Reconfigurable antenna array schematic

(4) For a given frequency and radiation pattern, the following shifts are easily carried out: two polarization modes with orthogonal polarization directions [57], two circular polarization modes [58–60], and left- and right-hand circular polarization as well as linear polarization [61].

The Defense Advanced Research Projects Agency (DARPA) of the United States has been actively working on a variety of reconfigurable phased array antennas, including electronic reconfigurable antenna concepts, dynamically adaptively changing antenna patterns, and achieving bandwidth greater than octave. Results are planned for SAR, spaceborne radar, satellite communications and other fields [15].

1.5 Research Status of Large Spaceborne Deployable Antennas in China

China's research activities in the field of spaceborne antennas began in the 1970s. The study of large-diameter deployable antennas began in the 1990s. Except for the development of mesh reflector antennas (such as the existing 16 m diameter mesh reflector antennas in service), most of the current research work is still in the theoretical research and prototype development stage.

1.5.1 Solid Surface Deployable Antenna

Domestic research on solid surface deployable antennas began in the 1980s. China's first solid surface antenna was applied in the elliptical-cut paraboloid of the "Oriental Red" No. 2 OC star launched in 1986. The projection size was 0.62 m × 1.5 m, and the profile accuracy was 0.3 mm. Most antennas currently in operation are solid surface antennas. For example, "Oriental Red" No. 2, "Oriental Red" No. 3, and "Fengyun" No. 2 all adopted solid surface antennas of similar structure, but they are all non-deployable structures. The representative solid surface spaceborne deployable antenna is a 3 m-diameter self-rebounding solid surface antenna with certain application prospects.

1.5.2 Inflatable Deployable Antenna

Domestic research on inflatable deployable antennas is still in the stage of model testing and development. Prototypes of inflatable reflector antennas with diameters of 3–5 m has been developed, as shown in Figs. 1.14 and 1.15.

Fig. 1.14 Prototype test of a 3 m inflatable antenna

Fig. 1.15 Prototype of a 5 m inflatable antenna

1.5.3 Mesh Deployable Antenna

There are many studies on mesh deployable antennas in China, such as a 4.2 m aperture (UHF band) umbrella antenna (Fig. 1.16), a 6 m × 2.8 m aperture (S band) frame antenna (Fig. 1.17), a 4.2 m aperture (S/Ka band) umbrella antenna prototype (Fig. 1.18) and the 16 m aperture (L-Ku band) surrounding truss-type deployable antenna in service (Fig. 1.19).

Fig. 1.16 A 4.2 m radial rib antenna

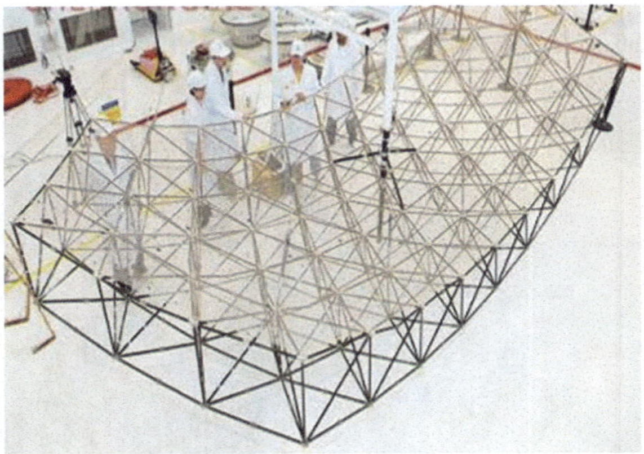

Fig. 1.17 A 6 m × 2.8 m diameter frame antenna

Fig. 1.18 A 4.2 m umbrella antenna prototype

Fig. 1.19 A 16 m diameter truss antenna

1.5.4 Electrostatic Forming Deployable Membrane Antenna

As for the deployable membrane antenna, research in China is still in its infancy. Investigation of the coupling theory of the deformation field and electrostatic field, forming technology, boundary cable and electrostatic force control technology of the membrane reflector antenna has been performed. Experimental models with diameters of 0.55–2 m were constructed with good results (Fig. 1.20a, b).

In China, universities have carried out in-depth and systematic research on key technologies and theoretical methods for deployable antennas and have made satisfactory progress, such as designing a variety of deployable antennas with tetrahedral and hexagonal prisms and ring trusses and carrying out research on space deployable structures and model testing. In fact, flexible structural dynamic analysis has been carried out since the 1980s, and further research on flexible structural multibody kinetic/dynamic analysis and control technology was carried out in the 1990s [62–70]. In the 21st century, with the support of several national projects, electromechanical coupling theory and method of deployable antenna, flexible multi-body deployment kinetic/dynamic analysis, support mesh reflector surface profile analysis, design and adjustment, and two state (deployed and stowed) dynamic optimization [71–85]. China's first large-scale spaceborne deployable antenna design software platform named SDADS (spaceborne deployable antenna design software) has been developed by research institute on mechatronics, Xidian University, China [86] with integrating electromagnetic, structural and thermal technologies.

1.6 Deployable Antenna Design Theory and Methods

In recent years, research on antenna design and manufacturing technology around the world has focused on the design of flexible structures and deployment mechanisms, the analysis and adjustment of reflector surface profiles, electromagnetic performance

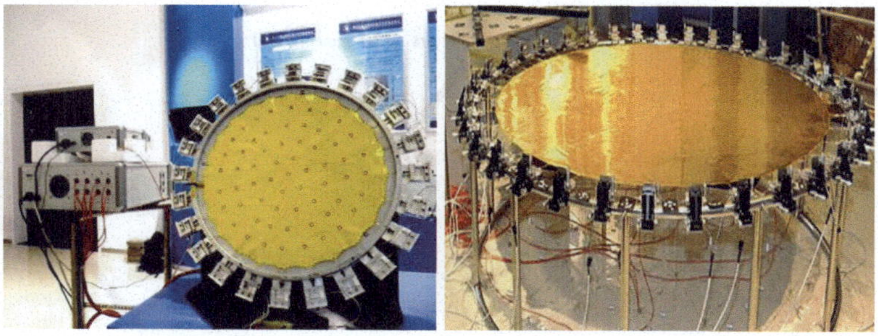

(a) 0.55m diameter (b) 2m diameter

Fig. 1.20 Prototypes of electrostatic forming membrane antennas

analysis and reliability analysis. Due to the increasing demand for large-diameter deployable antennas, exploration in related technical fields will continue to be carried out in depth.

1.6.1 Theory and Approach of Comprehensive Optimization Design with Electromagnetic, Mechanical and Thermal Techniques

Spaceborne deployable antennas have the characteristics of nonlinearity (in terms of material and geometry), great flexibility, performing in a harsh environment and high reliability requirements. The structural optimization design problem is a multidisciplinary, large-scale, multi-objective design problem. Optimal results will determine the antenna structure performance (such as quantity, furled rate, and repeatability). In a high-precision large-scale spaceborne deployable antenna, interactions exist among the structural displacement field, electromagnetic field and temperature field. As shown in Fig. 1.21, specific interaction types include (1) the influence of structural parameters and serving environmental loads on the structural displacement field (Fig. 1.21①); (2) the influence of the displacement field on the electromagnetic field (Fig. 1.21②); (3) the influence of the temperature field on the performance of the electronic device and thus the electromagnetic field (Fig. 1.21③); and (4) the influence of changes in temperature on the displacement field and, in turn, on the electromagnetic field (Fig. 1.21④). These interactions will ultimately affect the electromagnetic performance of the deployable antenna. To objectively describe this multi-physical coupling problem, it is necessary to establish a corresponding field coupling theory model. That is, we should try to find the inner links among three differential governing equations, such as elastic differential equation of structural deformation, Maxwell's equations of the electromagnetic field, and the differential equations of the temperature field (for instance, Navier-Stokes for heat convection). At the same time, based

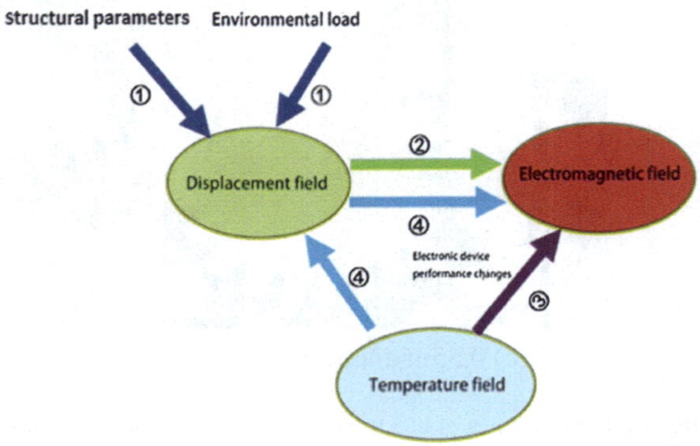

Fig. 1.21 Multi-field coupling problem of spaceborne antenna

on the research on antenna structural deployment dynamics, profile precision adjustment and maintenance, and on-orbit thermal analysis, an electromechanical thermal analysis and optimization design method for spaceborne deployable antennas is studied to overcome the difficulties of multidisciplinary, multi-objective coupling and computational complexity.

1.6.2 *Reflector Antenna Design Technology with Shaped Beam*

In a specific application case, a generated pattern is required to cover a certain terrain. In this case, the satellite antenna needs to be designed in terms of a shaped beam, as shown in Fig. 1.22. There are two implementing methods: (1) An array feed; that is, the feed of the reflector surface is an array, and the array generates the required shaped beam. For this purpose, the beamforming network is necessary to control the excitation coefficient of each individual feed [87]; (2) The reflector surface is contoured. For instance, a single feed may illuminate one shapeable reflector surface. This technique is based on a paraboloid (or elliptical surface, hyperboloid, etc.) on which an adjustable surface is superimposed to cover the required area [88–90]. However, due to the complexity of a beamforming network of antennas, the

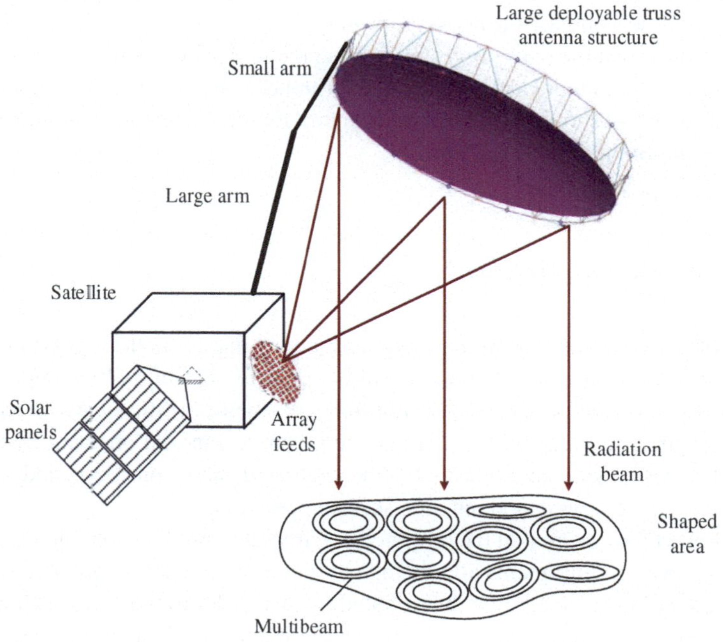

Fig. 1.22 Antenna beam shaping design

multi-feed array will introduce RF loss. To overcome this shortcoming, an additional method and devices must be utilized. This will result in an increase in the system weight. Another problem of shaped beams is the large curvature of the reflector, which creates manufacturing difficulty. Fortunately, a cable-net structure can solve the problem.

1.6.3 Performance Test and Measurement Technology

The deployable antenna works in a microgravity (or gravity-free) environment in space. It is, however, manufactured and tested on the ground. Clearly, finding a way to eliminate the influence of ground gravity and accurately measure the structure and electromagnetic performance of large-diameter spaceborne antennas becomes particularly important. There are two possible ways to eliminate the influence of gravity: (1) Eliminating gravity by changing the antenna measurement environment, such as by liquid buoyancy and free fall. (2) Integrating the ground-based gravity environment model test with simulation software to estimate the performance of the antenna in space. The latter method can be divided further into two types: (a) The antenna surface error estimation method based on antenna gain analysis [91]. That is, the performance is estimated with the Ruze formula based on the electronic performance change corresponding to the artificial structural deformation. (b) Utilizing a mathematical model to compensate for the effect of ground assembly and adjustment [92]. That is, the comparison between prototype test and the mathematical model simulation, and then the correction strategy and method are obtained. To summarize, the testing of future deployable antenna performance cannot rely solely on measurement equipment but should include effective and economical testing techniques based on a combination of "soft" and "hard".

1.6.4 Evaluation Method

The performance evaluation of deployable antennas mainly includes two tasks: (1) Meeting specific diameter requirements and evaluating feasibility by comparing the performance and cost of several different design schemes (ratio of performance to cost and engineering feasibility); (2) For the given design, the physical realization limit of the deployable size is asked to be evaluated based on the furled volume, deployment driving force, and antenna surface precision.

As is known, to ensure that the designed and manufactured antenna on the ground satisfies behavior requirements after being put into orbit, much fundamental research and key technologies need to be investigated further. Additionally, the evaluation of technology is necessary. Based on the research on deployable antenna testing techniques and performance prediction methods, evaluation includes both correctness and effectiveness assessments. Regarding effectiveness, it can be measured from two perspectives: (1) for the same structural parameters, how great a benefit, in terms of the electric performance of the antenna, can be obtained with and without the application of the proposed method; (2) for the same performance requirements, how much the mechanical error allowance can be relaxed.

1.6.5 Passive Inter-modulation (PIM) of Mesh Antenna

In communication and reconnaissance systems, under the condition of a multicarrier high-power signal, the intermodulation effect caused by the nonlinearity of the passive component is called PIM. PIM is mainly caused by contact and material nonlinearities. The former refers to interference caused by contact with nonlinear current and voltage behavior, such as looseness of the waveguide flange and rivet joint. The latter refers to interference caused by electromagnetic nonlinear materials, e.g., ferromagnetic materials and carbon fiber materials. With the continuously increasing index of communication and shortage of frequency spectrum resources due to the increase of telecommunication, PIM has received increasing attention. It is necessary to carry out passive cross-over analysis to check whether the cross-tuned wave that exits from the transmitted signal, will be dropped in the receiving frequency band. If there are high-energy harmonics in the receiving frequency band, communication might fail because the satellite receiver is damaged and system oscillation occurs. The mesh reflector is composed of a wire mesh (Fig. 1.23), which is complicated in terms of preparation, unevenness in the metal plating, inconsistent processes, the contact friction between wires, and the magnetic material itself. Nonlinear problems caused by, for example, magnetic hysteresis characteristics might result in a higher passive intermodulation of the antenna. For every 1 dB increment in the

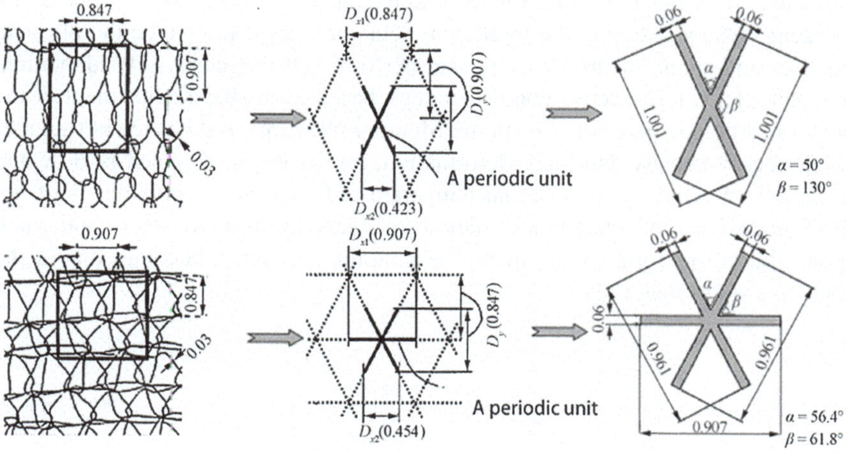

Fig. 1.23 Wire mesh structure surface-patch model

communication signal input, the passive intermodulation will increase by 3 dB [93]. Since there is currently a lack of mature mathematical models for PIM analysis and design in engineering, the only way is depended on preventive measures, experience and testing. Clearly, the investigation of theoretical modeling based on the PIM generation mechanism is one of the main basic problems.

In order to quantitatively analyze the electrical properties of the reflector surface of the support mesh, an approximate analytical model and calculation method are described in Ref. [88]. Figure 1.23 shows two types of mesh weaves. For each one, three subfigures (left, middle and right) are shown. The left subfigure is the practical case with a small window, the middle one is the detailed wire connection of the window, and the right subfigure is the magnified view of the middle one.

1.6.6 Application of New Materials

New material development is very important for satisfying the special requirements of spaceborne deployable antennas. Currently, ultra-light materials and smart materials are the two main strategies for the urgent development of ultra-light, high-modulus, and tough materials to reduce antenna weight. Both carbon nanotube and graphene materials have excellent mechanical properties, dielectric constants, electrical conductivities and thermal conductivities. Dot matrix material is another kind of material, developed from metal foam modeling and metal superposition modeling. It has a large porosity inside and good thermal conductivity, and is suitable for large-scale structures in space.

The combination of intelligent materials and structural technology enables structures to have not only the carrying function of ordinary materials but also special sensing and response functions. The applications on spaceborne antennas mainly include the following: (1) In sensing the internal and external environment based on intelligent sensing material, the intelligent actuator is used to change the physical properties and shape of the structure and perform self-diagnosis, self-adaptation, self-repair, etc. (2) The active control strategy-based local shape adjustment of the reflector can help realize beam forming. Memory alloy material has the advantages of high energy density and large deformation, but its response speed is slow and it is greatly affected by the external temperature. Electroactive polymer is of not only characteristics of a large deformation (recovery strain up to 380%) and quick response but also strong environmental adaptability, so that it becomes one of the prospective smart materials.

1.7 Next Generation of Large Spaceborne Deployable Antennas

The developing trend of deployable antennas is toward large-scale and high-precision antennas for high-frequency communication [92, 94, 95]. Deployable antennas so far are subject to certain limitations of technological feasibility and cost when expanding to large apertures. Based on the current research and applications around the world, large-scale deployable antennas have the following developing trends.

1.7.1 Deployable Frame Reflector Antenna

The modular design concept of the frame deployable antenna is to balance, by selecting the appropriate module size and number, two main characteristics. One is the diameter and surface precision in the deployed state, and the other is the height and diameter of the furled state. Since the topology of the structure itself allows the reflector surface to be easily deployed in the form of a honeycomb structure, it satisfies the requirements of a large diameter, high precision, and a light weight. At present, the design concept of deployable antennas of several tens of meters or even hundreds of meters in developed countries adopts a frame-like deployable form of multi-cell units. For example, the frame-type mesh deployable antenna designed by General Dynamics in the United States is composed of multiple combined tetrahedral modules with a surface precision of 1.5 mm in terms of the root of mean square (RMS) value and a frequency range of 20–30 GHz [96]. Another representative of this type of antenna is the Japanese engineering test satellite ETS-8, with a size of 17 m × 19 m [97].

1.7.2 Inflatable Deployable Reflector Antenna

The obvious advantage of the inflatable reflector antenna is that the furled ratio is high. The disadvantage is that it is difficult to achieve high precision, and if space debris is impacted after inflation, the surface is easily damaged, causing the satellite to be scrapped. After the required shape is inflated, if the material is hardened by light, the compressor can be thrown away. In this case, the difficulty will become the development of new materials. As a result, the aerated hardening materials have naturally become a popular research topic.

NASA is developing next-generation weather radar (NEXRAD) in geostationary orbit with a diameter of 35 m at the Ka-band (35 GHz) [98]. The design concept is an inflatable hardened deployable antenna (Fig. 1.24), and the reflector is a gas-hardened shape memory polymer (SMP). This type of antenna combines the advantages of the mesh antenna and inflatable antenna, but it is worthwhile to note that when the

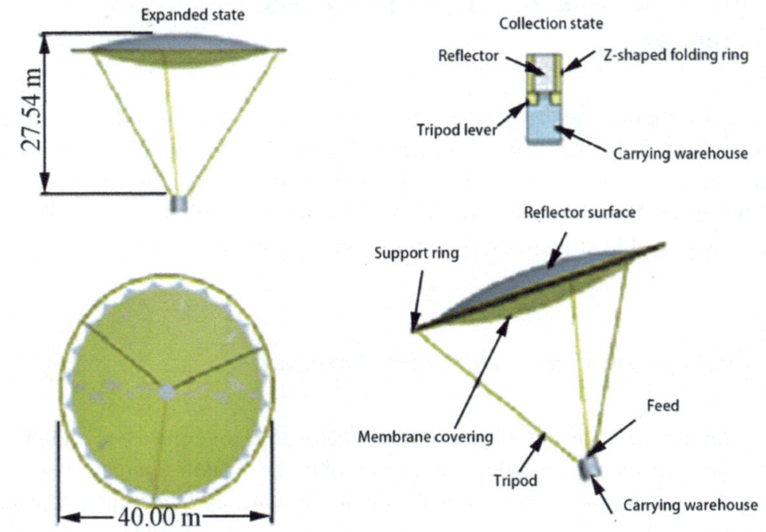

Fig. 1.24 SMP Reflector deployable antenna scheme

antenna aperture is further increased, the SMP material requires a large amount of thermal energy to be deployed. Thus, shape memory aerated hardened composites have become the key enabling technology for such large, lightweight deployable antennas.

1.7.3 Space Assembled Large Antenna

In order to have space antenna with a very large diameter, a possible method is to send the modules of the antenna to space, and then, in space, assemble the modules into a large-diameter antenna by robot (or hand). In 2007, for instance, JPL proposed a space based automated assembly optical telescope design with a diameter of 150 m, as shown in Fig. 1.25 [99]. Based on China's space station "Tiangong No. 1", such large antennas will have immeasurable development potential. The key technologies involve the control of space robots (or hands) and virtual/advanced reality (VR/AR) for space assembly. The specific techniques include assembly modeling in an immersive virtual reality environment, operational positioning based on geometric constraints, interactive assembly planning and evaluation by virtual reality, and factor analysis of humans and machines in the assembly process.

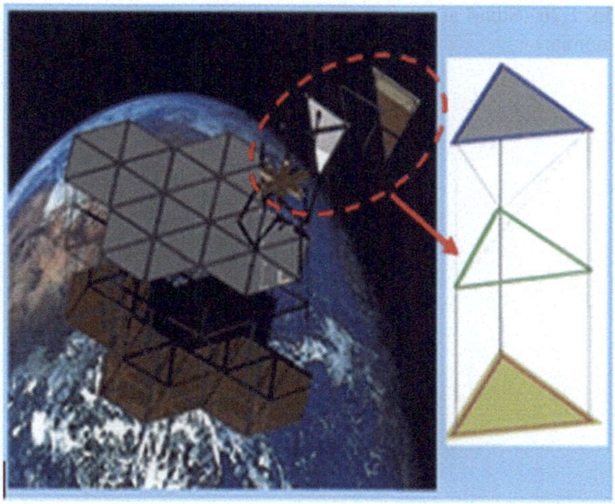

Fig. 1.25 Space assembly optical telescope

1.7.4 Smart Array Antenna

With the development of space technology, phased array antennas have been gradually applied in aerospace fields due to the application requirements of satellite communication anti-jamming, space microwave imaging radar, beam on-orbit reconstruction and fast change, multi-target and beam scanning. Combining antenna arrays with signal processing and MEMS technology is expected to form a new type of deployable smart array antenna. In this kind of antenna, signal processing technology is utilized to control and improve the performance of the antenna for the purpose of receiving the maximum amount of information. Clearly, this type of intelligence-based adaptive array technology antenna will play a positive role in aerospace applications. Figure 1.26 shows the concept of a deployable planar phased array patch smart antenna (20–40 m) proposed in France for the application of mobile communication satellites. It works in the L/S band and can perform autonomous control of the beam and its direction through software processing. Mechanical adjustment can compensate for surface errors automatically [100].

1.8 Humans Entering Aerospace Era

We are now in an era of great development in space science and technology [101]. It is a general trend to explore the ocean and space, especially the deep sea and deep space. As in the great geographical discoveries of the 15th century, the universe contains infinite resources and wealth for human beings to find. The great voyage era represented by Columbus in 1492 when the American continent was found and when Magellan completed the round-the-world voyage in 1522 greatly promoted the

Fig. 1.26 Smart array antenna

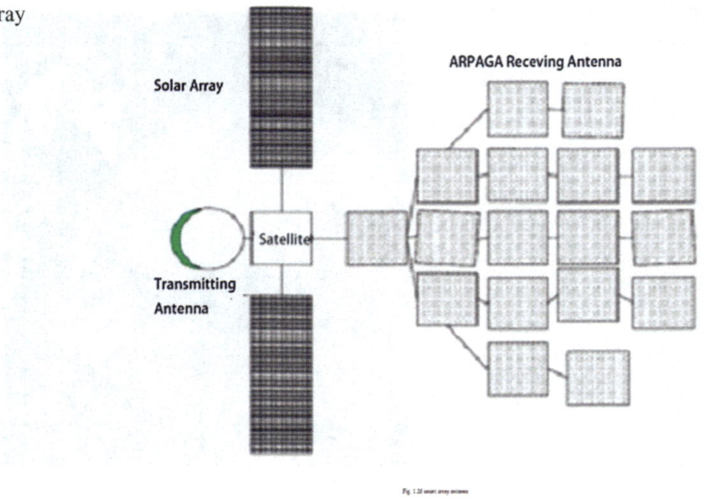

development of world shipbuilding, machinery, navigation and other technologies, and now we are entering the era of Great Aerospace, which will promote the development of various aerospace technologies. The design and manufacturing technology of spaceborne antennas with ultra large diameters, high precision, light weights and large furled ratios is one of the most urgent and critical technologies. Although China started in the field several decades later than other developed countries, the antennas with rapid progress and diameters of more than 10 m have been delivered to space. Antennas of 20–30 m diameters are under development. New concepts and new antennas are making progress continually. It can be anticipated that new concepts and new technologies for spaceborne antennas with great progress will be realized soon.

References

1. Kunkee D, Poe GA, Boucher D et al (2008) Design and evaluation of the first special sensor microwave imager/sounder (SSMIS). IEEE Transa Geosci Remote Sens 46(4):863–883
2. Cherny IV, Chernyavsky GM, Nakonechny VP et al (2002) Spacecraft 'Meteor-3 M' microwave imager/sounder MTVZA: First results. In: Proceedings of IGARSS'02 symposium, Toronto, Canada, pp 2660–2662
3. Im E, Thomson M, Fang HF (2007) Prospects of large deployable reflector antennas for a new generation of geostationary Doppler weather radar satellite. In: Proceedings of AIAA space 2007 conference & exposition, Long Beach, California, USA, pp 66–174
4. Takano T (1999) Large deployable antennas concepts and realization. Proceedings of IEEE Antennas and Propagation Society International Symposium, Orlando, Florida, USA, pp 1512–1515
5. Freeland RE (1983) Survey of deployable antenna concepts. NASA Langley Research Center, Hampton, Virginia, USA, pp 613–652

6. Misawa M (1998) Stiffness design of deployable satellite antennas in deployed configuration. J Spacecraft Rockets 35(3):380–386
7. Soykasap O, Watt AM, Pellegrino S (2004) New deployable reflector concept. In: Proceedings of the 45th AIAA/ASME/ASCE/AHS/ASC structures, structural dynamics and materials conference, Palm Spring, California, USA
8. Barer H, Datashvili L, Gogava Z et al (2001) Building blocks of large deployable precision membrane reflectors. In: Proceedings of the 42nd AIAA/ASME/ASCE/AHS/ASC structures, structural dynamics and materials conference and exhibit, Seattle, Washington, USA
9. Mikulas J, Collins TJ, Hedgepeth JM (1991) Preliminary design considerations for 10–40 meter-diameter precision truss reflectors. J Spacecraft Rockets 28(4):439–447
10. Miyasaki A, Homma M, Tsujigate A et al (2001) Design and ground verification of large deployable reflector. In: Proceedings of the 42nd AIAA/ASME/ASCE/AHS/ASC structures, structural dynamics and materials conference and exhibit, Seattle, Washington, USA
11. Natori MC, Takano T, Noda T et al (1998) Ground adjustment procedure of a deployable high accuracy mesh antenna for space VLBI mission. In: Proceedings of the 39th AIAA/ASME/ASCE/AHS/ASC structures, structural dynamics and materials conference and exhibit and AIAA/ASME/AHS adaptive structures forum, Long Beach, California, USA
12. Rogers CA, Stutzman WL, Campbell TG et al (1993) Technology assessment and development of large deployable antennas. J Aerosp Eng 6(1):34–55
13. Duan B, Li T (2007) Large deployable antenna and its application in aerospace. In: Proceedings of the 2nd aerospace electronic information supporting development strategy forum, Nanjing, China, pp 245–253. (in Chinese)
14. Zhang G, Zhao Y (2004) Application of micro electro mechanical system in phased array antenna. J Electromechan Eng 20(6):1–13 (in Chinese)
15. Sieracki V (2000) Advances in MEMs for RF technology. In: Proceedings of AOC radar and EW conference session 2: technology developments & impact on Radar/ESM
16. Escrig F (1985) Expandable space structures. Int J Space Struct 1(2):79–91
17. Guest SD, Pelegrino S (1996) A new concept for solid surface deployable antennas. Acta Astronaut 38(2):103–113
18. Liu R, Tian D, Deng Z (2010) Research status and prospects of spaceborne deployable antenna structure. Mech Design 27(9):1–21 (in Chinese)
19. Li T, Ma X (2012) Large spaceborne expandable antenna technology. Space Electron 9(3):35–43 (in Chinese)
20. Lichod Z (2003) Inflatable deployed membrane waveguide array antennaforspace. In: Proceedings of AIAA/ASME/ASCE/AHS/ASC structures, structural dynamics and materials conference, Norfolk, USA
21. Guan F, Li G, Xia J (2003) Inflatable expandable space structure. In: Proceedings of the symposium on satellite structure and institutional technology progress, Xi'an, China. (in Chinese)
22. Di J (2005) Research on optimization and adjustment technology of reflecting surface accuracy of cable net deployable antenna structure. Ph.D. Thesis, Xidian University, Xi'an, China. (in Chinese)
23. Huang J (2001) The development of inflatable array antennas. IEEE Antennas Propag Mag 43(4):44–50
24. Huang J, Lou M, Feria A et al (1998) An inflatable L band microstrip SAR array. In: Proceedings of IEEE antennas and propagation society international symposium. Atlanta, Georgia, USA, pp 2100–2103
25. Cassapakis CG, Love AW, Palisoc AL (1998) Inflate spaceborne antennas: a brief overview. In: Proceedings of 1998 IEEE aerospace conference, Snowmass, Colombia, USA, pp 453–459
26. Wang Y (2003) Inflatable antenna structure technology overview. Telecommun Technol 43(2):6–11 (in Chinese)
27. Akira M, Satoshi H, Mitsunobu W (2003) Key technologies for high-accuracy large mesh antenna reflectors. Acta Astronaut 53(11):899–908

28. Love AW (1976) Some highlights in reflector antenna development. Radio Sci 11(8-9):671–684
29. Miura K, Miyazaki Y (1990) Concept of the tension truss antenna. AIAA Journal 28(6):1098–1104
30. Mikulas MM, Collins TJ, Hedgepeth JM (1990) Preliminary design approach for large high precision segmented reflectors. NASA Technical Memorandum, TM 261050
31. Semler D, Tulintseff A, Sorrell R et al (2010) Design, integration and deployment of the TerreStar 18-meter reflector. In: Proceedings of the 28th AIAA international communications satellite systems conference, California, USA
32. Akira M, Kyoji S, Motofumi U et al (2009) In-orbit deployment characteristics of large deployable antenna reflector onboard Engineering Test Satellite VIII. Acta Astronaut 65(9):1306–1316
33. Canada, USA, SkyTerra 1, 2 (MSV1, 2, SA) http://space.skyrocket.de/docsdat/skyterra-1.htm, 2016-10-02
34. Zhang P, Jin G, Shi G (2009) Research and development status of space film mirrors. China Opt Appl Optics 2(2):91–101 (in Chinese)
35. Geoff P, Randall J (2002) Large aperture holographically corrected membrane telescope. Opt Eng 41(7):1603–1607
36. Palisoc L (2000) Large telescope using a holographically-corrected membrane mirror-final report to the NASA Institute for advanced concepts. L'Garde, Inc., Technical Report, LTR00-AP-021
37. Errico S, Angle R, Stamper B et al (2002) Stretched membrane with electrostatic curvature (SMEC) mirrors: A new technology for large lightweight spaceborne telescopes. In: International optical design conference, Tucson, Arizona, United States, pp 356–364
38. Angle R, Burge J, Hege K et al (2000) Stretched membrane with electrostatic curvature (SMEC) mirrors: a new technology for ultra-lightweight spaceborne telescopes. In: Proceedings of SPIE, pp 699–705
39. Burley GS, Stilburm JR, Walker GAH (1998) Membrane mirror and bias electronics. Appl Opt 37(21):4649–4655
40. Soh M, Lee JH, Youn SK (2005) An inflatable circular membrane mirror for spaceborne telescopes. Proc Opt Design Testing II, Proc SPIE 56(38):262–271
41. Rai E, Nishimoto S, Katada T et al (1996) Historical overview of phased array antennas for defense application in Japan. In: Proceedings of IEEE international symposium on phased array systems and technology, Tokyo, Japan
42. Brejcha AG, Keeler LH, Sanford GG (1978) The SEASAT–a synthetic aperture radar antenna. In: Proceedings of synthetic aperture radar technology conference, Las Cruces, New Mexico, USA
43. Ward JC (1979) "Large space systems technology-1979", NASA Conference Publication 2118, First Annual Program Technical Review, NASA LRC, Hampton, USA, pp 157–171
44. Kopriver F (1980) "Large space systems technology-1980", NASA Conference Publication 2168, Second Annual Technical Review, NASA LRC, Hampton, USA
45. Larson TR (1981) A microstrip honeycomb array for the low altitude space-based radar mission. Ball Aerosp Syst Div, F81-06
46. Tauno VH, Koen VC (2007) RF MEMS impedance tuners for 6-24 GHz applications. Int J RF Microwave Comput Aided Eng 17(3):265–278
47. Ramadoss R, Lee S, Lee YC et al (2007) MEMS capacitive series switch fabricated using PCB technology. Int J RF Microwave Comput Aided Eng 17(4):387–397
48. Ramadoss R, Lee S, Lee YC et al (2006) RF-MEMS capacitive switch fabricated using printed circuit processing techniques. J Microelectromech Syst 15(6):1595–1604
49. Malczewski A, Eshelman S (1999) X-band RF MEMS phase shifters for phased array applications. IEEE Microwave Guided Wave Letters 9(12):517–518
50. Alastalo A (2006) Microelectromechanical resonator-based components for wireless communications: filters and transmission lines. VTT Publications 616, Espoo, Finland

51. Yu L, Andrea B (2001) Distributed MEMS transmission lines for tunable filter applications. John Wiley & Sons Incorporated, New York, USA
52. Erdil E, Topalli K, Unlu M et al (2007) Frequency tunable microstrip patch antenna using RF MEMS technology. IEEE Trans Antennas Propag 55(4):1193–1196
53. Li S (2012) Review of RF MEMS technology and main problems. Space Electronic Technol 2012(4):6–13 (in Chinese)
54. Maciel JJ, Slocum JF, Smith JK et al (2007) MEMS electronically steerable antennas for fire control radars. Aerosp Electron Syst Mag 22(11):17–20
55. Chiao JC, Cheng SY, Jeff JL et al (2001) MEMS reconfigurable antennas. Int J RF Microwave Comput Aided Eng 11(5):301–309
56. Coleman CM, Rothwell EJ, Ross JE et al (2002) Self-structuring antennas. IEEE Trans Antennas Propag 44(3):11–23
57. Weily AR, Guo YJ (2007) An aperture coupled patch antenna system with MEMS-based reconfigurable polarization. In: Proceedings of IEEE international symposium on communications and information technologies, Sydney, Australia, pp 325–328
58. Khaleghi A, Kamyab M (2009) Reconfigurable single port antenna with circular polarization diversity. IEEE Trans Antennas Propag 57(2):555–559
59. Kovitz J, Rajagopalan H, Rahmat-Samii Y (2015) Design and implementation of broadband MEMS RHCP/LHCP reconfigurable arrays using rotated E-shaped patch elements. IEEE Trans Antennas Propag 63(6):2497–2507
60. Hsu SH, Chang K (2007) A novel reconfigurable microstrip antenna with switchable circular polarization. Antennas Wirel Propag Lett IEEE 6(11):160–162
61. Hum SV, Okoniewski M, Davies RJ (2007) Modeling and design of electronically tunable reflect arrays. IEEE Trans Antennas Propag 55(8):2200–2210
62. Qiu Y, Ye S, Liu M (1992) Dynamic analysis of elastic mechanism considering hinge flexibility and friction. J Electromech Eng 1992(5):11–17 (in Chinese)
63. Liu M, Wang Y (2002) Variable structural dynamic analysis and control. Mech Sci Technol 21(11):23–24 (in Chinese)
64. Feng L, Ye S, Liu M (2000) Research on symbolic calculus of multi-flexible system dynamics. Math Rese Comment 20(1):143–148 (in Chinese)
65. Tan Z, Ye S, Liu M (1994) Dynamic multibody system dynamics research and existing problems. Progress Mech 24(2):248–256 (in Chinese)
66. Tan Z, Ye S, Liu M (1994) Dynamics of flexible multibody system dynamics, Mechanics and practice. Mech Eng 16(5):14–19 (in Chinese)
67. Qiu Y, Liu M (1992) Expansion dynamics of large spaceborne deployable antennas. Chinese Space Sci Technol 1992(1):1–7 (in Chinese)
68. Wang J, Liu M (2009) Large-scale spaceborne deployable antenna dynamics modeling and simulation. J Syst Simul 21(6):1730–1733 (in Chinese)
69. Wang J, Liu M, Zhao Y (2009) Research on combined control of large spaceborne deployable antenna deployment process. China Mech Eng 20(6):728–732 (in Chinese)
70. Liu M, Li W, Zheng F (1998) Sub-ribbed mesh shaped reflective surface shape accuracy adjustment. J Xidian Univ 25(4):506–509
71. Li TJ, Wang Y (2009) Performance relationships between ground model and spaceborne prototype of deployable spaceborne antennas. Acta Astronaut 65(9):1383–1392
72. Li TJ, Wang Y (2009) Deployment dynamic analysis of deployable antennas considering thermal effect. Aerosp Sci Technol 13(4-5):210–215
73. Li TJ, Su JG (2011) Electrical properties analysis of wire mesh for mesh reflectors. Acta Astronaut 69(1-2):109–117
74. Li TJ, Su JG, CAO YY (2011) Dynamic characteristics analysis of deployable spaceborne structures considering joint clearance. Acta Astronautica 68(7–8): 974–9832011
75. Li TJ (2012) Deployment analysis and control of deployable spaceborne antenna. Aerosp Sci Technol 18(1):42–47
76. Du JL, Bao H, Yang DW et al (2012) Initial equilibrium configuration determination and shape adjustment of cable network structures. Mech Based Design Struct Mach 40(3):277–291

77. Zhang YQ, Duan BY, Li TJ (2012) A controlled deployment method for deployable flexible spaceborne antennas. Acta Aerosp 81(1):19–29
78. Zhang YQ, Ru WR, Xu HR (2015) A deployment trajectory design method based on the Bezier curves. J Comput Theor Nanosci 12(12):5288–5296
79. Zhang YQ, Ru WR, Yang GG et al (2016) Deployment analysis considering the cable-net tension effect for deployable antennas. Aerosp Sci Technol 48(1):193–202
80. Zhiqiang Dong and BaoyanDuan (2001) Study on mechanical properties of spaceborne antenna wound ribs. J Xidian Univ 28(6):755–758 (in Chinese)
81. Yang D (2010) Structural design and profile adjustment of spaceborne large expandable cable antenna. PhD Thesis, Xidian University, Xi'an, China. (in Chinese)
82. You G (2013) Morphological analysis and optimization of cable net deployable antenna. PhD Thesis, Xidian University, Xi'an, China. (in Chinese)
83. Zhang Y (2013) Integrated design of flexible spaceborne deployable antenna with structural and control technologies. Large Spaceborne Deployable Antennas (LSDAs)—A Comprehensive Summary 15 PhD Thesis, Xidian University, Xi'an, China. (in Chinese)
84. Zong Y (2015) Influence of structural random factors on electronic performance of spaceborne mesh reflector antennas and optimum design. PhD Thesis, Xidian University, Xi'an, China. (in Chinese)
85. Zhang S (2015) On integrated optimum design of reflector antennas with mechanical and electronic technologies. PhD Thesis, Xidian University, Xi'an, China. (in Chinese)
86. Hao J, Duan B et al (2015) The computer program of synthetic design and digital modeling for large deployable satellite antennas. Space Electron Technol 12(3):35–42 (in Chinese)
87. Pontoppidan K (2005) Technical description of GRASP9. Denmark, Copenhagen
88. Miura A, Rahmat-Samii Y (2007) Space-borne mesh reflector antennas with complex weaves: Extended PO/periodic-MoM analysis. IEEE Trans Antenna Propag 55(4):1022–1029
89. Gonzalez-Valdes B, Martinez-Lorenzo JA, Rappaport C et al (2008) Generating contoured beams with single-shaped reflector using an iterative field-matrix approach. Antennas Wirel Propag Lett IEEE 2008(7):697–700
90. Bergmann JR, Moreira FJS (2009) Omnidirectional ADE antenna with a GO-shaped main reflector for an arbitrary far-field pattern in the elevation plane. IET Microw Anten Propag 2009(3):1028–1035
91. Hiroaki T (2011) Surface error estimation and correction of a spaceborne antenna based on antenna gain analysis. Acta Astronaut 68(7):1062–1069
92. Natora MC, Hirabayashi H, Okuizumi N (2002) A structure concept of high precision mesh antenna for spaceborne VLBI observation. In: Proceedings of the 43rd AIAA/ASME/ASCE/AHS/ASC structures, structural dynamics and materials conference, Denver, Colorado, USA
93. Duan B (2005) Analysis, optimization and control of flexible microwave antenna structures. Science Press of China, Beijing, China. (in Chinese)
94. Surya PC, James DM, Jennie O (2006) Design evaluation of a large aperture deployable antenna. In: Proceedings of the 47th AIAA/ASME/ASCE/AHS/ASC structures, structural dynamics and materials conference, Newport, Kentucky, USA
95. Duan B (2004) Review of the development of electromechanical engineering. Electro-Mechan Eng 20(6):14–30 (in Chinese)
96. Bolli P (2002) Passive intermodulation on large reflector antennas. Antennas Propag Maga 44(5):13–19
97. Delaune D, Tanaka T, Onishi T et al (2004) Simple satellite tracking stacked patch array antenna for mobile communications experiments aiming at ETS-VIII applications. IEE Proc Microw Anten Propag 151(2):173–179
98. Lin JK, Fang HF, Im E et al (2006) Concept study of a 35 m spherical reflector system for NEXRAD in space application. Proceedings of the 47th AIAA structures, structural dynamics and materials conference, Newport, USA, 2006
99. Wilkie WK, Williams RB, Agnes GS et al (2007) Structural feasibility analysis of a robotically assembled very large aperture optical spaceborne telescope. Proceedings of the 48th AIAA structures, structural dynamics and materials conference, Honolulu, USA

100. Gayrard JD (2004) A very large patch receiving antenna for mobile communication satellites. In: Proceedings of the 22nd AIAA international communications satellite systems conference & exhibit, Monterey, USA
101. Baoyan D (2020) Large Spaceborne deployable antennas (LSDAs)—a comprehensive summary. Chinese J Electron 29(1):1–5

Chapter 2
Space Service Environment Analysis

2.1 Overview

For satellites to operate reliably in space, it is necessary to understand and solve special service environment problems such as vacuum, extremely high and low temperatures, microgravity, and solar irradiation. These factors will bring unprecedented challenges to the design, manufacture and maintenance of spaceborne deployable antennas [1].

2.2 Satellite Orbit and the Requirements for Antenna Design

The orbit of the satellite is obviously inseparable from the three orbits of spacecraft and the speed of universe. We owe this discovery to the astrophysical scientist Johannes Kepler (Fig. 2.1), who proposed three laws of planetary movement. As shown in Fig. 2.2, the first orbit around the Earth is the intercontinental ballistic missile zone, with a circular trajectory and a velocity of 7.9 km/s. The second is the elliptical orbital zone, and craft overcome the gravity of Earth at 11.2 km/s. The third is the hyperbolic orbital zone in the solar system. The speed at which aircraft can escape from the solar system is 14.6 km/s.

As the "eyes" and "ears" of space equipment, to complete tasks such as communication, reconnaissance, remote sensing, and meteorological observation, spaceborne antennas need to operate in various specific orbits. Generally, orbits can be roughly divided into the following four kinds: low, middle, high and large elliptical orbits. There are two main kinds of classification methods (Fig. 2.3). One is continuous orbit; in this case, low orbit refers to an orbit height less than 1000 km (for the specific low orbit of near-Earth orbit, the distance is 300–600 km). Middle orbit is 1000–20,000 km in height, high orbit is greater than 20,000 km, and geosynchronous

B. Duan et al., *Large Deployable Satellite Antennas*,
Springer Tracts in Mechanical Engineering,
https://doi.org/10.1007/978-981-15-6033-0_2

Fig. 2.1 Johannes Kepler (1571–630) was a German astronomer and mathematician. He discovered the three laws of planetary motion and was the founder of modern experimental optics

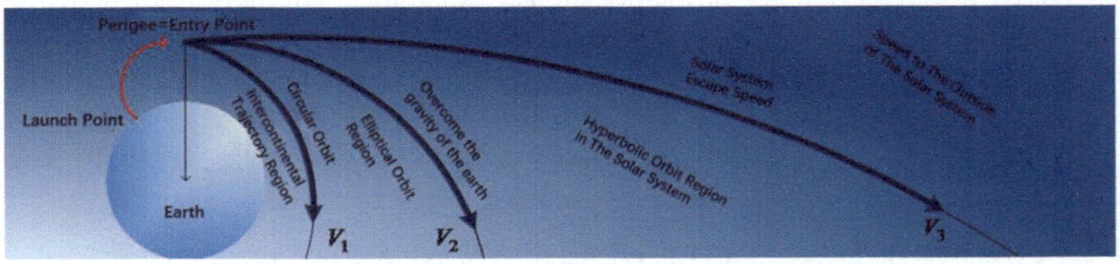

Fig. 2.2 Three cosmic speed diagrams

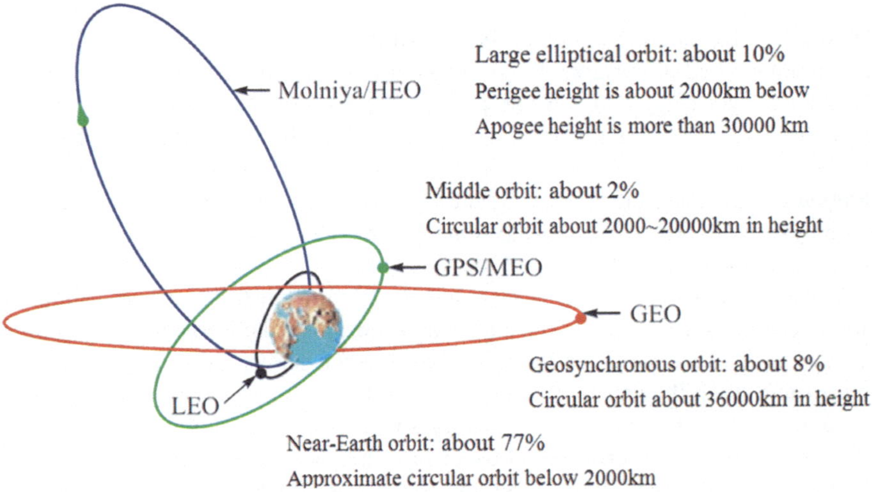

Fig. 2.3 Four kinds of satellite orbits

orbit, in the high orbit category, is 36,000 km. A large elliptical orbit means that the perigee from the Earth is several hundred to several thousand km, and the apogee is greater than 30,000 km. The other type is discontinuous orbit. Low orbit refers to an orbit less than 1000 km, middle orbit refers to an orbit between 18,000 and 25,000 km, and high orbit refers to an orbit of more than 36,000 km. The perigee of a large elliptical orbit may be approximately 300 km or hundreds to thousands of kilometers, and the apogee is from 10,000 km to more than 30,000 km.

Although the distance is so great, the requirements for signal strength and image sharpness are still high, which is reflected in the performance indexes, such as the positioning accuracy, resolution, sensitivity, and code rate. For example, the BeiDou Navigation Satellite System (BDS) has a ground positioning accuracy better than 10 m, a speed measurement accuracy of 0.2 m/s, and a timing accuracy better than 10 ns.

2.3 Very High and Low Temperature Environment

When an antenna operates in various orbits, whether it is directly irradiated by the sun will cause a significant temperature change. For example, if it is irradiated, the temperature can reach tens or even one hundred degrees centigrade, and when entering the shadow area of the Earth, the temperature may drop to negative one hundred degrees centigrade [2]. In addition, even if operating in the illuminated area of an orbit, although the ambient temperature does not change, the antenna may be blocked by a satellite or solar panel, and some part of the antenna itself may be blocked by another part, causing a change in the temperature gradient of the antenna. This temperature difference environment will create new requirements for the design and manufacture of antennas.

To understand the temperature changes in different orbits, the above five kinds of orbits are numerically simulated. The results of the on-orbit thermal analysis of a 16 m antenna are shown in Table 2.1.

Table 2.1 Results of on-orbit thermal analysis of a 16 m antenna

Temperature	Low orbit	Middle orbit	High orbit	GEO orbit	Large elliptical orbit
Highest/°C (Sunlight)	42.7	39.3	38.7	38.3	39.4
Lowest/°C (Shadow)	−106.5	−139.6	−159.1	−159.5	−159.7
Maximum gradient/°C (Side lighting)	72.7	87.6	96.4	100.6	104.7

2.4 Microgravity

Even though a spaceborne antenna operates in a microgravity environment in space, its design, manufacture and even measurement have to be done on the ground. The question is that how can the antenna's performance be confirmed after it is launched into orbit [3]. For the measurement of a large diameter antenna surface, creating a microgravity (or non-gravity) environment is essential and difficult. Another example is the large mesh reflector antenna. When the locking device is released, the support cable net may float randomly due to micro- or non-gravity. Intertwining may cause deployment failure, which is also difficult to simulate on the ground.

2.5 Vacuum

The very thin atmosphere in space will give rise to the following problems of analysis, design and manufacture of deployable satellite antennas:

(1) Because of the high vacuum, the internal molecular and solid surface states of the material will change, greatly affecting the damping characteristics of various parts of the antenna system. The determination of material's damping coefficient commonly used on the ground is no longer applicable. As a result, the vibration performance, such as the dominant modes of the antenna, will be different from that on the ground. If this occurs on the supporting arms, it will take time to stabilize the antenna.

(2) The high vacuum changes the solid surface state and evaporates liquid and solid materials with high vapor pressure, causing dry friction on the surface and thus leading to motion damping or jamming of the mechanism movement. Therefore, the determination of the hinge gap will be different from that on the ground.

(3) Discharge between the electrodes may damage materials. Adjacent surfaces with voltage differences, such as electrostatic forming membrane antennas, may generate inter-electrode discharge between the membrane and base electrode when the vacuum reaches a certain level and the voltage is too high, damaging the aluminized polyimide serving as a reflective surface.

References

1. Natori MC, Takano T, Nora T et al (1998) Ground adjustment procedure of a deployable high accuracy mesh antenna for space VLBI mission, A98-25353, AIAA/ASME/AHS adaptive structures forum, April 20–23, 1998, Long beach, California, USA
2. Min G 1991) Satellite thermal control technology. Aerospace Press, Beijing (in Chinese)
3. Song J (2007) Aerospace vertical and horizontal—aerospace pulls on basic science. Higher education press, Beijing (in Chinese)

Chapter 3
Cable Net Design and Form-Finding of Cable and Truss Combined Structures

3.1 Introduction

To commemorate the great contribution of Galileo Galilei (Fig. 3.1) in the area of space, NASA launched a Jupiter probe named Galileo on October 11, 1989. The probe is equipped with a camera and a lower gain antenna that can measure Jupiter's atmospheric, cloud and magnetic information. There are two kinds of spaceborne reflector antennas: axisymmetric (prime-feed) and non-axisymmetric (offset-feed) reflector antennas. The non-axisymmetric reflector antenna is mainly a cutting parabolic antenna. The cutting antenna here is generally not a ground warning or search antenna that is cut by two planes parallel to the axis but an antenna obtained from a cylinder intersecting the axisymmetric reflector. As a result, an elliptical reflector surface is obtained.

Spaceborne reflector antennas are mostly offset feed reflectors, since they can reduce occlusion and save space by placing the feed on the satellite. Therefore, the following discussion will be focused on the offset-feed reflector antenna because the axisymmetric prime feed antenna is just a special case of the offset-feed reflector antenna.

3.1.1 Coordinate Systems

To facilitate the following discussion, coordinate systems are shown in Figs. 3.2, 3.3, 3.4 and 3.5. The meaning of each coordinate system in Fig. 3.2 is as follows,

(1) $O_1X_1Y_1Z_1$, $O_2X_2Y_2Z_2$, $O_3X_3Y_3Z_3$ and $O_4X_4Y_4Z_4$ are the coordinate systems of the world, satellite, parent paraboloid and cutting paraboloid, respectively. The coordinate system $O_3'X_3'Y_3'Z_3'$ is the transition coordinate system utilized to transform the parent paraboloid coordinate system to the cutting paraboloid coordinate system $O_4X_4Y_4Z_4$.

B. Duan et al., *Large Deployable Satellite Antennas*,
Springer Tracts in Mechanical Engineering,
https://doi.org/10.1007/978-981-15-6033-0_3

Fig. 3.1 Galileo Galilei, 1564–1642, Italian mathematician, physicist, astronomer, and one of the founders of modern experimental science. The refractive telescope was invented and its astronomical observations strongly supported the heliocentric theory

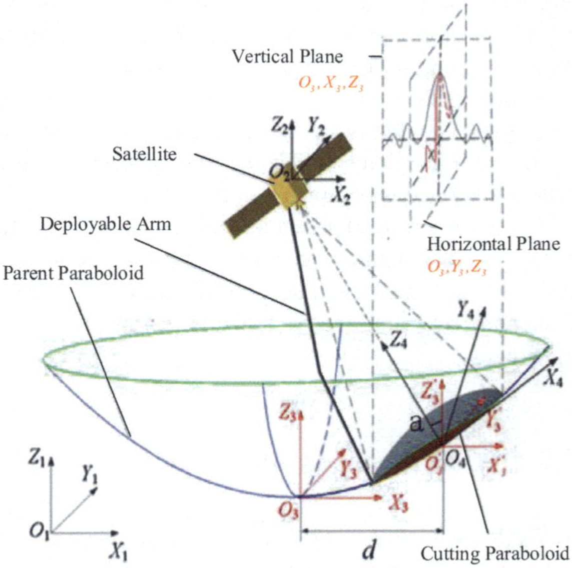

Fig. 3.2 Satellite-borne offset reflector antenna

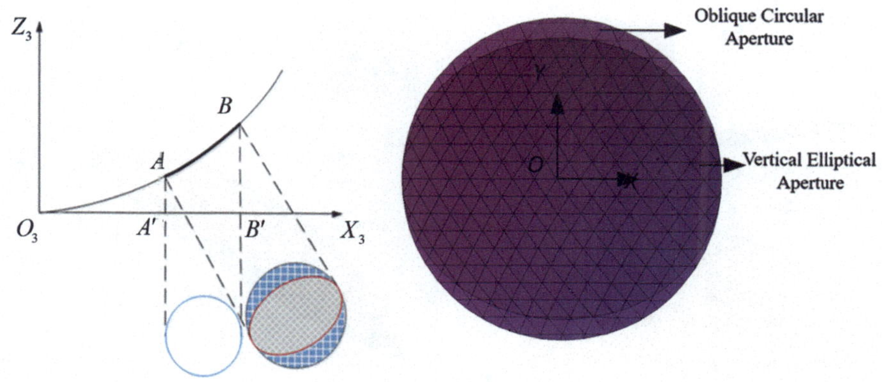

Fig. 3.3 Offset paraboloid formed by different cutting methods

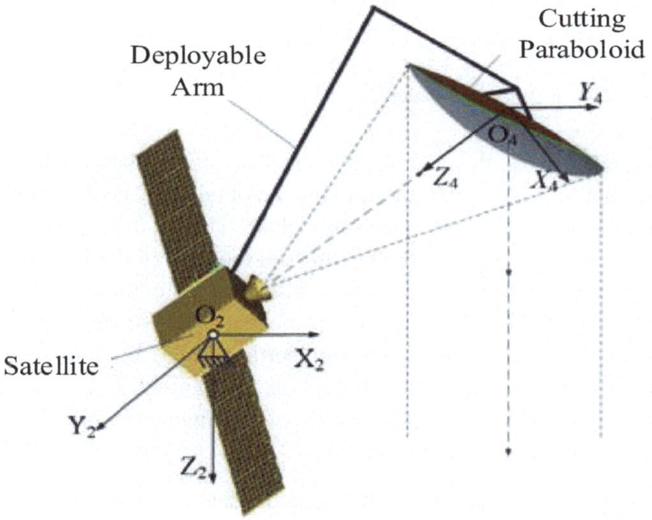

Fig. 3.4 Integration of satellite and antenna

(2) The vertical plane parallel to plane $O_3X_3Z_3$ is the 0-degree plane of the far field radiation pattern, and that parallel to $O_3Y_3Z_3$ is the 90° plane of the far field radiation pattern. The parameter α is the offset angle between axis O_3Z_3 $(O_3'Z_3')$ and O_4Z_4. The parameter d is the offset distance.

(3) The cutting paraboloid (offset parabolic reflector) is obtained by cutting a paraboloid with a cylinder. There are two ways of cutting, namely vertical cutting (parallel to the axis) and oblique cutting (along the normal). An ellipse can be obtained by vertical cutting, as shown by the red line in Fig. 3.3a; while a circle is obtained by oblique cutting, as shown in Fig. 3.3b. For the

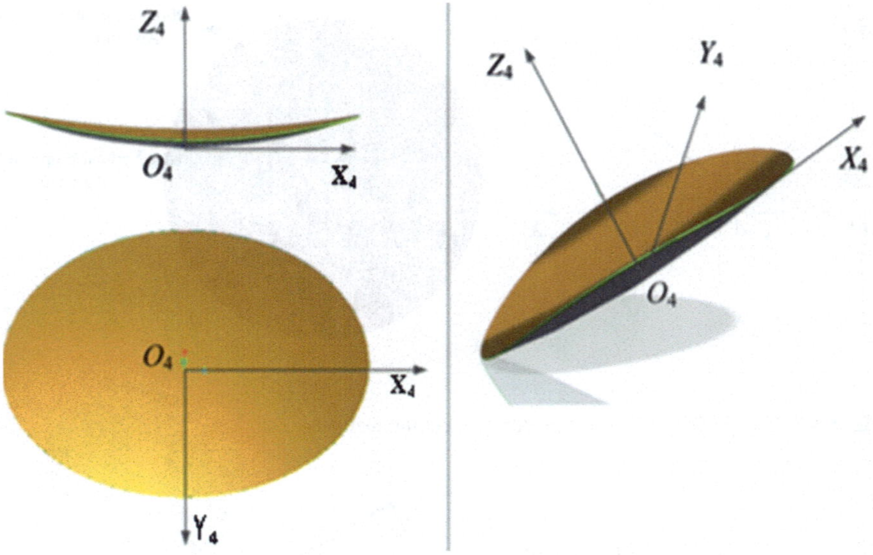

Fig. 3.5 Three views of cutting paraboloid

oblique cutting case, the advantage is that not only a relatively large electronic aperture, but also a circular aperture can be obtained. The disadvantage is that the obtained aperture is not in the same plane. That is, the Z_4 coordinates at the ends of the X_4 axis are smaller than those at the ends of the Y_4 axis. In contrast, the obtained aperture is at the same level in vertical cutting, so vertical cutting is usually adopted. In this case, the projected circle of the cylinder on plane $O_3X_3Y_3$ is regarded as the electronic aperture of the antenna, denoted as D_e, and the projection of the cut paraboloid on the plane $O_4X_4Y_4$ is an ellipse. The short axis is the electronic aperture D_e of the antenna, while the long axis is the physical aperture D of the antenna.

Figure 3.4 shows the relationship between the satellite and antenna. The feed is placed on the satellite. Electromagnetic waves, emitted from the feed, can be reflected from the offset reflector and then emitted to the Earth or other specified targets. The reverse is also true, that is, the reflector concentrates the received electromagnetic wave to the feed.

The picture on the right in Fig. 3.5 shows a three-dimensional solid model of the cutting paraboloid in the parent paraboloid, and the left picture shows the main and top views of the cutting paraboloid. The projection of the cutting paraboloid on plane $O_4X_4Y_4$ is an ellipse with the long axis of the ellipse on the X_4 axis and the short axis on the Y_4 axis.

3.1.2 Description of the Common Parameters

The specific structural implementation of the offset reflector described above can vary, such as by using rigid surface, inflatable, membrane, and mesh reflectors as presented in Chap. 1.

At present, there are three main types of large space mesh reflector antennas, namely, AstroMesh antennas, modular antennas, and rigid-rib antennas, which will be discussed separately in this chapter. Since these types of antennas have the advantages of a large aperture, high precision, a light weight and a high furled rate [1], they have received extensive attention and application in the aerospace field.

To avoid confusion, two specific terms related to the mesh reflector are defined, mesh and support cable net. The size of the mesh openings is generally smaller than the operating wavelength. Since the mesh tension is small, the mesh configuration is mainly determined by the support cable net.

To properly design the support cable net and the corresponding support truss, it is necessary to first determine the support cable net nodes and preliminarily design their coordinate values. It is assumed that the operating frequency f, wavelength λ, focal length F and the allowable approximation error of the antenna are given. The reason it is called the approximation error is that there are other types of errors, such as manufacturing and assembly errors.

There are two ways to perform mesh generation for the support cable net. One is to mesh the paraboloid surface directly and the other is to find the best surrogate surface of the paraboloid reflector and mesh the surrogate surface to obtain the mesh of the paraboloid reflector.

Because the curvature of the parabolic surface along the generatrix is different from point to point, the surface is inconvenient for mesh generation, so the second method is often used. Clearly, the key to solving the problem is to find a suitable surrogate surface. The idea is that the surrogate surface can not only minimize the error of paraboloid approximation but also make it easy to mesh.

Fortunately, an approximate surface to meet this requirement is found, namely, the sphere surface (see Fig. 3.6a). The sphere surface coincides with the paraboloid at the vertex and opening points of the paraboloid. The radius of the sphere surface is

$$R = 2F + \frac{D^2}{32F} \tag{3.1}$$

When the focal length F is relatively large, if the focal length ratio $F/D \in (0.5, 1.0)$, the second term in the above equation takes a value in ($D/32$, $D/16$). Since the spaceborne antenna generally operates in an orbit of several hundred kilometers or even tens of thousands of kilometers away from the Earth, the focal length is relatively large. In this case, the second item is very small compared with the first item, so that it can be ignored in the initial design of the support cable net.

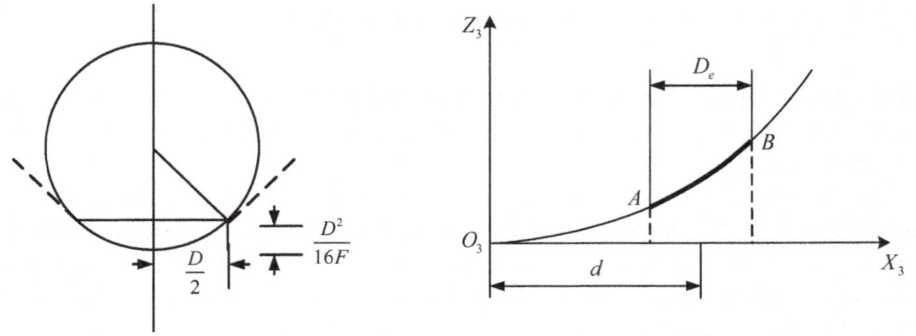

(a) Approximation of spherical surface to paraboloid

(b) Offset reflector antenna

Fig. 3.6 Approximation sphere and offset reflector of parabolic reflector

The paraboloid can be mathematically described in $O_3X_3Y_3Z_3$ as

$$X_3^2 + Y_3^2 = 4FZ_3 \tag{3.2}$$

The length of the offset parabolic generatrix *AB* is S_l in Fig. 3.6b, and it can be expressed as,

$$\begin{aligned} S_l &= \int_{d-D_e/2}^{d+D_e/2} \sqrt{1 + (\frac{dz_3}{dx_3})^2}dx = \frac{1}{2F}\int_{d-D_e/2}^{d+D_e/2} \sqrt{(2F)^2 + (x_3)^2}dx \\ &= \frac{1}{2F}\left[\frac{x_3}{2}\sqrt{(2F)^2 + x_3^2} + \frac{(2F)^2}{2}\ln(x + \sqrt{(2F)^2 + x_3^2})\right]_{d-D_e/2}^{d+D_e/2} \\ &= \frac{1}{4F}(D_B D'_B - D_A D'_A) + F\ln\frac{D_B + D'_B}{D_A + D'_A} \end{aligned} \tag{3.3}$$

where $D_A = d - \frac{D_e}{2}$; $D'_A = \sqrt{4F^2 + D_A^2}$; $D_B = d + \frac{D_e}{2}$; $D'_B = \sqrt{4F^2 + D_B^2}$.

If the antenna is a prime-feed antenna, we just set the offset distance $d = 0$ in formula (3.3).

The spherical reflector can be approximated by a series of planar polygons (see Fig. 3.7). To approximate the spherical reflector as well as possible, the length of the polygons should be short, which in turn leads to considerable computational work and a complex structure. Therefore, the problem is summarized in terms of the quantitative relationship between the length of the polygons l and the spherical approximation error. The determination of length l can be considered in the following situations.

1. Without considering pillow distortion of the support cable net

For the given approximation error ε_{rms} (the area between the straight line segment and the corresponding arc) with the regular triangle in Fig. 3.7a, the side length l and approximation error ε_{rms} will obey the following relationship [2]

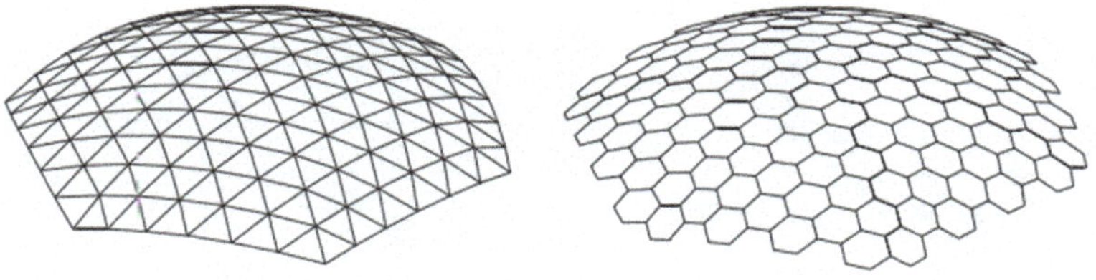

(a) Plane triangle (b) Plane hexagon

Fig. 3.7 Using a series of planes to approximate the spherical reflector surface

$$\varepsilon_{rms} = \frac{l^2}{8\sqrt{15}R} \tag{3.4}$$

Dividing Eq. (3.4) by D and ignoring the second term on the right hand side of Eq. (3.1) gives

$$\frac{\varepsilon_{rms}}{D} = \frac{1}{16\sqrt{15}} \frac{(l/D)^2}{F/D} = 0.01614 \frac{(l/D)^2}{F/D} \tag{3.5}$$

The side length l can be directly expressed as a function of the approximation error. Equation (3.5) can be simply expressed as

$$\frac{l}{D} = \alpha \sqrt{\frac{F}{D} \frac{\varepsilon_{rms}}{D}} \tag{3.6}$$

in which the coefficient α is

$$\alpha = \begin{cases} 7.872 & \text{for a regular triangle} \\ 6.160 & \text{for a regular quadrilateral} \\ 4.046 & \text{for a regular hexagon} \end{cases} \tag{3.7}$$

Now the question is which situation is best. The electronic performance was analyzed in [1]. If the aperture and focal length of the antenna are taken as 660 m and 533 m, respectively, the operating frequency is 1 GHz (wavelength 30 cm) and the allowable surface error is 12 mm ($\lambda/25$), the corresponding electronic performance (such as the far-field pattern) shows that the triangle is in the middle and upper levels, the quadrilateral is the worst one, and the hexagon is the best case. At the same time, it is observed that the far field pattern of the triangle is closer to that of the hexagon. As a result, the triangles are commonly used in engineering because they are much simpler than hexagons.

2. Considering pillow distortion of the support cable net

Considering the pillow distortion of the support cable net can lead us to a slight reduction of the side length of the polygon. For a known support cable net thickness l and distribution tension p, the relationship between ε_{rms} and l is [3]

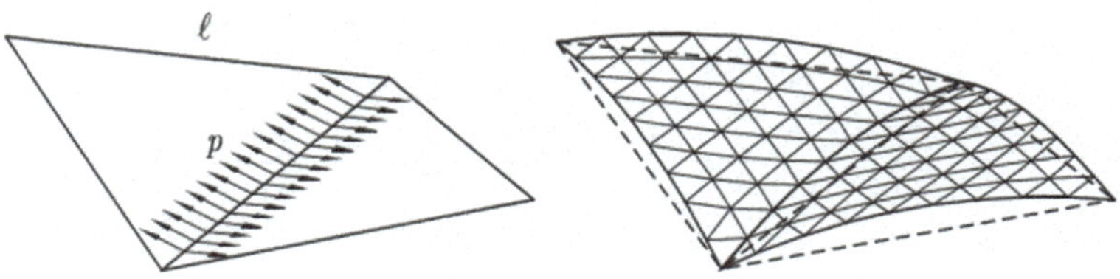

(a) The straight side of triangle (b) The "bending" phenomenon of the straight side of triangle

Fig. 3.8 Saddle characteristics of triangular patches on a codirectional surface

$$\frac{\varepsilon_{rms}}{D} = 0.01614\frac{(l/D)^2}{F/D}(1 + 0.33\frac{pl}{t}) \tag{3.8}$$

Equation (3.8) is obtained by expanding the error $\frac{\varepsilon_{rms}}{D}$ as a Taylor series with respect to the parameter $\frac{pl}{t}$ by retaining the first two terms only.

The so-called pillow distortion of the support cable net means that if there is no lateral tension for the isotropic uniform tension support net, when it is divided into triangular meshes, the tension p will be reversed at the collinear point of two triangles (Fig. 3.8a), which will lead to the rectangular bending of the triangles (Fig. 3.8b). Otherwise, their balance will be lost.

If the physical quantity $\frac{\varepsilon_{rms}}{D}$ is expanded as a Fourier (non-Taylor) series with respect to the parameter $\frac{pl}{t}$, we obtain [3]

$$\frac{\varepsilon_{rms}}{D} = 0.01614\frac{(l/D)^2}{F/D}\sqrt{1 + 0.660\frac{pl}{t} + 0.133(\frac{pl}{t})^2} \tag{3.9}$$

At first glance, Eq. (3.8) seems different from Eq. (3.9). It is however not true; both equations are based on the same condition that parameter $\frac{pl}{t}$ is small. This condition is in accordance with practical engineering. Since the parameter is a small quantity, it is advisable to expand the root part of Eq. (3.9) as a Taylor series, i.e.

$$\sqrt{1 + 0.660\frac{pl}{t} + 0.133(\frac{pl}{t})^2} = 1 + 0.33\frac{pl}{t} + O[(\frac{pl}{t})^2] \tag{3.10}$$

Clearly, if the small (second-order and above) parts of Eq. (3.10) are omitted and substituted into Eq. (3.9), the corresponding parts of Eqs. (3.9) and (3.8) are the same.

3. Determination of rib number of rigid-rib antenna

There are two cases here. One is that the mesh is supported directly with rigid ribs, and the other is that there is a cable net between the two rigid ribs, and the mesh is then laid on the support cable net.

For the second case, instead of connecting the mesh directly with the rigid ribs, a support cable net is designed between the two rigid ribs, which can obviously increase the stiffness of the whole structure. The mesh on the support cable net is used in the electronic performance.

For the first case, the relationship between the distance l_{ra} (the outermost ends of two adjacent ribs) and the error ε_{rms} becomes [2]

$$\frac{\varepsilon_{rms}}{D} = 0.01076\frac{(l_{ra}/D)^2}{F/D}(1 + \frac{p_1}{p_2}) \tag{3.11}$$

where p_1 and p_2 are the radial and circumferential tensions of the support cable net, respectively.

If $p_1 = p_2$ (isotropic mesh tension), Eq. (3.11) becomes [2]

$$\frac{l_{ra}}{D} = 6.817\sqrt{\frac{F}{D}\frac{\varepsilon_{rms}}{D}} \tag{3.12}$$

Supposing $D = 5$ m, $F = 2$ m and $\varepsilon_{rms}^{G+Sa} = 3$ mm in (3.12), we have $l_{ra} = 0.528$ m. This means that the total number of rigid ribs must be greater than 29.75 ($n_{rb} = \pi D/l_{ra}$), that is, at least 30 ribs, which corresponds to a support rigid rib every 12 degrees. This will inevitably increase the areal density of the entire antenna and is generally unacceptable.

As far as the second case is concerned, the mesh is laid on the support cable net between adjacent radial ribs (see Fig. 3.9).

In Fig. 3.9, the black lines indicate two adjacent radial ribs, the red and green lines are the upper and lower support cable net between the two adjacent radial ribs, the blue lines are the vertical adjustment cables, and the mesh is indicated in yellow. The rigid rib structure corresponding to Fig. 3.9d may be a truss (a), a perforated stiffened plate (b) or another structural form as shown in Fig. 3.10.

It should be noted that regardless of the type of mesh reflector antenna it is, the coordinates of the surface nodes can be determined from the side length l of the equilateral triangle (derived from Eqs. (3.6) or (3.8)) and the curve length S_l of parabola AB. The total number of the support cable nets to be divided along $S_l/2$ is (where Int means the integer function)

$$n_{rbi} = Int(\frac{S_l}{2l}) \tag{3.13a}$$

For the total number of rigid ribs n_{rb}, if the distance between the outermost ends of the two adjacent ribs is taken as $\bar{\zeta}$ times the side length l of the triangle, we obtain

$$n_{rb} = Int(\frac{\pi D}{\bar{\zeta} l}) \tag{3.13b}$$

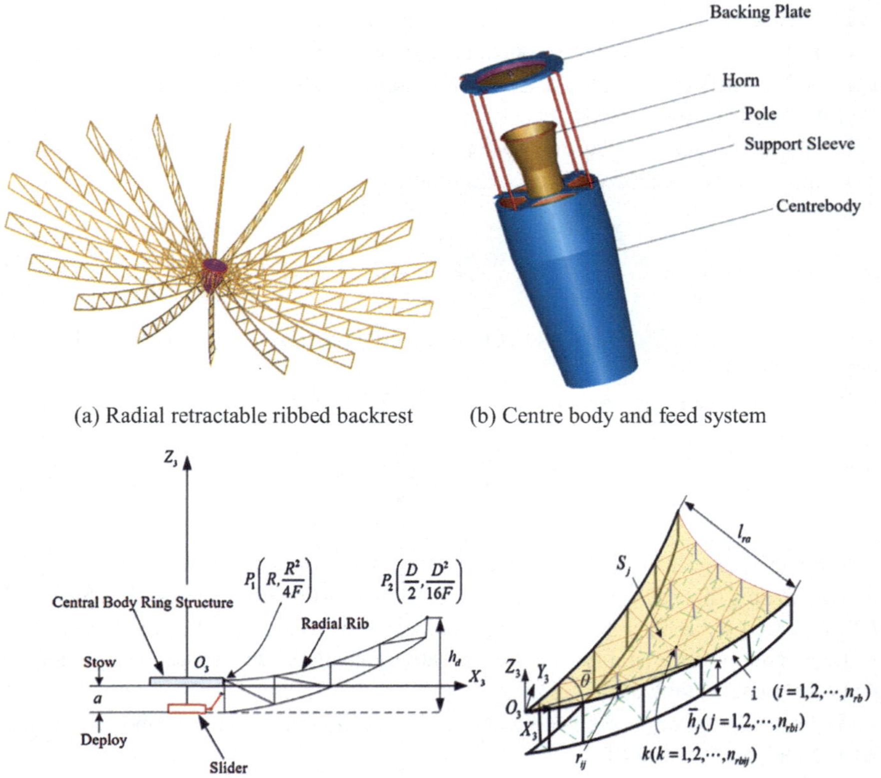

(a) Radial retractable ribbed backrest
(b) Centre body and feed system
(c) Radial rib deployment mechanism
(d) Support net structure

Fig. 3.9 Rigid rib antenna structure

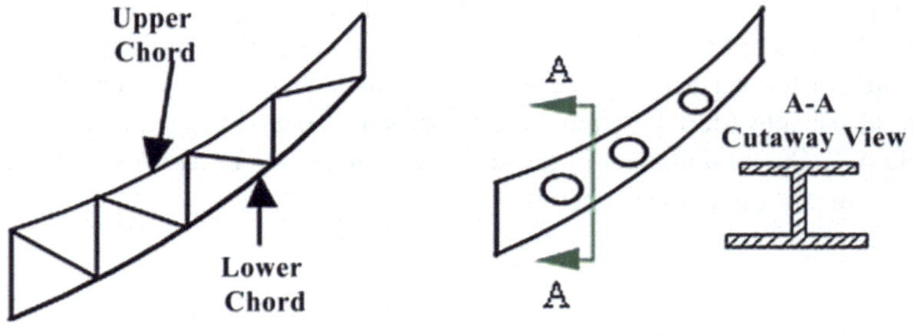

(a) Truss structure
(b) Rib plate structure

Fig. 3.10 Two typical basic forms of a single radial rib

Initially, $\bar{\zeta} = 3 - 5$ can be used, and then it can be determined from the detailed design of the support cable net and rigid ribs.

The surface approximation error ε_{rms} mentioned above is related to the operating wavelength λ of the antenna. The conceptual design can be based on the Ruze formula and is initially taken as one-thirtieth to one-sixtieth of the wavelength.

3.2 Reflective Surface and Support Structure Design

As mentioned above, large space reflector antennas mainly include three types: AstroMesh antenna, modular antenna and rigid-rib antenna. The design for these antennas mainly includes determining the support cable net and support truss, deployable status and stowed states. The detailed design methodologies are given below.

3.2.1 AstroMesh Antenna

The AstroMesh antenna has the advantages of a simple structure, high furled rate (small stowed radius for a large diameter in the deployed state), large aperture and low areal density [4, 5], but with the increasing antenna diameter, two problems are caused. First, the height of the antenna in the stowed state is raised, and it is difficult to fit into the rocket fairing. Second, with the increasing number of tension cables, it is difficult to adjust the accuracy of the surface. For these reasons, the design of the support cable net, the support truss and the connection between them are discussed below.

1. Basic composition of AstroMesh antenna

As shown in Fig. 3.11a, this type of antenna is offset in most cases; the feed is placed at the satellite, and the antenna is connected to the satellite with extension arms. The antenna itself mainly includes the following parts (Fig. 3.11b): a deployable truss consisting of expandable parallelograms, a front and rear support cable net, adjustable tension cables, and a wire mesh attached to the front support cable net. The support truss is connected with the front and rear cable net. To make the front or rear support cable net form the required shape of the reflective surface at the same time, the structural member size and cable tension can be optimized. It should not only meet the shape requirements of the support cable net, but also achieve the optimal tension distribution. The mesh connected to the front support cable net serves as the reflective surface for reflecting electromagnetic waves, and in practice the mesh is woven with soft alloy metal filaments.

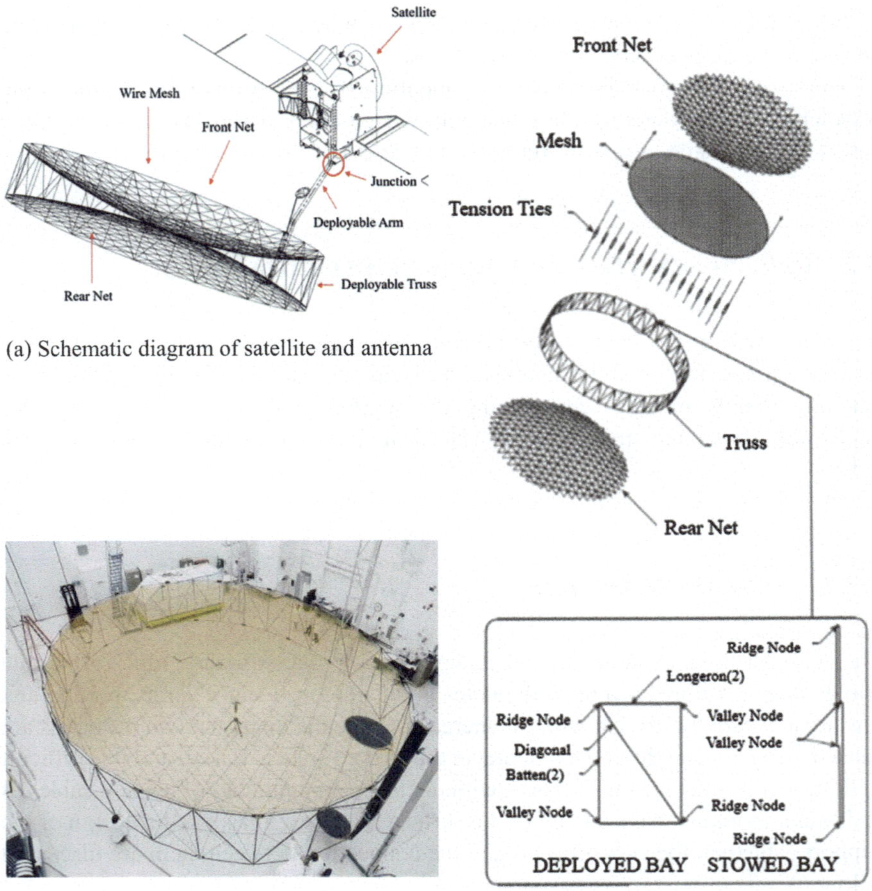

(a) Schematic diagram of satellite and antenna

(b) Flight hardware of the antenna

(c) Components of the antenna

Fig. 3.11 Large space deployable AstroMesh reflector antenna

2. Design of deployable AstroMesh antenna structures

(1) Cable net

There are three types of supporting cable nets for the AstroMesh antenna, including radial cable net, three-directional cable net and quasi-geodesic cable net. The mechanical properties of the three-directional cable net are the most reasonable among these. Therefore, the structure of the three-directional cable net is the focus of the discussion below. According to the different connection types between three-directional cable nets and their support truss structures, they can be further divided into six-ring three-directional cable net, six-ring three-directional cable net with multiple hanging nodes, and quasi-geodesic six-ring three-directional cable net (Fig. 3.12).

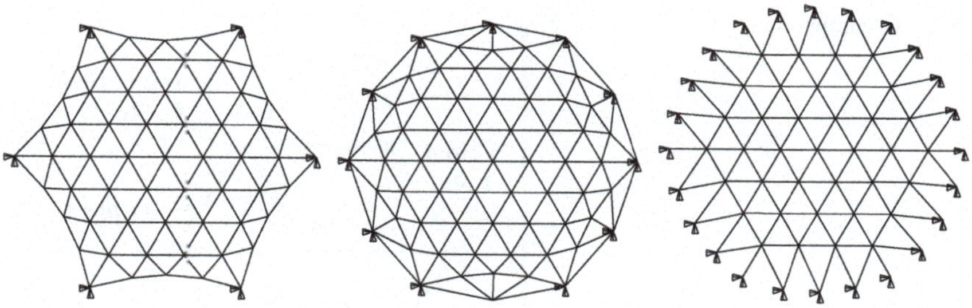

(a) Six-ring tension net (b) Six-ring tension net with multiple hanging nodes (c) Six-ring quasi-geodesic net

Fig. 3.12 Types of triangular cable net

The design of the support cable net has the goal of generating the cable net corresponding to the working paraboloid to meet the requirements of the antenna operating frequency. The working paraboloid is formed by cutting the parent paraboloid with a cylinder (Fig. 3.3). Since the curvature of the paraboloid along the z-axis is not equal everywhere, triangular meshes are usually employed first to divide the horizontal projection plane of the paraboloid, and then triangular meshes are generated by function mapping. This method has difficulty ensuring the accuracy of the generated triangular mesh and even causes singularities, making the support cable net tension analysis and design accuracy difficult to guarantee. For this reason, we can first find a sphere closest to the paraboloid and then project the equilateral triangular mesh of the sphere onto the paraboloid. At the same time, because the aperture surface of the working paraboloid of the offset antenna is elliptical and the aperture of the sphere is circular, they do not match on the aperture surface, so it is necessary to further project the meshed sphere to the ellipsoid to make the final mesh consistent with the shape of the aperture surface of the actual working paraboloid.

To achieve this goal, the basic idea is to first define a spherical surface that approximates the working paraboloid and use the spherical surface as the reference spherical surface for mesh division. Second, we perform triangular division of the reference spherical surface to obtain mesh nodes and connection relationships. Then, we map the nodes on the reference spherical surface to the reference ellipsoid and then project the nodes on the ellipsoid surface radially onto the working paraboloid. In the process of node mapping and projection, the topological relationship of the element connections remains unchanged. Third, the coordinates of the rear cable net nodes are calculated, and an axis-symmetrical paraboloid is usually employed for the rear cable net. It is assumed that the X and Y coordinates of the corresponding nodes of the working paraboloid and the rear cable net are the same in the initial design. The details of the mesh generation process are given in Sect. 3.2.1, Point 3.

(2) Support perimeter truss

The perimeter truss is the support structure of the antenna, which plays a vital role in antenna performance such as deployment performance (stationarity and reliability) and surface accuracy. The basic design idea is as follows. First, the side length of

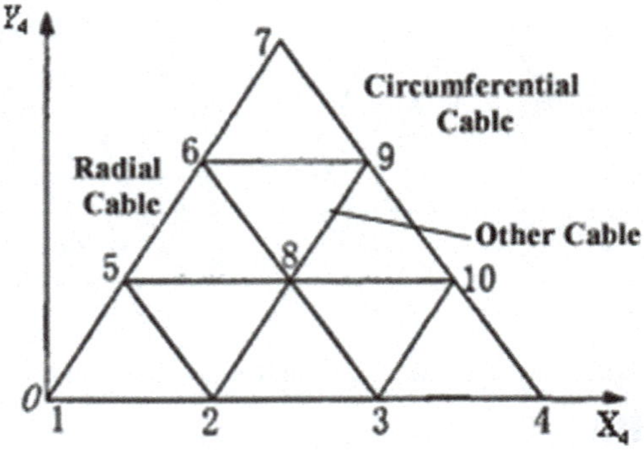

Fig. 3.13 Grouping of cable types

each short side of a quadrilateral in the perimeter truss is determined according to the antenna aperture requirement. Then, the heights h_1 and h_2 of the front and rear support cable net and the spacing Δh between the bottom points of the two supporting cable nets are calculated from geometric relationships. After that, the length of the longeron, batten and diagonal of the quadrilateral truss are determined. Finally, the height of the corresponding stowed state is calculated; refer to Point 5 of Sect. 3.2.1.

3. Design of support cable net of the offset reflector antenna

(1) Determination of cables in three-directional cable net

The cables of the equilateral triangle cable net structure can be divided into radial cable, circumferential cable, other cable connecting the inner and outer circumferential cables (Fig. 3.13), and vertical tension cables between the front and rear support cable net [6].

(2) Shape of working paraboloid of offset antenna

The geometric design of support cable nets has been explored by scholars [7, 8]. The following is discussed in detail in connection with Fig. 3.2. For ease of discussion, an offset reflector is shown in Fig. 3.14. The diameter and focal length of the parent paraboloid are D_p and F_p, respectively.

The intersection points between the cylindrical surface and parent paraboloid in the X_3Z_3 plane are P_r and P_s. The intersection point between chord P_rP_s and the

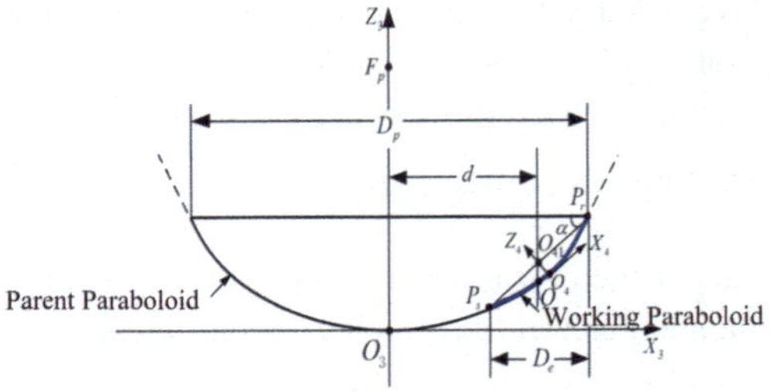

Fig. 3.14 Projection of working paraboloid in X_3Z_3 plane

Z_4 axis is O_{41}, and the radius of the cylinder is R_c, then the electronic aperture of the working paraboloid $D_e = 2R_c$ and the physical aperture $D = P_r P_s$. In the coordinate system $O_3X_3Y_3Z_3$, the coordinates $(x_{3p_r}, 0, z_{3p_r})$ and $(x_{3p_s}, 0, z_{3p_s})$ of points P_r and P_s can be found as follows:

$$\begin{cases} x_{3p_s} = d - R_c, x_{3p_r} = d + R_c \\ z_{3p_r} = \frac{x_{3pr}^2}{4F_p}, z_{3p_s} = \frac{x_{3ps}^2}{4F_p} \end{cases} \tag{3.14}$$

According to the geometric relationship, we have

$$\tan\alpha = \frac{z_{3p_r} - z_{3p_s}}{x_{3p_r} - x_{3p_s}} \tag{3.15}$$

$$\tan\alpha = \frac{\sqrt{D^2 - D_e^2}}{D_e} \tag{3.16}$$

The relationship between the electronic and physical apertures of the antenna can be be obtained by Eqs. (3.14)–(3.16) as

$$D_e = \frac{2F_p D}{\sqrt{4F_p^2 + d^2}} \tag{3.17}$$

Assuming that the height of the parent paraboloid is h_g, the distance from point P_s to the X_3 axis is e_z, the distance from the lowest point O_4 to the line $P_r P_s$ is h_1, and the rotation matrix of $O_3X_3Y_3Z_3$ with a counterclockwise rotation of α around Y_3 is $R_{y_4}(\alpha) = \begin{pmatrix} \cos\alpha & 0 & -\sin\alpha \\ 0 & 1 & 0 \\ \sin\alpha & 0 & \cos\alpha \end{pmatrix}$, then the coordinates of point O_{41} of the working paraboloid aperture become

$$\begin{Bmatrix} x_{3o_{41}} \\ y_{3o_{41}} \\ z_{3o_{41}} \end{Bmatrix} = \begin{Bmatrix} d \\ 0 \\ \frac{1}{2}(e_z + h_g) \end{Bmatrix} \tag{3.18}$$

where e_Z and h_g are respectively,

$$e_z = \frac{(d - R_c)^2}{4F_p} \tag{3.19}$$

$$h_g = \frac{D_p^2}{16F_p} \tag{3.20}$$

$$R_c = \frac{1}{2}D_e = \frac{F_p D}{\sqrt{4F^2 + d^2}} \tag{3.21}$$

$$D_p = 2(d + R_c) = 2d + \frac{2F_p D}{\sqrt{4F_p^2 + d^2}} \tag{3.22}$$

In general, if the coordinates of a given point are (x_3, y_3, z_3) in $O_3X_3Y_3Z_3$ and (x_4, y_4, z_4) in $O_4X_4Y_4Z_4$, the transformation from $O_4X_4Y_4Z_4$ to $O_3X_3Y_3Z_3$ will be

$$\begin{Bmatrix} x_3 \\ y_3 \\ z_3 \end{Bmatrix} = R_{y_4}(\alpha)\begin{Bmatrix} x_4 \\ y_4 \\ z_4 \end{Bmatrix} + \begin{Bmatrix} x_{3o_4} \\ y_{3o_4} \\ z_{3o_4} \end{Bmatrix} \tag{3.23}$$

where

$$\begin{Bmatrix} x_{3o_4} \\ y_{3o_4} \\ z_{3o_4} \end{Bmatrix} = \begin{Bmatrix} \frac{D}{2}\cdot\cos\alpha + \frac{b_1+\sqrt{b_1^2-4a_1\cdot c_1}}{2a_1}\cdot\sin\alpha + d \\ 0 \\ \frac{D}{2}\cdot\sin\alpha - \frac{b_1+\sqrt{b_1^2-4a_1\cdot c_1}}{2a_1}\cdot\cos\alpha + \frac{1}{2}(e_z + h_g) \end{Bmatrix}$$

Therefore, it can be seen from Eqs. (3.2) and (3.23) that the working paraboloid in the local coordinate system $O_4X_4Y_4Z_4$ will satisfy

$$x_4\sin\alpha + z_4\cos\alpha + z_{3o_4} = \frac{1}{4F_p}\left[\left(x_4\cos\alpha - z_4\sin\alpha + x_{3o_4}\right)^2 + y_4^2\right] \tag{3.24}$$

This shows that the aperture surface of the working paraboloid is an ellipse in plane $O_4X_4Y_4$ as

$$\frac{x_4^2}{a_e'^2} + \frac{y_4^2}{b_e'^2} = 1 \tag{3.25}$$

where $a_e' = \frac{D_e}{2\cos\alpha}$, $b_e' = \frac{D_e}{2}$.

In particular, if $d = 0$, we have $D_e = D$, $\alpha = 0$, and $a_e' = b_e' = D_e/2$, and the paraboloid becomes a circle. If $x_4 = D/2$ and $y_4 = 0$, the height of the working paraboloid becomes

$$h_1 = \frac{F_p D^2}{16F_p^2 + 4d^2}\cos\alpha \tag{3.26}$$

(3) Nodal coordinate generation on the reference sphere surface

As shown in Fig. 3.15, the connected points of the reference sphere and the working paraboloid are points P_r, P_s and O_4, and the equation of the reference sphere is

$$x_4^2 + y_4^2 + (z_4 - R_s)^2 = R_s^2 \tag{3.27}$$

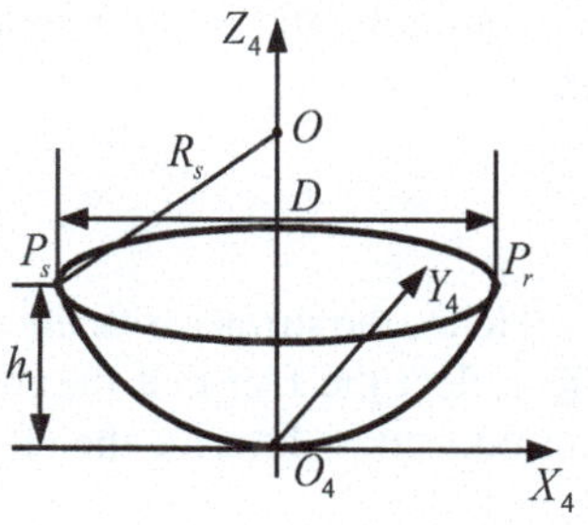

Fig. 3.15 Reference sphere

where $R_s = 2F_p$.

(4) Division of reference sphere

The mesh of the reference surface is built upon the rings and subdivisions. The subdivision lines are geodesic curves passing through the vertex O_4 and all the subdivisions are axis-symmetric with respect to the z_4 axis, as shown in Fig. 3.16.

Let n denote the total number of subdivisions, which is commonly chosen as 4, 6 or 8. Each subdivision is further divided by rings as shown in Fig. 3.17. The portion of each ring inside every subdivision has two intersection nodes with two subdivision lines and is a geodesic line of the reference surface passing through those intersections. Since the facet size is bounded by distances between each pair of neighboring rings, the facet edges with the ends located at two neighboring intersection nodes on a subdivision line must be shorter than the allowable facet length (Eqs. (3.6) or (3.12)). Thus, from Eq. (3.13a), the minimum number of rings obeys

$$\bar{n}_r = n \tag{3.28}$$

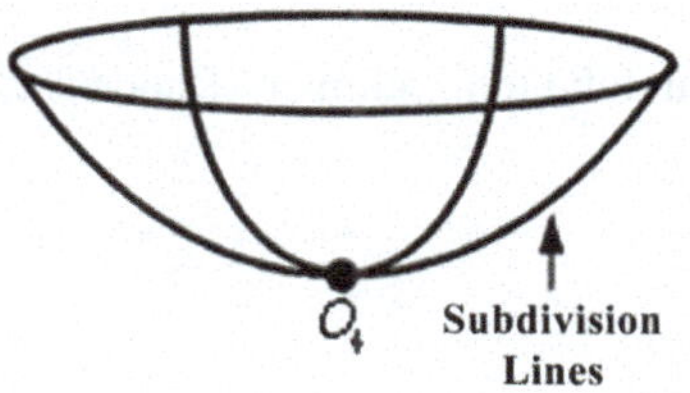

Fig. 3.16 The subdivision line on the reference sphere

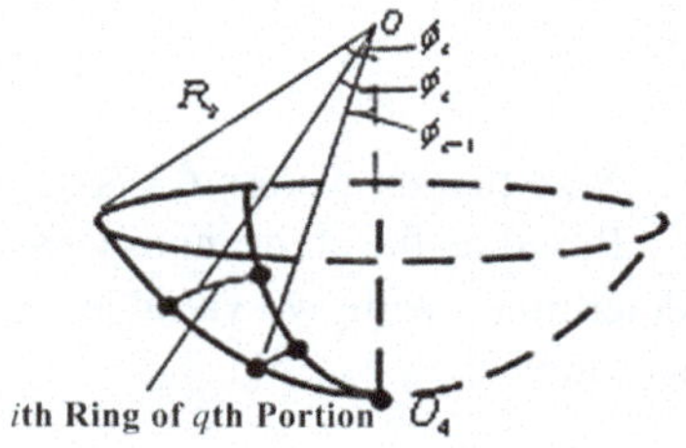

Fig. 3.17 The rings and subdivisions

Naturally, the actual number of rings n_r of a sphere should not be greater than $\bar{n}_r$, i.e.,

$$n_r \le \bar{n}_r \tag{3.29}$$

Hence the surface is divided into n_r layers from the center to the outside boundary. To reflect the fact that the thickness of each layer is different, the weight $w(i)$ is introduced to describe the relative thickness of each layer.

$$w(i) = \begin{cases} 1 & i = 1, 2, \ldots, n_r - 1 \\ w_B & i = n_r \end{cases} \tag{3.30}$$

The radian corresponding to the ith layer arc is

$$\phi(i) = \frac{\phi_t}{\sum\limits_{i=1}^{n_r} w(i)} w(i) + \phi(i-1) \begin{cases} i = 1, 2, \ldots, n_r \\ \phi(0) = 0 \end{cases} \tag{3.31}$$

where $\phi_t = \arcsin(D/(2R_s))$.

(5) Nodal number and coordinate determination

Within each subdivision, the nodes including the two ending nodes at the ith ring are the nodes that divide the ith ring equally. The corresponding chords or facet edges are also equal. Due to the symmetry of subdivisions, all the facet edges with vertices on the same ring have the same length and all the nodes of the surface are on geodesics. In each ring the total number n_i of nodes is

$$n_i = i n_s \ (i = 1, 2, \ldots, n_r) \tag{3.32}$$

and the total number of nodes n_{total} of the reflector is

$$n_{total} = 1 + \sum_{i=1}^{n_r} i n_s = \frac{1 + n_r}{2} n_r n_s + 1 \tag{3.33}$$

Then the jth node on the ith ring can be represented as

$$N_{i,j} = \{i, j\} \tag{3.34}$$

Note that the vertex O_4 is defined as $N_{0,1}$.

Based on the above number system, the coordinates of nodes on geodesics can be determined with two variables $\theta_{i,j}$ and $\alpha_{i,j}$, according to the geometry in Fig. 3.18, as follows:

$$\begin{cases} \bar{x}_{4,(i,j)} = R_s \sin\theta_{i,j} \cos\alpha_{i,j} \\ \bar{y}_{4,(i,j)} = R_s \sin\theta_{i,j} \sin\alpha_{i,j} \\ \bar{z}_{4,(i,j)} = R_s - R_s \cos\theta_{i,j} \end{cases}, \quad (i = 1, 2, \ldots, n_r - 1; \; j = 1, 2, \ldots, n_i) \tag{3.35}$$

For $i = 1$, $j = 1$,

$$\left(\bar{x}_{4,(0,1)}, \bar{y}_{4,(0,1)}, \bar{z}_{4,(0,1)}\right) = (0, 0, 0) \tag{3.36}$$

For $i = n_r$,

$$\begin{cases} \bar{x}_{4,(n_r,j)} = \frac{D}{2} \cos\alpha_{n_r,j} \\ \bar{y}_{4,(n_r,j)} = \frac{D}{2} \sin\alpha_{n_r,j} \\ \bar{z}_{4,(n_r,j)} = h_1 \end{cases}, \quad (j = 1, 2, \ldots, n_r n_s) \tag{3.37}$$

where

$$\alpha_{n_r,j} = (j-1)\frac{2\pi}{n_r n_s} \tag{3.38}$$

To obtain variables $\theta_{i,j}$ and $\alpha_{i,j}$, two different cases should be considered, that is, whether or not the nodes are located on the subdivision lines.

1) The nodes on the subdivision lines

The nodes on subdivision lines satisfy the following equation

$$\bar{j} = 1 + (q-1)i \tag{3.39}$$

where $q = 1, 2, \ldots, n_s$ and $i = 1, 2, \ldots, n_r - 1$.

For convenience of discussion, the nodes on subdivision lines are denoted as $N_{i,\bar{j}}$. Figures 3.18 and 3.19 show that

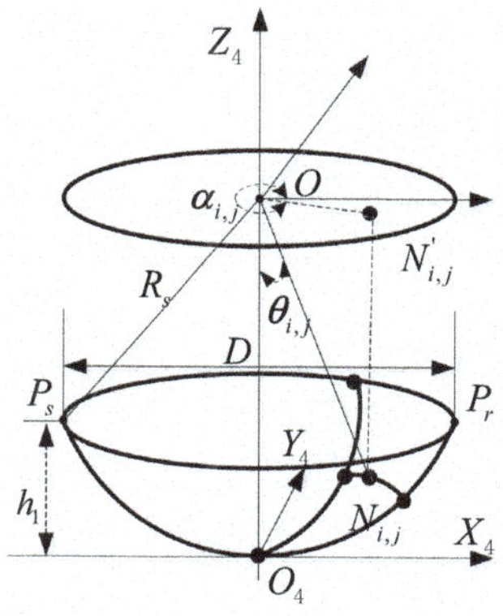

Fig. 3.18 Nodes in reference spherical subdivision

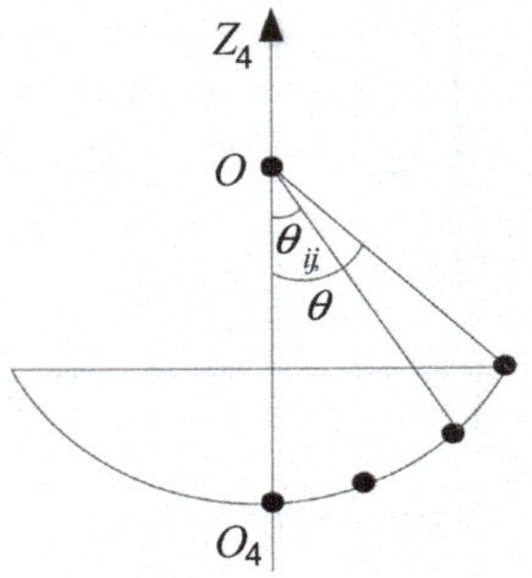

Fig. 3.19 Nodes on the subdivision line

$$\theta_{i,\bar{j}} = \phi(i) \tag{3.40}$$

$$\alpha_{i,\bar{j}} = (q-1)\frac{2\pi}{n_s} \tag{3.41}$$

2) Nodes not on the subdivision lines

For this case, $\theta_{i,\bar{j}}$ can still be determined by Eq. (3.40). The question is how to determine $\alpha_{i,\bar{j}}$. Because of the symmetry of the subdivision, just considering one subdivision, called the ith one, is fine. Projecting the ith ring of the subdivision onto the X_4Y_4 plane which passes through the origin O of the reference sphere yields Fig. 3.20. According to the previous discussion, there are $i + 1$ nodes in a portion of this ring and the nodes at the ends of the portion are on the subdivision lines. Denote one end node of the horizontal line as $N_{i,\bar{j}}$ and the other node as $N_{i,\bar{j}+i}$.

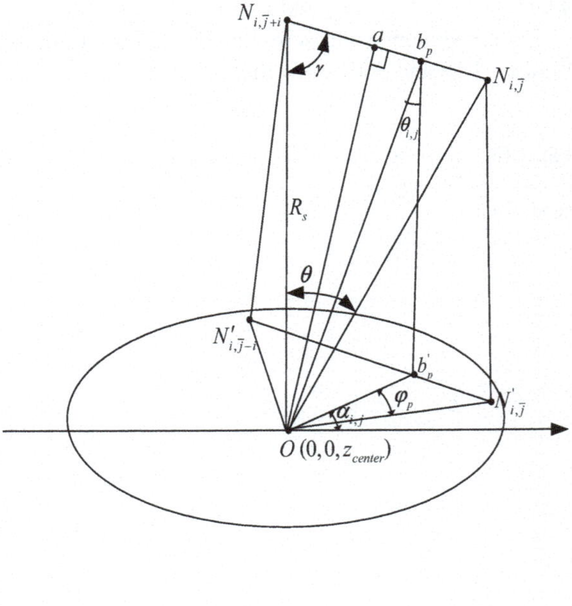

Fig. 3.20 Nodes in a subdivision

By the symmetry of subdivision lines, the two end nodes are at the same height and the chord $\overline{N_{i,\bar{j}}N_{i,\bar{j}+i}}$ is parallel to the X_4Y_4 plane. The nodes $N_{i,\bar{j}}$ and $N_{i,\bar{j}+i}$ together with their projected nodes $N'_{i,\bar{j}}$ and $N'_{i,\bar{j}+i}$ form a rectangle.

Figure 3.20 shows that

$$\overline{ON'_{i,\bar{j}}} = \sqrt{\bar{x}^2_{4,(i,\bar{j})} + \bar{y}^2_{4,(i,\bar{j})}} \tag{3.42}$$

$$\overline{N_{i,\bar{j}}N_{i,\bar{j}+i}} = \overline{N'_{i,\bar{j}}N'_{i,\bar{j}+i}} = \sqrt{\left(\bar{x}_{4,(i,\bar{j})} - \bar{x}_{4,(i,\bar{j}+i)}\right)^2 + \left(\bar{y}_{4,(i,\bar{j})} - \bar{y}_{4,(i,\bar{j}+i)}\right)^2} \tag{3.43}$$

Since a is the midpoint of edge $N_{i,\bar{j}}N_{i,\bar{j}+i}$ of the isosceles triangle $ON_{i,\bar{j}}N_{i,\bar{j}+i}$, we obtain

$$\cos\gamma = \frac{\overline{N_{i,\bar{j}}N_{i,\bar{j}+i}}}{2R_s} \tag{3.44}$$

$$\overline{Oa} = R_s \sin\gamma \tag{3.45}$$

$$\beta = \pi - 2\gamma \tag{3.46}$$

The angle (acute angle) between segments Ob_p and $ON_{i,\bar{j}}$ is

$$\beta_p = p\beta/i, \ (p = 1, 2, \ldots, i-1) \tag{3.47}$$

The length of the line segment Ob_p in the right triangle Oab_p is

$$\overline{Ob_p} = \frac{\overline{Oa}}{\sin(\gamma + \beta_p)} \tag{3.48}$$

Letting b'_p be the projection of b_p on plane OX_4Y_4, we have

$$\overline{b'_pN'_{i,\bar{j}}} = \overline{b_pN_{i,\bar{j}}} = \frac{\overline{N_{i,\bar{j}}N_{i,\bar{j}+i}}}{2} - \overline{Ob_p}\cos(\gamma + \beta_p) \tag{3.49}$$

$$\overline{b_pb'_p} = \overline{N_{i,\bar{j}}N'_{i,\bar{j}}} = R_s - z_{4,(i,\bar{j})} \tag{3.50}$$

In the right triangle $Ob_{p'}b_p$

$$\cos\theta_{i,\bar{j}+p} = \frac{\overline{b_pb'_p}}{\overline{Ob_p}} \tag{3.51}$$

$$\overline{Ob'_p} = \overline{Ob_p}\sin\theta_{i,\bar{j}+p} \tag{3.52}$$

Since

$$\cos\varphi_p = \frac{\left(\overline{Ob'_p}\right)^2 + \left(\overline{ON'_{i,\bar{j}}}\right)^2 - \left(\overline{b'_p N'_{i,\bar{j}}}\right)^2}{2\overline{Ob'_p}\,\overline{ON'_{i,\bar{j}}}} \tag{3.53}$$

the angle of the node on the ith ring is

$$\alpha_{i,\bar{j}+p} = \varphi_p + (q-1)\frac{2\pi}{n_s} \quad (p = 0, 1, \cdots, i-1) \tag{3.54}$$

(6) Boundary nodes

Although all nodes are located on geodesics, the curvatures of the geodesic rings are different from the curvature of the outside aperture, which commonly results in a larger surface RMS error in the outer region of the reference sphere surface than that in the inner region. To make the inner and outer surface RMS errors of the reference sphere as consistent as possible, the following two aspects need to be considered.

1) Adjust the curvatures of the rings

For rings close to the aperture which are defined as converging rings, their curvatures gradually increase from inside to outside and converge to the aperture curvature. Set the largest geodesic ring as the $(n_c - 1)$th ring. Set all the outer rings except for the boundary ring as the converging rings. On the n_cth ring which is the first converging ring from the inside, the coordinates of the middle point on the geodesic arc are

$$\begin{cases} \bar{x}_{4,n_c} = R_s \sin\phi(n_c)\cos\alpha_{n_c,n_c/2+1} \\ \bar{y}_{4,n_c} = R_s \sin\phi(n_c)\sin\alpha_{n_c,n_c/2+1} \end{cases} \tag{3.55}$$

For the case of $i = n_c$, $q = 1$ and $p = n_c/2$, the parameters $\alpha_{n_c,n_c/2+1}$ and $\varphi_{n_c/2}$ can be obtained from Eqs. (3.40)–(3.55). The distance between the point $\left(\bar{x}_{n_c}, \bar{y}_{n_c}\right)$ and line $N'_{n_c,1}N'_{n_c,n_c+1}$ is

$$d_{n_c} = \frac{\left|\varsigma_1 \bar{x}_{4,n_c} - \bar{y}_{4,n_c} + \varsigma_2\right|}{\sqrt{\varsigma_1^2 + 1}} \tag{3.56}$$

$$\varsigma_1 = \frac{\bar{y}_{4,(n_c,n_c+1)} - \bar{y}_{4,(n_c,1)}}{\bar{x}_{4,(n_c,n_c+1)} - \bar{x}_{4,(n_c,1)}}, \quad \varsigma_2 = \bar{y}_{4,(n_c,1)} - \varsigma_1 \bar{x}_{4,(n_c,1)} \tag{3.57}$$

Denote the curvature of the n_cth ring in the XY plane as $1/r_{n_c}$ where the radius r_{n_c} satisfies

$$\left(\frac{\overline{N'_{n_c,1}N'_{n_c,n_c+1}}}{2}\right)^2 + \left(r_{n_c} - h_{n_c}\right)^2 = r_{n_c}^2 \tag{3.58}$$

and we have

$$r_{n_c} = \frac{\left(\frac{\overline{N'_{nc,1}N'_{nc,nc+1}}}{2}\right)^2 + h_{n_c}^2}{2h_{n_c}} \tag{3.59}$$

Suppose the curvatures in the X_4Y_4 plane are increased in terms of the power function, then the radius of each converging ring r_i in the X_4Y_4 plane will become

$$\frac{1}{r_i} = \left(\frac{1}{R_s \sin\phi(i)} - \frac{1}{r_{n_c}}\right)\left(\frac{i - n_c}{n_r - n_c - 1}\right)^{\rho} + \frac{1}{r_{n_c}} \begin{cases} i = n_c, n_c + 1, \ldots, n_r - 1 \\ 1 \le n_c < n_r - 1 \end{cases} \tag{3.60}$$

$$r_i = R_s \sin\phi(i),\ i = n_c,\ n_c = n_r - 1 \tag{3.61}$$

where ρ is a proper positive factor.

Combining Eqs. (3.35), (3.40) and (3.41) yields the coordinates $(\overline{x}_{4,(i,j)}, \overline{y}_{4,(i,j)}, \overline{z}_{4,(i,j)})$ of the nodes $N'_{i,j+p}$ on the convergence ring as

$$\begin{cases} \bar{x}_{4,(i,j)} = \overline{A'N'_{i,j}} \cos\alpha_{i,j} \\ \bar{y}_{4,(i,j)} = \overline{A'N'_{i,j}} \sin\alpha_{i,j} \\ \bar{z}_{4,(i,j)} = R_s - \sqrt{R_s^2 - \bar{x}_{4,(i,j)}^2 - \bar{y}_{4,(i,j)}^2} \end{cases} \quad (i = n_c, n_c + 1, \ldots, n_r - 1;\ j = 1, 2, \ldots, in_s) \tag{3.62}$$

This converging function is especially important when the task is to design the effective surface (main working surface) of the diameter D_e and does not put requirements on the overall diameter including the boundary layer. In this case, the $(n_r - 1)$th circle will be the outer ring of the effective surface and its projection in the XY plane needs to converge to an exact circle, which can be achieved by the above converging procedure.

To calculate Eq. (3.62), we need to know the parameter α_{ij}, which is derived below.

As seen from Fig. 3.21, in the isosceles triangle $O_1N'_{i,\bar{j}+i}N'_{i,\bar{j}}$, point a' is the midpoint of line segment $N'_{i,\bar{j}+i}N'_{i,\bar{j}}$, so

$$\overline{a'N'_{i,\bar{j}}} = \frac{\overline{N'_{i,\bar{j}}N'_{i,\bar{j}+i}}}{2} \tag{3.63}$$

$$\sin\frac{\mu}{2} = \frac{\overline{a'N'_{i,\bar{j}}}}{r_i} \tag{3.64}$$

Point A' is the projection of the spherical vertex O_4 in the X_4Y_4 plane and the coordinates of point A' are (0, 0), so

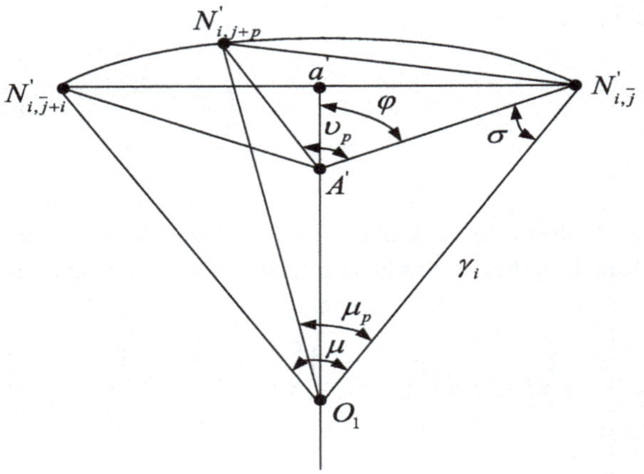

Fig. 3.21 Convergence ring points in the projection plane of XY

$$\overline{A'N'_{i,\bar{j}}} = \sqrt{\bar{x}^2_{4,(i,\bar{j})} + \bar{y}^2_{4,(i,\bar{j})}} \tag{3.65}$$

In the right triangle $A'a'N'_{i,\bar{j}}$, we have

$$\sin\varphi = \frac{\overline{a'N'_{i,\bar{j}}}}{A'N'_{i,\bar{j}}} \tag{3.66}$$

The angle corresponding to point $N'_{i,\bar{j}}$ is

$$\mu_p = p\frac{\mu}{i} \tag{3.67}$$

In triangle $O_1N'_{i,\bar{j}}N'_{i,\bar{j}+p}$, the well-known triangle formula yields

$$\overline{N'_{i,\bar{j}}N'_{i,\bar{j}+p}} = \sqrt{2r_i^2 - 2r_i^2\cos\mu_p} \tag{3.68}$$

Similarly, according to the relationship between the inner and outer angles of a triangle, we have

$$\sigma = \varphi - \frac{\mu}{2} \tag{3.69}$$

In the triangle, it can be obtained by the cosine theorem

$$\overline{A'N'_{i,\bar{j}+p}} = \sqrt{\left(\overline{N'_{i,\bar{j}}N'_{i,\bar{j}+p}}\right)^2 + \left(\overline{A'N'_{i,\bar{j}}}\right)^2 - 2\overline{N'_{i,\bar{j}}N'_{i,\bar{j}+p}}\,\overline{A'N'_{i,\bar{j}}}\cos\left(\frac{\pi-\mu_p}{2} - \sigma\right)} \tag{3.70}$$

$$\cos \upsilon_p = \frac{\left(\overline{A'N'_{i,\bar{j}+p}}\right)^2 + \left(\overline{A'N'_{i,\bar{j}}}\right)^2 - \left(\overline{N'_{i,\bar{j}}N'_{i,\bar{j}+p}}\right)^2}{2\overline{A'N'_{i,\bar{j}+p}}\,\overline{A'N'_{i,\bar{j}}}} \tag{3.71}$$

Therefore

$$\alpha_{i,\bar{j}+p} = \upsilon_p + (q-1)\frac{2\pi}{n_s} \quad (p = 0, 1, \cdots, i-1) \tag{3.72}$$

2) Improve the non-uniform of the facets

Without the converging rings, all nodes will be exactly on the geodesics of the reference sphere, and all the facets will be quite uniform. Unfortunately, if the converging rings are introduced, the nodal positions will be slightly changed and facets in the outer layer will tend to possess a larger size than those in the inner layer. Compensation to reject the increasing trend in facet length is necessary and therefore the following treatment is presented. Defining the ratio of the maximum facet length in the $(n_r - 1)$th layer to the maximum facet length in the first layer as ς, another power function is defined to describe the converging thickness weight of each layer and the weight function (3.30) is transformed into Eq. (3.73). The two power functions in Eqs. (3.60) and (3.73) share the same exponent ρ.

$$w(i) = \begin{cases} 1 + (1-\varsigma)\left(\frac{i-1}{n_r-2}\right)^{\frac{\rho}{2}} + \frac{\varsigma-1}{2} & (i = 1, 2, \ldots, n_r - 1) \\ w_B & (i = n_r) \end{cases} \tag{3.73}$$

(7) Cable member connections

The cable members of the reflector support cable net can be defined by the connections of the nodes. According to the previous discussion, the rings divide the reference sphere into several layers. The cable members are categorized into two cases:

1) Members inside the layers.

Without repeatedly counting the members, each node has three members connected inside the layer if it is on the subdivision lines and two members when it is not.

Since the first node in the ith ring is always on the subdivision lines, its connection can be defined by the two nodal numbers of the ends as follows,

$$M_k = \begin{cases} \{N_{i,1}, N_{i+1,(i+1)n_s}\} \\ \{N_{i,1}, N_{i+1,1}\} \\ \{N_{i,1}, N_{i+1,2}\} \end{cases} \quad (i = 1, 2, \ldots n_r - 1) \tag{3.74}$$

Every node satisfying Eq. (3.39) will connect with the three members in Eq. (3.75), while other nodes will connect with the two members in Eq. (3.76).

$$M_k = \begin{cases} \{N_{i,\bar{j}}, N_{i+1,s}\} \\ \{N_{i,\bar{j}}, N_{i+1,s+1}\} \\ \{N_{i,\bar{j}}, N_{i+1,s+2}\} \end{cases} \quad \begin{cases} i = 1, 2, \ldots n_r - 1 \\ j \in Eq.(3.36) \end{cases} \tag{3.75}$$

$$M_k = \begin{cases} \{N_{i,j}, N_{i+1,s}\} \\ \{N_{i,j}, N_{i+1,s+1}\} \end{cases} \quad \begin{cases} i = 1, 2, \ldots n_r - 1 \\ j \notin Eq.(3.36) \end{cases} \tag{3.76}$$

where $N_{i+1,s}$ is the number of the second end from the last member connection of the previous node on the same ring. For example, when considering the member connections of the second node in the ith ring, the last member connection of the first node is $\{N_{i,1}, N_{i+1,2}\}$ and hence $s = 2$. Specifically, in the first layer, the members are defined by

$$M_k = \{N_{0,1}, N_{1,j}\} \; j = 1, 2, \ldots, n_s \tag{3.77}$$

2) Members on the rings

On each ring, the members are connected between every pair of neighboring nodes. Their connections can be described by

$$M_k = \begin{cases} \{N_{i,j}, N_{i,j+1}\} \\ \{N_{i,n_i}, N_{i,1}\} \end{cases}, \; (i = 1, 2, \ldots n_r; \; j = 1, 2, \ldots, n_i - 1) \tag{3.78}$$

To calculate the number of members, they should be counted by layers. For the ith layer, the members include the members inside this layer and those on the ith ring. The numbers of members and triangular patches of the ith layer are $(3i - 1)n_s$ and $(2i - 1)n_s$, respectively. The total number of members m_{total} and triangular patches f_{total} can be determined as

$$m_{total} = \sum_{i=1}^{n_r} (3i - 1)n_s = \frac{3n_r + 1}{2} n_r^2 n_s \tag{3.79}$$

$$f_{total} = \sum_{i=1}^{n_r} (2i - 1)n_s = n_r^2 n_s \tag{3.80}$$

(8) Reduce the number of nodes in the boundary layer

In most applications of reflectors, the boundary layer is not used as the working surface. The number of nodes on the aperture rim may be less than the number of facets of the boundary layer. Assume that the number of nodes at the aperture rim is n_i and that each node is connected with c_B nodes of the inner ring. Note that c_B is an odd number and larger than 1. These connections follow two rules: first, along all the connected nodes on the $(n_r - 1)$th ring, the node positioned in the middle is the only one that connects one node on the aperture rim or the n_rth ring; second, the

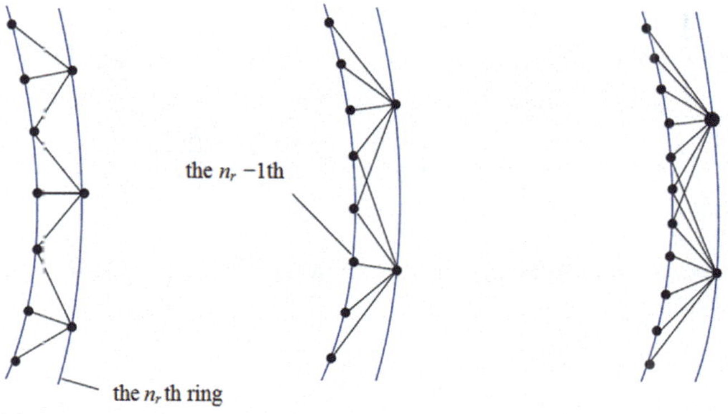

Fig. 3.22 Topology of connection inside boundary layer

other $c_B - 1$ nodes must connect two nodes from the $(n_r - 1)$th ring. The topology of the connection inside the boundary layer is illustrated in Fig. 3.22. The number of nodes can be reduced to $2/(c_B + 1)$ of the number of nodes on the $(n_r - 1)$th ring, or mathematically

$$n_i = \frac{2(n_r - 1)n_s}{(c_B + 1)}, \ i = n_r \tag{3.81}$$

Hence the total number $n_{total,B}$ of nodes on the entire surface is

$$n_{total,B} = \left(\frac{n_r}{2} + \frac{2}{(c_B + 1)}\right)(n_r - 1)n_s + 1 \tag{3.82}$$

Similar to the derivation in Eqs. (3.37) and (3.38), the nodal coordinates are obtained by the following equation

$$\begin{cases} \bar{x}_{4,(n_r,j)} = \frac{D}{2}\cos\alpha_{n_r,j} \\ \bar{y}_{4,(n_r,j)} = \frac{D}{2}\sin\alpha_{n_r,j} \\ \bar{z}_{4,(n_r,j)} = h_1 \end{cases}, \ j = 1, 2, \ldots, \frac{2(n_r - 1)n_s}{(c_B + 1)} \tag{3.83}$$

where

$$\alpha_{n_r,j} = (j - 1)\frac{\pi(c_B + 1)}{(n_r - 1)n_s} \tag{3.84}$$

The boundary cable connections can be described as

$$M_k = \begin{cases} \left\{N_{n_r,1}, N_{n_r-1,(n_r-1)n_s-k+1}\right\} \\ \left\{N_{n_r,1}, N_{n_r-1,k+1}\right\} \\ \left\{N_{n_r,1}, N_{n_r-1,1}\right\} \end{cases}, \ k = 1, 2, \ldots, (c_B - 1)/2 \tag{3.85}$$

$$M_k = \left\{N_{n_r,j}, N_{n_r-1,\frac{c_B+1}{2}(j-1)\frac{c_B-1}{2}+k}\right\}, \ (j = 2, 3, \ldots, \frac{2(n_r-1)n_s}{(c_B+1)}, k = 1, 2, \ldots, c_B) \tag{3.86}$$

The number of members m_B in the boundary layer is obtained below and is independent of c_B.

$$m_B = 2(n_r - 1)n_s \tag{3.87}$$

The number of facets in the boundary layer is $f_B = \frac{3n_r+2}{2}(n_r - 1)n_s$, and it is c_B times the number of nodes on the aperture rim. From Eqs. (3.79) and (3.87), the total number of members of the entire reflector is then

$$m_{total} = \frac{3n_r + 2}{2}(n_r - 1)n_s \tag{3.88}$$

(9) Mapping of the support net nodes from reference sphere to reference ellipsoid

As shown in Fig. 3.23, to generate the support net nodes on the working paraboloid (black), it is necessary to map the nodes on the reference sphere (blue) to the reference ellipsoid (red), which requires a set of mapping functions (corresponding to different horizontal heights z_4, i.e., different coordinates). Then a set of mapping functions is designed to perform two tasks. One is to, in the same horizontal plane, uniquely project the points from the circle of the spherical reference to the ellipse of the secondary reference; the other is that if all pairs of neighboring points on the circle result in equal chords (equally separate the circle), the mapped points on the ellipse should do the same. The nodal coordinate z_4 remains unchanged during the mapping process. Note that each ellipse is inscribed on the long axis with the corresponding circle of the reference sphere. The mapping function can be written as

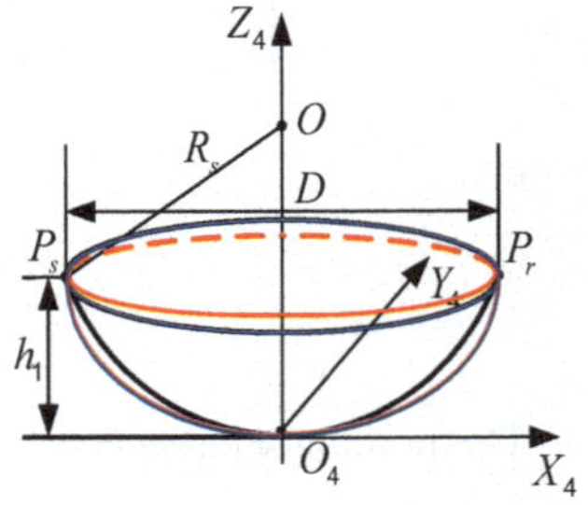

Fig. 3.23 Reference sphere, ellipsoid and working paraboloid

$$\begin{cases} \frac{x_4^2}{a_e^2} + \frac{y_4^2}{b_e^2} + \frac{(z_4 - R_s)^2}{a_e^2} = 1 \\ a_e = R_s, b_e = \frac{2R_s}{D} b'_e \end{cases} \tag{3.89}$$

where a_e and b_e are the long and short axes of the reference ellipsoid, respectively.

The reference ellipse with the same ellipticity b_e/a_e and a different size can be obtained by setting different values z_4 in Eq. (3.89), and then the points on the reference sphere can be mapped to the ellipsoid one by one.

If the mapping point of node $\overline{N}_{i,j}$ $(i = 1, 2, \ldots, n_s;\ j = 1, 2, \ldots, \mathrm{nr_i})$ at the reference sphere on the reference ellipsoid is $\widetilde{N}_{i,j}$, its orthogonal coordinate $(u_{i,j}, v_{i,j})$ in the X_4Y_4 plane will be

$$\begin{cases} \mathrm{u}_{i,1} = 1, v_{\mathrm{i},1} = 0 \\ \mathrm{u}_{i,\mathrm{j}} = \cos(2\pi \eta_{\mathrm{i,j}}), \mathrm{v_{i,j}} = \frac{\mathrm{b_e}}{\mathrm{a_e}} \sin(2\pi \eta_{\mathrm{i,j}}), \ j = 2, \ldots, in_s \\ \eta_{i,\mathrm{j}} = \sum_{\mathrm{k}=1}^{\mathrm{j}-1} \varepsilon_{\mathrm{ik}} \end{cases} \tag{3.90}$$

in which $\varepsilon_{i,k}$ is the normalized polar angle increment at the kth chord and $\eta_{i,j}$ is the dimensionless polar angle of the jth mapped node on the ith ring. Applying the equal-chords assumption, $\varepsilon_{i,k}$ and $\eta_{i,j}$ can be obtained from the following equations:

$$\begin{cases} (u_{i,j+1} - u_{i,j})^2 + (v_{i,j+1} - v_{i,j})^2 = (u_{i,j+2} - u_{i,j+1})^2 + (v_{i,j+2} - v_{i,j+1})^2 \\ (u_{i,in_s} - u_{i,in_s-1})^2 + (v_{i,in_s} - v_{i,in_s-1})^2 = (u_{i,in_s} - u_{i,1})^2 + (v_{i,in_s} - v_{i,1})^2 \end{cases}, \quad (j = 1, \ldots in_s) \tag{3.91}$$

The mapping function of the ith ring is denoted as $\eta_{i,j} = f_{m,i}((j-1)/in_s)$, and the mapping function of the ith ring $f_{m,i}$ is expressed as a cubic spline. Denote $\overline{\alpha}_{i,j}$ as the polar angle of the nodes on the ellipsoid reference in the X_4Y_4 plane, we have

$$\overline{\alpha}_{i,j} = 2\pi f_{m,i}\left(\frac{\alpha_{i,j}}{2\pi}\right) \tag{3.92}$$

where $\alpha_{i,j}$ is the X_4Y_4 plane polar angle of the jth node at the ith ring on the reference sphere.

The mapped node $\widetilde{N}_{i,j}(\tilde{x}_{4,(i,j)}, \tilde{y}_{4,(i,j)}, \tilde{z}_{4,(i,j)})$ on the ellipsoid is at the same height of the node on the spherical reference, so

$$\begin{cases} \tilde{x}_{4,(i,j)} = \sqrt{a_e^2 - (\tilde{z}_{i,j} - R_s)^2} \cos(\overline{\alpha}_{i,j}) \\ \tilde{y}_{4,(i,j)} = \frac{b_e}{a_e} \sqrt{a_e^2 - (\tilde{z}_{i,j} - R_s)^2} \sin(\overline{\alpha}_{i,j}) \\ \tilde{z}_{4,(i,j)} = \overline{z}_{4,(i,j)} \end{cases} \tag{3.93}$$

$$(\tilde{x}_{4,(0,1)}, \tilde{y}_{4,(0,1)}, \tilde{z}_{4,(0,1)}) = (0, 0, 0) \tag{3.94}$$

(10) Project nodes of the reference sphere onto the working paraboloid surface

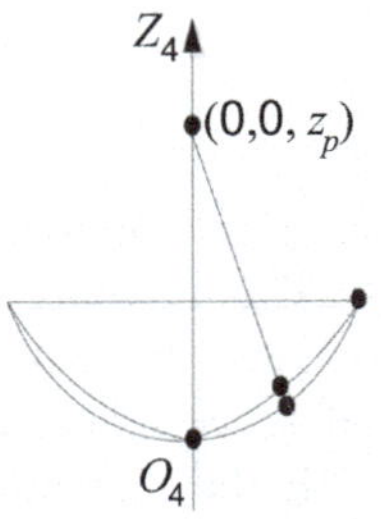

Fig. 3.24 Node mapping

Once node $\widetilde{N}_{i,j}(\tilde{x}_{4,(i,j)}, \tilde{y}_{4,(i,j)}, \tilde{z}_{4,(i,j)})$ is generated on the reference ellipsoid, node $(x_{4,(i,j)}, y_{4,(i,j)}, z_{4,(i,j)})$ of the working parabola (Fig. 3.24) can be generated by mapping. In this process, the member connections remain unchanged. Assume the projection center is $(0, 0, z_p)$ with $z_p > h_1$; from the line equation, we have

$$\frac{x_{4,(i,j)}}{\tilde{x}_{4,(i,j)}} = \frac{y_{4,(i,j)}}{\tilde{y}_{4,(i,j)}} = \frac{z_p - z_{4,(i,j)}}{z_p - \tilde{z}_{4,(i,j)}} = \xi_{i,j} \tag{3.95}$$

Substituting Eq. (3.95) into Eq. (3.24) leads to

$$\widetilde{b}_2\xi_{i,j}^2 + \widetilde{b}_1\xi_{i,j} + \widetilde{b}_0 = 0 \tag{3.96}$$

where

$$\begin{cases} \tilde{b}_2 = \frac{1}{4F_p}\left(\left(\tilde{z}_{4,(i,j)}\cos\alpha - \left(z_p - \tilde{z}_{4,(i,j)}\right)\sin\alpha\right)^2 + \tilde{y}_{4,(i,j)}^2 + y_{3o_4}^2\right) \\ \widetilde{b}_1 = \tilde{x}_{4,(i,j)}\sin\alpha + \left(z_p - \tilde{z}_{4,(i,j)}\right)\cos\alpha + \frac{1}{2F_p}\left(\tilde{x}_{4,(i,j)}\cos\alpha - \left(z_p - \tilde{z}_{4,(i,j)}\right)\sin\alpha\right)\left(z_p\sin\alpha + x_{3o_4}\right) \\ \widetilde{b}_0 = \frac{1}{4F_p}\left(z_p\sin\alpha + x_{3o_4}\right)^2 - z_p\cos\alpha - z_{3o_4} \end{cases} \tag{3.97}$$

If the working paraboloid is a prime-feed paraboloid with $d = 0, \alpha = 0, x_{3o_4} = 0$, $y_{3o_4} = 0$ and $z_{3o_4} = 0$, Eq. (3.97) can be simplified as

$$\begin{aligned} \tilde{b}_2 &= \frac{1}{4F_p}\left(\tilde{x}_{4,(i,j)}^2 + \tilde{y}_{4,(i,j)}^2\right) \\ \tilde{b}_1 &= z_p - \tilde{z}_{4,(i,j)} \\ \tilde{b}_0 &= -z_p \end{aligned} \tag{3.98}$$

Two sets of solutions can be obtained. The plus or minus symbol of the nodal X_4 or Y_4 has to be unchanged before and after projection; as a result, $\xi_{i,j}$ must be positive. Considering the condition of $z_{4,(i,j)} \geq 0$ at the same time gives

$$\begin{cases} \xi_{i,j} > 0 \\ z_p - \xi_{i,j}\left(z_p - \widetilde{z}_{4,(i,j)}\right) \geq 0 \end{cases} \tag{3.99}$$

When the projection center $z_p \geq 0$ and $\xi_{i,j} > 0$, only the second condition of Eq. (3.99) needs to be satisfied. In this case, the projection center can be moved along that Z_4 axis, and the value of the variable z_p that minimizes the total cable length of the support cable net and meets the requirement of the surface root mean square (rms) error, can be selected.

Thus far, the node coordinates and connection relations of the working paraboloid in the local coordinate system can be obtained. The nodes can be indexed by $N_{i,j}$ and the coordinates are $\left(x_{4,(i,j)}, y_{4,(i,j)}, z_{4,(i,j)}\right)$. The member connections obey M_k.

(11) Determination of node coordinates of the rear cable net

In terms of simplicity, the rear cable net surface can be designed as an axisymmetric paraboloid. If its focal length is F_2, the equation can be written as follows

$$z_4 = -\Delta h - \frac{x_4^2 + y_4^2}{4F_2} \tag{3.100}$$

where Δh is the distance between the front and rear support cable net, as mentioned previously.

The rear support cable net nodes can be obtained by simply mapping the front support cable net nodes vertically. That is, the front and rear support cable net have the same form of topological connection, the corresponding nodes have the same x_4 and y_4 coordinate values, the difference is only in the z_4 value. Thus, the coordinates of the corresponding rear support cable net nodes $\left(x''_{4,(i,j)}, y''_{4,(i,j)}, z''_{4,(i,j)}\right)$ are

$$\begin{cases} x''_{4,(i,j)} = x_{4,(i,j)} \\ y''_{4,(i,j)} = y_{4,(i,j)} \\ z''_{4,(i,j)} = -\Delta h - \frac{x''^2_{4,(i,j)} + y''^2_{4,(i,j)}}{4F_2} \end{cases} \tag{3.101}$$

4. Antenna height design

As shown in Fig. 3.25, the height h of the perimeter truss includes the heights h_1 and h_2 of the front and rear support nets and the distance Δh between the bottoms of the two support nets.

$$h = h_1 + h_2 + \Delta h \tag{3.102}$$

where h_1 is determined by Eq. (3.26). If the focal length of the rear support net is F_2, the corresponding height is $h_2 = \frac{D^2}{16F_2}$. Δh is determined by the designer.

Take an AstroMesh reflector antenna with a diameter of 16 m in orbit as an example. If $F_p = 8.4$ m, $F_2 = 45$ m and $d = 9.3$ m, we have the following calculated parameters of $h_1 = 1.11$ m and $h_2 = 0.34$ m If $\Delta h = 0.65$ m, the deployable height of the antenna is $h = 2.1$ m.

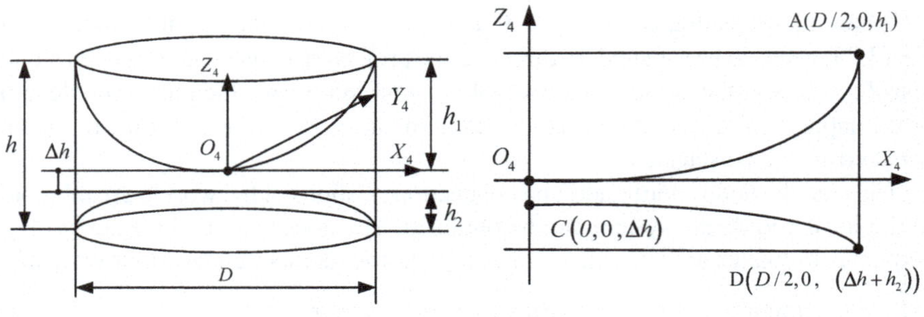

(a) Overall schematic diagram (b) In $X_4O_4Z_4$ plane

Fig. 3.25 Antenna height in deployed state

5. Design of antenna perimeter truss

(1) Design of perimeter truss in deployed state

Considering the stiffness of the structure, the perimeter truss is designed as a circle. From an engineering viewpoint, only a few quadrilateral trusses can be used to encircle the ring. The main body of the AstroMesh deployable reflector antenna is formed by the gradual outward expansion of a regular hexagon at the center, so the edge of the annular truss is selected as a multiple of 6. The center of the hexagon coincides with the center of the reflector, as shown in Fig. 3.26a–c, so that the regular connection between the reflector and annular truss can be ensured.

Suppose the total number of quadrilateral trusses is nuqt (number of quadrilateral trusses), and the side length (also the longeron length of the quadrilateral truss) is

$$l_t = \frac{D^2}{2}(1 - \cos\frac{360}{nuqt}) \tag{3.103}$$

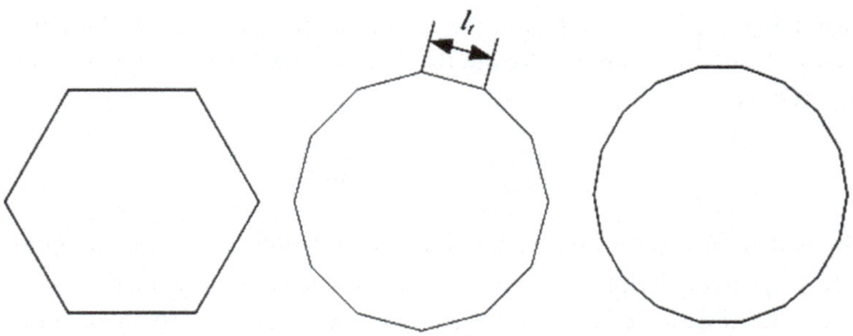

(a) Hexagon (b) Twelve sided shape (c) Eighteen sided shape

Fig. 3.26 Ring truss projection diagram

Specifically, for each quadrilateral truss, there will be locking and synchronization joints at the four corners (Fig. 3.27) because of the smooth transition from the stowed state to the deployed state. As shown in Fig. 3.27, parameters t_1, t_2 and t_h are the lengths of the synchronous joint, locking joint arm and horizontal bar, respectively. R_x and R_y are the projection lengths of the locking joints in the directions of X_4 and Y_4, respectively. h_v is the height of the vertical bar and the height of the deployed state is $h = h_v + t_1 + t_2$.

(2) Design of perimeter truss in stowed state

During launch, the antenna is fixed within the fairing of the rocket in the stowed state (see Fig. 3.28). A quadrilateral in the stowed state is shown in Fig. 3.28a. The height of the stowed state is $h_S = h + l_t = h + t_1 + t_2 + t_h$. Taking the 16 m antenna in orbit as an example, the horizontal bar has $l_t = 1.6\,\text{m}$ and the stowed height is $h_S = 3.7\,\text{m}$.

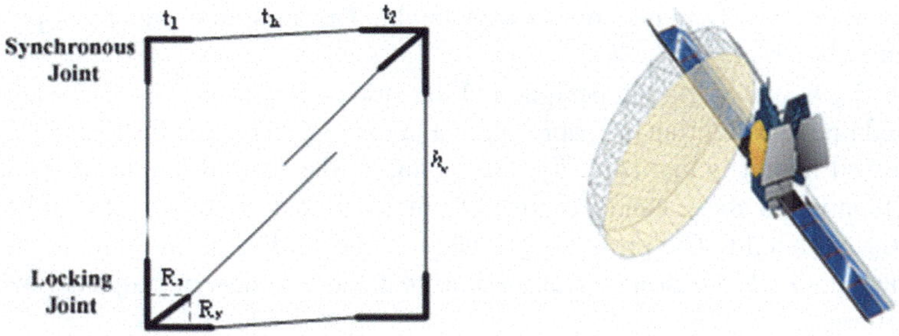

(a) Quadrilateral truss structure (b) The deployed state of truss

Fig. 3.27 Deployment of AstroMesh antenna

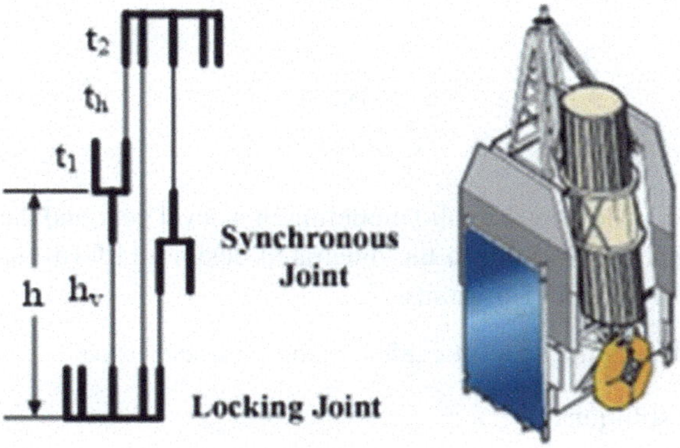

(a) A stowed quadrilateral element (b) The stowed state of truss

Fig. 3.28 The stowed state of AstroMesh antenna

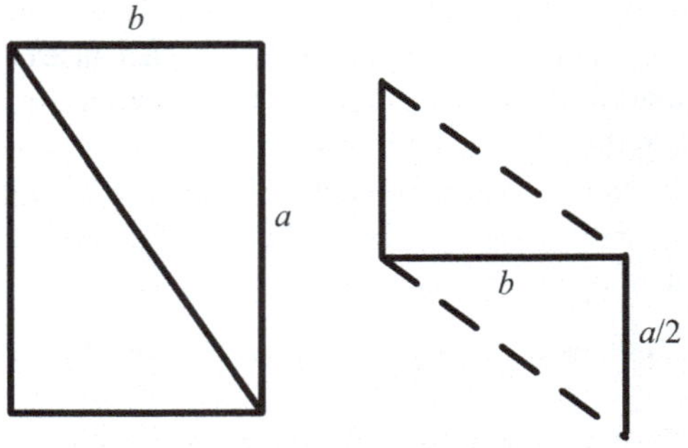

Fig. 3.29 High unfold/stow ratio scheme for AstroMesh deployable antenna

Note that when the aperture of the AstroMesh deployable antenna is required to be large enough, the stowed height becomes the restriction factor because the rocket fairing cannot accommodate it. For this reason, researchers have been working for a long time to effectively reduce the stowed height. The authors of this book propose the scheme shown in Fig. 3.29.

Figure 3.29a is the original design, and the stowed height is $a + b$. If scheme II (the dotted line representing the cable) shown in Fig. 3.29b is used instead of scheme I, the stowed height is $\max\{a/2, b\}$. For example, the stowed height of the 16 m antenna mentioned above can be reduced from 3.7 m to 1.05 m, which is only 28% of the original height. The problem that needs to be studied at this time is that the cable tension should be properly chosen so that the two design projects have the same stiffness.

(3) Design of bevel gear with synchronized joint for AstroMesh antenna

There is a pair of bevel gears at the synchronous joint to control the closure and expansion of the quadrilateral. The bevel gear is installed in the synchronous joint, and the relationship between the bevel angle γ and the number of edges of the perimeter truss is

$$\gamma = \frac{180}{nuqt} \tag{3.104}$$

Figures 3.30 and 3.31 show the solid modeling of a bevel gear and the bevel gears installed in the synchronous joint in the integrated design platform of the satellite antenna developed by Xidian University.

6. Connection between the support cable net and perimeter truss

(1) Problems and solutions

There are two kinds of connection modes between the support cable net and perimeter trusses, i.e., a circle and ellipse. If it is a circle, there is no problem in connecting the

Fig. 3.30 Bevel gears with a bevel angle of 30°

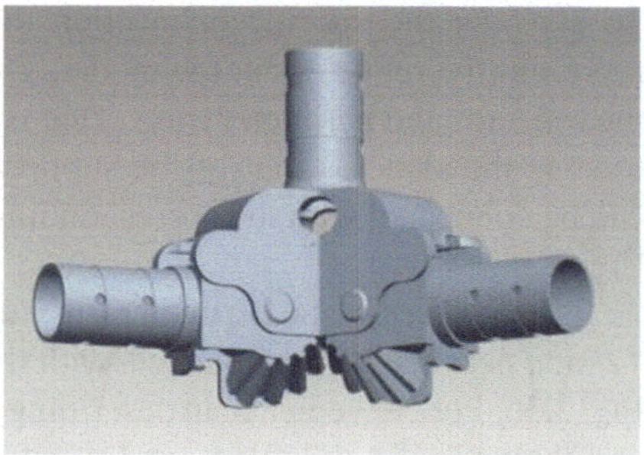

Fig. 3.31 Integrated model of synchronized joints

support cable net with the perimeter truss. If it is an ellipse, the connection between the support cable net and perimeter trusses should be addressed specifically, since the points of the support cable net are not on the same level as the upper ridge point of the truss. At this time, there is a need of increasing transition of the support cable net or new connection structure.

Figure 3.32 is a connection form. Sliders A and B are added to two adjacent quadrilateral cross bars, and rods are extended from A and B to intersect point C. Then rod CD is connected from point C. In this case, point D is connected to the nodes of the non-long axis of the upper support cable net.

Similarly, the latter is to make the surrounding truss higher (Fig. 3.33), that is, to increase the height of the truss up to the appropriate height of the short axis of the cable net according to the curvature of the original cable net to the circular truss.

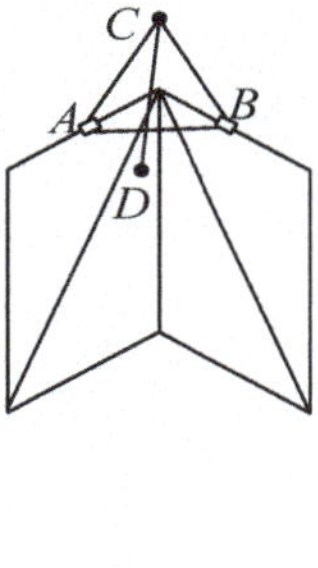

Fig. 3.32 Connection structure

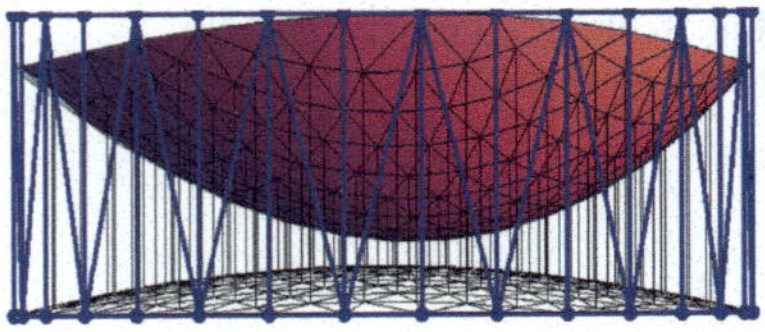

Fig. 3.33 Increasing the height of the peripheral truss

At this time, there is no problem with the short axis, while the nodes of the support cable net on the long axis can only be connected to the non-upper ridge of the truss.

(2) Connection of working parabolic support network with perimeter truss

To consider the specific connection model, we have to know how to divide and represent the reflectors in the model. There is a special case of connection between the reflector and perimeter truss. That is, when the number of edges of the perimeter truss is the same as that of the support cable net, the six vertices of the reflector can be connected with the corresponding six vertices of perimeter truss, as shown in Fig. 3.34.

If the vertices of the perimeter truss and reflector are inconsistent, for instance, the 12-side perimeter truss connects with the 6-side reflector, the geometry is shown in Fig. 3.35. For convenience in describing the problem, Zone 0 in Fig. 3.35 is separated and shown in Fig. 3.36. If n_1 and n_2 represent the total number of nodes on the outer edge of the reflector and on the annular truss respectively, we have

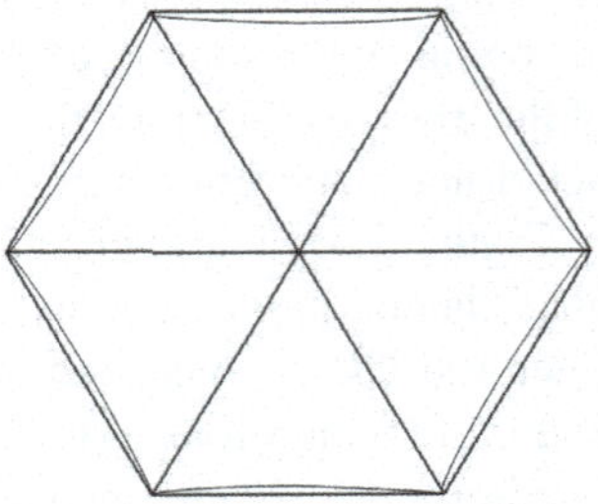

Fig. 3.34 Connection in special case

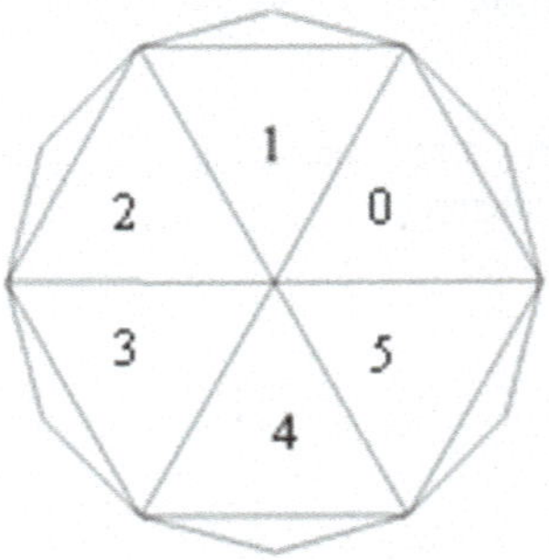

Fig. 3.35 Connection of 12 polygonal truss and 6 polygonal reflector

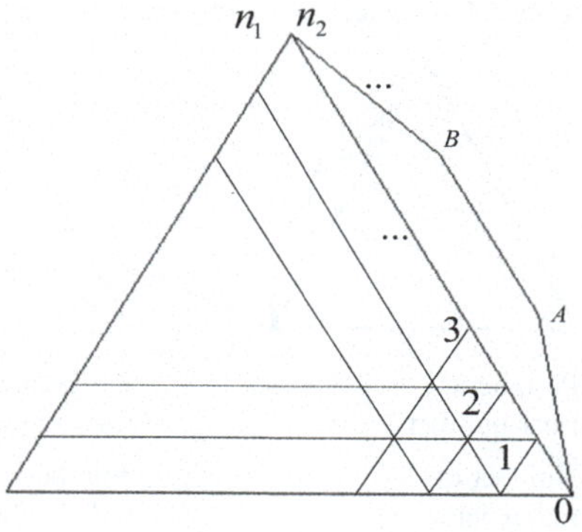

Fig. 3.36 Zeropartial reflector and peripheral truss

$$n_1 = n + 1 \tag{3.105}$$

$$n_2 = (\text{nuqt}/6) + 1 \tag{3.106}$$

As mentioned earlier, n is the number of radial cable segments and nuqt is the number of side of the perimeter truss.

In Fig. 3.36, nodes $(0, 1, 2, 3, \ldots, n_1)$ on the outermost edge of the support cable net and nodes $(0, A, B, C, \ldots, n_2)$ on perimeter truss need to be reasonably connected. Because nodes 0 and 0, n_1 and n_2 must be connected, it is necessary to study the connection between the outermost edge nodes χ_1 of the support cable net and the other nodes χ_2 of the perimeter truss. χ_1 and χ_2 are given as

$$\chi_1 = n_1 - 2 = n - 1 \tag{3.107}$$

$$\chi_2 = n_2 - 2 = nuqt/6 - 1 \tag{3.108}$$

with $\chi_1 \geq \chi_2$.

The connection between the perimeter truss and support cable net can be divided into the following situations (see Table 3.1 for several typical connection modes):

(1) If χ_1 is divisible by χ_2, let $\tau = \chi_1/\chi_2$.

① When $\chi_2 = 1$ (i.e. there is only one joint), there are χ_1 cables drawn from this joint;

② When $\chi_2 \neq 1$ (i.e. the number of joints is greater than 1), τ cables are drawn from each joint.

(2) If χ_1 is not divisible by χ_2, let $\tau = [\chi_1/\chi_2] + 1$.

① If χ_2 is an odd number, let $\eta = (\chi_2 + 1)/2$, there are τ cables drawn from the joints $(1, \eta)$ and (η, χ_2). The cables from joint η are connected to the nodes of the support cable net with the number in $[(\eta - 1)_{-1}, (\eta + 1)_1]$;

Table 3.1 Connection modes between peripheral trusses and support cable net

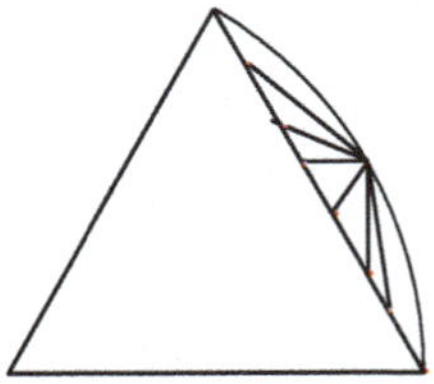		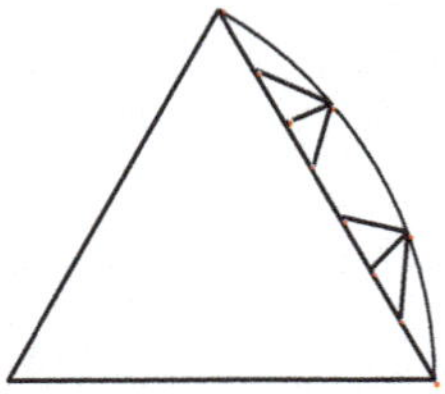		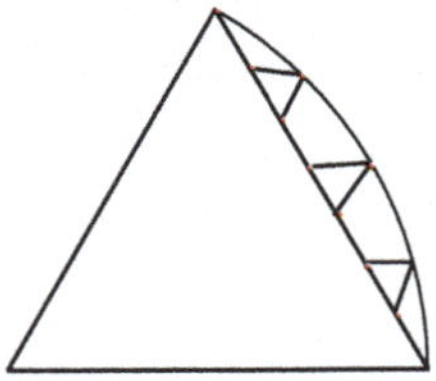	
Peripheral truss number	12	Peripheral truss number	18	Peripheral truss number	24
Number of radial cable segments	7	Number of radial cable segments	7	Number of radial cable segments	7
Categories	(1)①	Categories	(1)②	Categories	(1)②
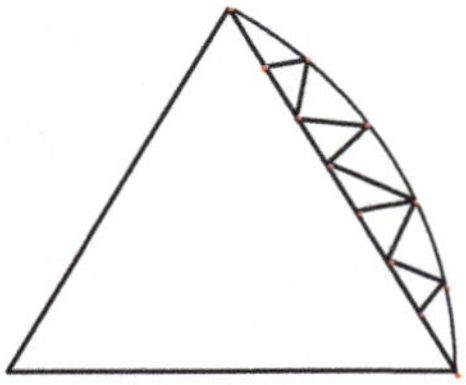		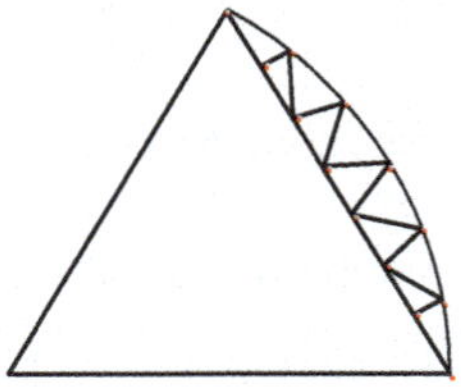		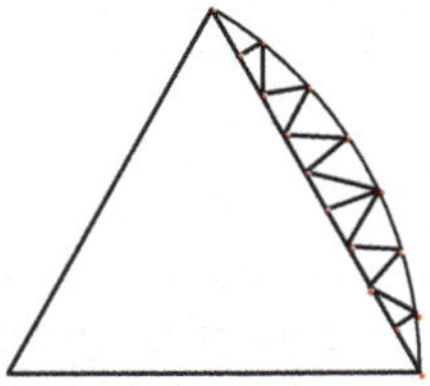	
Peripheral truss number	30	Peripheral truss number	36	Peripheral truss number	42
Number of radial cable segments	7	Number of radial cable segments	7	Number of radial cable segments	9
Categories	(2)②(i)	Categories	(2)①	Categories	(2)②(i)
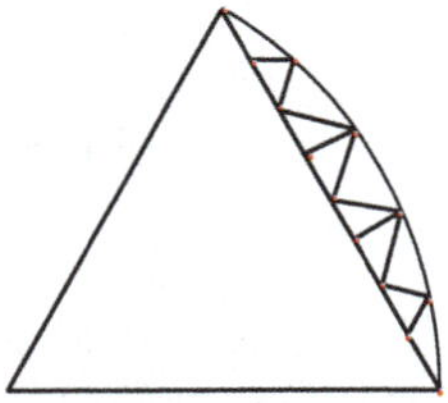		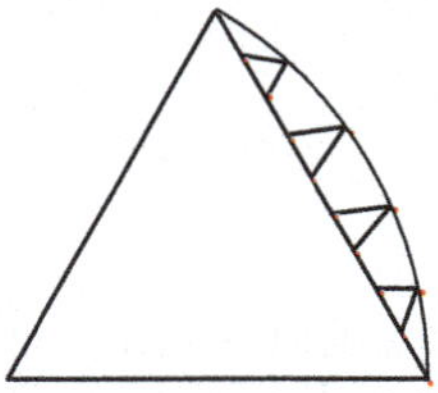		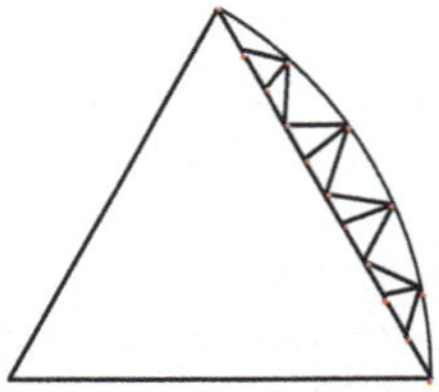	
Peripheral truss number	30	Peripheral truss number	30	Peripheral truss number	30
Number of radial cable segments	8	Number of radial cable segments	9	Number of radial cable segments	10
Categories	(2)②(ii)	Categories	(1)②	Categories	(2)②(ii)

② If χ_2 is an even number, let $\eta = \chi_2/2$; there are τ cables drawn from the joints whose numbers are in $(1, \eta)$ and $(\eta + 1, \chi_2)$.

(i) If χ_1 is an even number, let $middle = \chi_1/2$; the cables from the joint numbered η are connected to the nodes of the support cable net with the numbers in $[(\eta - 1)_{-1}, middle + 1]$, and the cables from the connection numbered $\eta + 1$ are connected to the nodes of the support cable net with the numbers in $[middle + 1, (\eta + 1)_1]$.

(ii) If χ_1 is an odd number, let $middle = (\chi_1 + 1)/2$; the cables from the joint numbered η are connected to the nodes of the support cable net with the numbers in $[(\eta - 1)_{-1}, middle]$, and the cables from the connection numbered $\eta + 1$ are connected to the nodes of the support cable net with the numbers in $[middle + 1, (\eta + 1)_1]$.

Note: $(\eta - 1)_{-1}$ indicates the nodal number of the last support cable net node connected to the $(\eta + 1)$th joint, and $(\eta + 1)_1$ indicates the number of the first support cable net node to which the $(\eta + 1)$th joint is connected.

3.2.2 *Modular Antenna*

The modular deployable antenna adopts a modular design concept, and its backup truss is a deployable truss system. NASA, NASDA, EADS (ESA) and RKA have produced such kind of deployable truss antennas in a variety of different configurations. Both the Japanese ETS-VIII satellite [9, 10] launched on December 18, 2006 and China's Environment-1 C satellite [11] launched on November 19, 2012 at the Taiyuan Satellite Launch Center all utilized modular deployable antennas.

Each module of the modular deployable antenna can be manufactured and assembled in parallel, so the development cycle can be shortened, and the scale and difficulty of testing can be significantly reduced because assembly can be completed after each unit test. Compared with AstroMesh and rigid-rib antennas, the stowed height can be reduced dramatically, and the deployable stiffness and stability are higher. The disadvantages are a large areal density, large diameter when stowed, and high manufacturing cost.

1. Basic composition of modular antenna

As shown in Fig. 3.37, the modular deployable antenna [9] consists of multi-layered hexagonal modules. Figure 3.37a shows a two-layer module with one module in the center layer and six modules in the surround (also called second) layer. It is conceivable that there are 18 modules in the third layer. Without loss of generality, it is assumed that there are N layers of modules. Each module consists of a reflective surface and support structure. The so-called reflective surface here includes the support cable net and mesh, and the support cable net further comprises front and rear support cable nets and adjustable tension cables. The support structure consists

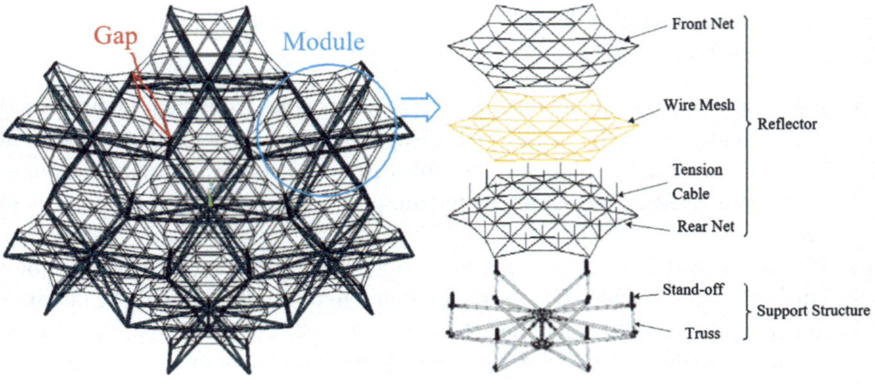

(a) Overall schematic diagram (b) Basic antenna composition

Fig. 3.37 Modular deployable antenna

of a support cable net stand-off and truss, and the truss is composed of six deployable units around the central bar along the circumferential array. Each module has a separate support cable net. The topology of the support cable net of each module is basically the same, but the curvature varies with the positions of the modules on the parent paraboloid. The nodes of the front support cable net of each module are located on the parabolic surface, and the rear support cable net forms a parabolic shape with its opening downward. The support cable net is installed on the support cable net stand-off, and the adjustable cable is parallel to the center bar of the module, connecting the front and rear support cable nets.

Figure 3.38 shows the deployed state (a) and stowed state (b) of a modular deployable antenna support structure. The unit consists of a central bar, an upper bar, a lower

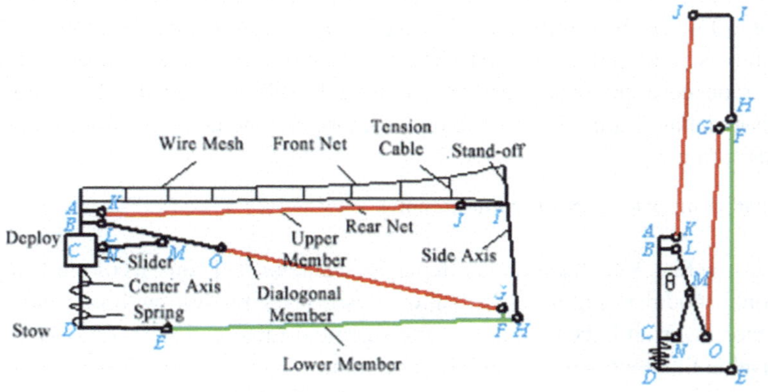

(a) Deployed state (b) Stowed state

Fig. 3.38 Deployable unit form

bar, a diagonal bar, a side bar, a driving spring and a slider. The support cable net is fixed at the end of the stand-off. The front and rear support cable net are stretched to the desired shape by adjusting the length of the cables, and the surface accuracy of the support cable net can be changed by altering the length of the adjustable cables. The driving spring can drive the slider to move up and down along the central bar and control the deployment and folding of the mechanism. Each module is connected with other modules through the side bar *IH*.

2. Fundamental design of modular deployable antenna

(1) Support cable net

The design of the modular deployable antenna support cable net can be characterized as a separate design for each module. Therefore, as long as the support cable net of each module is determined, the whole antenna support cable net can be obtained. The design of the support cable net of each module is the design of the support cable net of the offset feed antenna (Sect. 3.2.1). For this reason, the local coordinate system $O_4^{u,v}X_4^{u,v}Y_4^{u,v}Z_4^{u,v}$ of the module is established. After obtaining the nodal coordinates of the support cable net of the module in its coordinate system, the nodal coordinates in the local modular coordinate system are converted to the local antenna coordinate system through coordinate transformation, and then the overall support cable net of the antenna in the antenna local coordinate system $O_4X_4Y_4Z_4$ can be obtained.

(2) Module

The support cable net of a modular deployable antenna is a paraboloid. To minimize the size of each module, a spherical approximation paraboloid is used to make the position of each node on the framework as close as possible to the paraboloid of the reflector. Therefore, when designing the structure, a sphere is defined to approximate the parent paraboloid of the antenna, and the upper and lower structure nodes are located on the concentric sphere, so that the dimensions of the modules in each ring are the same. The difference in radius between the concentric spheres is the length of the stand-off and the side bars. Then, the coordinates of the truss nodes can be obtained by projecting the outermost circle nodes of the support cable net to the two concentric spheres.

(3) Symbol description

(1) Coordinate systems

As defined in Sect. 3.1, $O_4X_4Y_4Z_4$ is the local coordinate system of the paraboloid. For convenience in the following discussion, a coordinate system $O_4^{u,v}X_4^{u,v}Y_4^{u,v}Z_4^{u,v}$ is purposely defined for the *uv*th module.

(2) Coordinate symbols

Since coordinate system conversion is necessary in the design process of a modular deployable antenna support cable net, the following description is given. For the nodes of the support cable net, the superscript *uv* indicates the coordinates of the parent paraboloid or in the local coordinate system of the *uv*th module. Coordinates without the superscript *uv* represent the coordinates of the antenna parent paraboloid or in the local coordinate system. A number in a subscript indicates the parent paraboloid coordinate system (number "3") or the local coordinate system (number "4"). The subscripts u, v, i, j indicate the specific position of the node (the jth node of the ith ring of the *uv*th module). In particular, for the coordinates in the coordinate system of the *uv*th module, if the superscript *uv* is present, only *i,j* are included in the subscript. Here are some examples:

$(X_{4i,j}^{u,v}, Y_{4i,j}^{u,v}, Z_{4i,j}^{u,v})^{\mathrm{T}}$ represents the coordinates of the jth node in the ith ring of the *uv*th module in the local coordinate system of module $O_4^{u,v}X_4^{u,v}Y_4^{u,v}Z_4^{u,v}$.

$(X_{4u,v,i,j}, Y_{4u,v,i,j}, Z_{4u,v,i,j})^{\mathrm{T}}$ represents the coordinates of the jth node in the ith ring of the *uv*th module in the local coordinate system of module $O_4X_4Y_4Z_4$;

$(Xc_{u,v}, Yc_{u,v}, Zc_{u,v})^{\mathrm{T}}$ represents the coordinates of the center node of the *uv*th module in the parent paraboloid coordinate system $O_3X_3Y_3Z_3$.

Since truss nodes are obtained by projection of the support cable net nodes, coordinate conversion and superscript are not necessary. Because the truss has only upper and lower layers, and each layer has only one ring point, the subscript is not necessary to indicate the number of rings, which is marked as *u*, *v*, *U*(D), j, representing the jth key node of the upper (lower) truss of the uvth module. For example:

$(X_{4u,v,U,j}, Y_{4u,v,U,j}, Z_{4u,v,U,j})^{\mathrm{T}}$ represents the coordinates of the jth key node of the upper truss of the uvth module in the local coordinate system of the antenna $O_4X_4Y_4Z_4$.

$(X_{4u,v,D,j}, Y_{4u,v,D,j}, Z_{4u,v,D,j})^{\mathrm{T}}$ represents the coordinates of the jth key node of the lower truss of the uvth module in the local coordinate system of the antenna $O_4X_4Y_4Z_4$.

In this section, the offset reflector antenna is discussed as an example and the prime feed antenna is only a special case.

3. Design process of modular deployable antenna

(1) Determination of module number and truss size

If the physical aperture of the antenna is given, based on the analysis of Sect. 3.1, the number of segments n can be obtained from Eq. (3.13a) according to the antenna operating frequency and other parameters. The relationship between the number of cable net segments n' and n in the module is subject to

$$n' = n/(2N - 1) \tag{3.109}$$

Because n' and N must be positive integers, they should satisfy

$$n' \cdot (2N - 1) \geq n \tag{3.110}$$

For a given n, n' and N, there exists a finite group solution. By comparing the performance of the antenna below, it can be reasonably determined. After determining the total layers of the antenna modules, the support cable net of each module can be designed.

After determining the number of module layers N, it is necessary to consider the cases of both the deployed state and the stowed state because the number of layers affects the stiffness of the antenna; that is, it affects the reflector precision. The number of layers also affects the stowed height and diameter. In addition, there are two options for the wire mesh that forms the reflecting surface. One is to separately lay the mesh for each module, and the other is to use a whole mesh. The former will produce a leaf-like gap between the modules (Fig. 3.37), which will affect the electronic performance of the antenna, such as the gain. The latter must consider the structural difficulties of deployment from the stowed state to the deployed state.

(1) Antenna weight

From Fig. 3.38, it can be seen that the total number of bars in one deployable unit is

$$\mathrm{N}^{e}_{\mathrm{truss}} = i\left(1 + 6\left((N-1) + \frac{(N-1)(N-2)}{2}\right)\right) \tag{3.111}$$

where $\mathrm{N}^{e}_{\mathrm{truss}}$ represents the total number of the eth type of deployable bars. For the center unit, i = 1, and for the other parts, i = 6.

Thus, the weight of the modular antenna can be mathematically expressed as

$$\mathrm{W} = \mathrm{W}_{\mathrm{truss}} + \mathrm{W}_{\mathrm{cablenet}} + \mathrm{W}_{\mathrm{mesh}} = \rho \sum_{k=1}^{e} A_k \sum_{l \in k} l_l + \mathrm{W}_{\mathrm{cablenet}} + \mathrm{W}_{\mathrm{mesh}} \tag{3.112}$$

where ρ is the truss material density, A_k is the cross-sectional area of the kth member, $\sum_{l \in k} l_l$ is the sum of the lengths of all the bars which are controlled by A_k, $\mathrm{W}_{\mathrm{cablenet}}$ and $\mathrm{W}_{\mathrm{mesh}}$ are the weights of the support cable net and metal mesh, respectively.

Equations (3.111) and (3.112) tell us that an increasing number of layers in the module will lead to an increasing number of bars, which will increase the weight of the antenna. It is desirable to increase the module size as much as possible within the allowable range and to reduce the number of layers so the antenna's weight can be decreased.

(2) Stowed height and diameter

Changes in the module number of the modular antenna will directly affect the stowed volume of the antenna. Therefore, at the beginning of design, the stowed volume of the modular antenna with the designed layer must be guaranteed to meet the requirements of rocket fairing.

After the antenna is stowed, it consists of a number of basic deployable units (Fig. 3.38). The modules are precisely connected with the bar IH. Since the parabolic curvature is increased with the module outward, the length of the bottom chord and joint of the module will be increased. When the antenna is stowed, the side bars of each module are seamlessly connected, while the height of the lower bar is gradually increased. This will bring the stowed antenna to a "step-like ascending" state. Consequently, the height of the stowed antenna is determined by the length of the lower bar EF, joint FH and side bar HI of the outermost module.

$$H_{\text{stow}} = \overline{EF} + \overline{FH} + \overline{HI} = L - \overline{DE} + \overline{HI} \tag{3.113}$$

The stowed diameter is

$$D_{\text{stow}} = 2(2N - 1)\overline{DE} \tag{3.114}$$

It can be seen that a change in the number of module layers directly affects the size of the bar and joint, thus affecting the stowed height and diameter. In this case, the antenna becomes "short and fat". In contrast, if the number of module layers is reduced, the size of the modules increases, the length of the lower member and axis side increase, the stowed height increases while the diameter of the stowed antenna decreases, and the antenna becomes "thin and tall".

(3) Influence of the gap on antenna gain

There is a certain sag-to-span ratio (8–12%) [10] at the support cable net boundary of each module of the deployable antenna, which creates gaps between the modules. These gaps will affect the electronic performance of the antenna, mainly by reducing the gain. Next, the influence of the gap on the reflector gain is considered based on the area loss of the gap.

To determine the difference, the area of each gap is weighted, and the loss ratio of all gaps on the efficiency of the antenna will be

$$\eta_1 = \frac{\sum_{i=1}^{\text{NUMi}} \int_{S_i} Q_i dS_i}{\int_S Q dS} \tag{3.115}$$

where $\int_{S_i} Q_i dS_i$ and $\int_S Q dS$ are the integrals of the illumination factor $Q_i = 0.7 + 0.3(1 - r_i^2/R_0^2)$ [11] on the projection area S_i of the ith gap and the projection area S of the whole reflector and r_i is the distance of a point on the gap to the focal axis of the reflector. R_0 is the radius of the aperture surface, and NUMi is the total number of gaps.

Substituting Eq. (3.115) into the antenna's gain formula yields

$$G = D_1 \eta_1 \tag{3.116}$$

The effect of the gap on antenna gain can now be analyzed. Here, $D_1 = \frac{4\pi r^2}{\eta P_{rad}} \overrightarrow{E} * \overrightarrow{E}$ represents the directional coefficient of the reflector without gaps, r is the distance from the far field observation point to the coordinate center, $\eta = 120\pi$ is the free space wave impedance, and P_{rad} is the radiation power of the feed. The far-field radiated electric field $\overrightarrow{E}$ of the antenna can be obtained by the physical optics method. The specific method is to integrate every facet of the reflector first. After obtaining the far-field radiated electric field of each element, the far-field pattern of the whole antenna can be obtained by superposition.

It can be seen that in the case where the direction coefficients are the same, the larger η_1 is, the higher the gain is. Therefore, the gap in the reflective surface introduced by the module splicing will degrade the antenna gain, and the larger the area of the gap is, the greater the antenna gain loss will be. As a result, the sag-to-span ratio of the support cable net should be designed as small as possible.

(2) Determination of support cable net node coordinates

Figure 3.39 shows the projection of the modular deployable antenna on the parent paraboloid coordinate plane $O_3X_3Y_3$. It can be seen that the six vertices of the support cable net on each module form a regular hexagon in the projection plane, and the side length of the regular hexagon is L. For convenience in the following discussion, the modules are classified; the module located at the center is called the first layer, the peripheral modules outside of the first layer module are the second layer, the peripheral modules outside of the second layer modules are the third layer, and so on. The center of the vth module of the uth layer (hereinafter referred to as the uvth module) is marked as $c_{u,v}$. The numbering rules are shown in Fig. 3.39. The

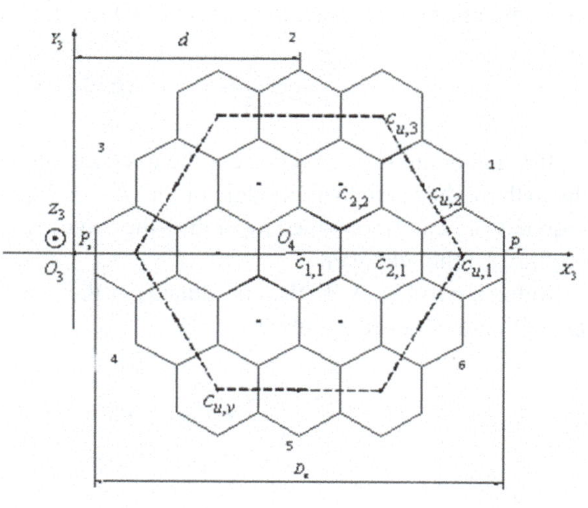

Fig. 3.39 Topology of modular deployable antenna

forward module of coordinate axis O_3X_3 in each layer is marked as the first module of each layer, and the other modules of the same layer are numbered counterclockwise. Clearly, the distance from the center $c_{1,1}$ to the axis of the parent paraboloid is the offset distance d of the whole reflector, and the coordinate of the center $c_{1,1}$ on the projection surface $O_3X_3Y_3$ is $(d, 0)$. The total width of all the modules along axis O_3X_3 is called the electronic aperture D_e of the deployable modular antenna. The two end points of the antenna on the O_3X_3 axis are, respectively, denoted as $P_r\left(d+\frac{D_e}{2}, 0, (d+\frac{D_e}{2})^2/(4F)\right)^{\mathrm{T}}$ and $P_s\left(d-\frac{D_e}{2}, 0, (d-\frac{D_e}{2})^2/(4F)\right)^{\mathrm{T}}$.

If there are a total of N layer modules, the total number of modules S is

$$S = 1 + 6\sum_{u=1}^{N}(u-1) \tag{3.117}$$

The relationship between D_e and the length L of the hexagon is

$$D_e = \sqrt{3}(2N-1)L \tag{3.118}$$

To obtain the coordinates of the center point of each module on the projection surface $O_3X_3Y_3$, the uvth module is taken as an example for illustration. First, the modules in each layer are subdivided into 6 partitions. As seen from Fig. 3.39, the center points of the same layer modules in the same partition are collinear, and the center points of all modules in the same layer connected in sequence will form a positive hexagon in the projection surface $O_3X_3Y_3$. The total number of modules in the uth layer is $6(u-1)$, $(u>1)$, and each partition in the uth layer contains u modules, where two modules located on the boundary of a partition also belong to its neighboring partition.

For modules in the same layer, if they are sorted according to the modules of the same partition counterclockwise, the central coordinates of the mth module in the kth partition can be obtained by rotating the central coordinates of the mth module of the first partition around the central point $c_{1,1}$ with angle φ_k. Angle φ_k is

$$\varphi_k = \frac{\pi}{3}(k-1) \quad (k = 2, 3, \ldots, 6) \tag{3.119}$$

If k and m are used to represent the quotient and remainder of v divided by u-1, the uvth module is the mth module of the $(k+1)$th partition. The coordinates of the required $c_{u,v}$ can be obtained by finding the center $c_{u,m}$ of the mth module in the first partition of the uth layer.

Referring to Fig. 3.39, the coordinates of the center $c_{u,m}$ on the projection plane $O_3X_3Y_3$ are obtained as follows

$$(T_1, T_2)^{\mathrm{T}} = \left(2\sqrt{3}(2u-m-1)L, \frac{3}{2}(m-1)L\right)^{\mathrm{T}} \tag{3.120}$$

where $m = 1, 2, \ldots u-2\ , u \geq 2$.

Therefore, the coordinates of $c_{u,v}$ are

$$c_{u,v} = \begin{bmatrix} \cos\varphi_k & -sin\varphi_k \\ \sin\varphi_k & cos\varphi_k \end{bmatrix} \begin{bmatrix} T_1 \\ T_2 \end{bmatrix} + \begin{bmatrix} d \\ 0 \end{bmatrix} \tag{3.121}$$

Substituting $Xc_{u,v}$ and $Yc_{u,v}$ into the parent paraboloid (3.2) gives the coordinate $Zc_{u,v}$ of the center of the *n*th module.

Next, we discuss the nodal coordinate calculation for the support cable net of an arbitrary module. After obtaining the node coordinates of the module in its coordinate system, the nodal coordinates in the antenna coordinate system can be obtained. For generality, the uvth module is taken as an example to illustrate.

When designing the support cable net for the *uv*th module, the paraboloid coordinate system $O_3^{u,v}X_3^{u,v}Y_3^{u,v}Z_3^{u,v}$ of the module is established at the origin O_3 (Fig. 3.40). The $O_3^{u,v}Z_3^{u,v}$ axis coincides with the O_3Z_3 axis. $O_3^{u,v}X_3^{u,v}$ points from the paraboloid center O_3 to the center $c_{u,v}$ of the uvth module. Then $O_3^{u,v}X_3^{u,v}Z_3^{u,v}$ is the plane containing the parent paraboloid of the module. The coordinate axis $O_3^{u,v}X_3^{u,v}$ coincides with the long axis of the ellipse, whose two end points are denoted as $P_s^{u,v}$ and $P_r^{u,v}$. Each module of the antenna can be formed by a parent paraboloid cut by a cylinder with diameter $D_e^{u,v}$. To obtain the nodal coordinates of the support cable net of a module, the module can be regarded as an independent offset reflector antenna.

Clearly the electronic aperture of the module is $D_e^{u,v}$, and the distance from O_3 to the center point $c_{u,v}$ of the module is the offset of the module, which is denoted

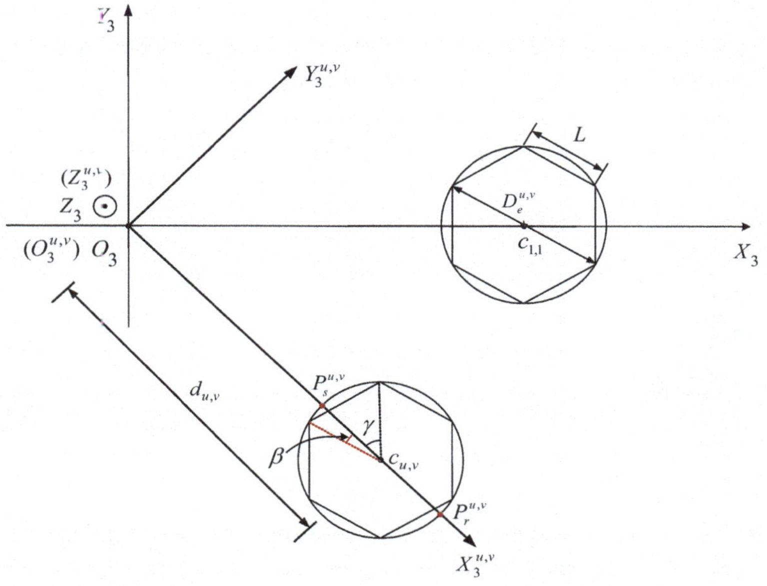

Fig. 3.40 Projection of the *uv*th module on $O_3^{u,v}X_3^{u,v}Z_3^{u,v}$

as $d^{u,v}$. As shown in Fig. 3.40, the angle β between the line from the lower vertex of the long axis of the module to the central point and the long axis of the ellipse is $\beta = \frac{\pi}{3} - \arctan(\left|Xc_{u,v}/Yc_{u,v}\right|)$.

As shown in Fig. 3.40, the electronic aperture $D_e^{u,v}$ of the module is

$$D_e^{u,v} = 2L \tag{3.122}$$

The offset distance $d^{u,v}$ of the *uv*th module is

$$d^{u,v} = \sqrt{(Xc_{u,v})^2 + (Yc_{u,v})^2} \tag{3.123}$$

The coordinates of the two end points of the elliptical long axis in coordinate system $O_3^{u,v}X_3^{u,v}Y_3^{u,v}Z_3^{u,v}$ are $P_s^{u,v}\left(d^{u,v} - D_e^{u,v}/2, 0, \frac{(d^{u,v}-D_e^{u,v}/2)^2}{4F}\right)^{\mathrm{T}}$ and $P_r^{u,v}\left(d^{u,v} + D_e^{u,v}/2, 0, \frac{(d^{u,v}+D_e^{u,v}/2)^2}{4F}\right)^{\mathrm{T}}$, respectively.

The process of mesh generation refers to the method in Point 3 of Sect. 3.2.1. The difference here is that the mesh generation begins with the long axis of the ellipse with the end points $P_s^{u,v}$ and $P_r^{u,v}$. Here we need to start with angle β; that is, the initial angle is different. Corresponding to $\alpha_{n_r,j} = (j-1)\frac{2\pi}{n_r n_s} - \beta$ in Eq. (3.39) of Point 3 of Sect. 3.2.1, the other derivation processes are similar. The coordinates of the mesh nodes of the uvth module in its local coordinate system $O_4^{u,v}X_4^{u,v}Y_4^{u,v}Z_4^{u,v}$ are

$$(X_{4i,j}^{u,v}, Y_{4i,j}^{u,v}, Z_{4i,j}^{u,v})^{\mathrm{T}} = (X_{i,j}, Y_{i,j}, Z_{i,j})^{\mathrm{T}} \quad i = 1, \ldots, n'; \; j = 1, 2, \ldots, 6i \tag{3.124}$$

Then, we convert the mesh node coordinates of the *uv*th module from its local coordinate system $O_4^{u,v}X_4^{u,v}Y_4^{u,v}Z_4^{u,v}$ to $O_3^{u,v}X_3^{u,v}Y_3^{u,v}Z_3^{u,v}$

$$(X_{3i,j}^{u,v}, Y_{3i,j}^{u,v}, Z_{3i,j}^{u,v}, 1)^{\mathrm{T}} = r_1^{u,v}(X_{4i,j}^{u,v}, Y_{4i,j}^{u,v}, Z_{4i,j}^{u,v}, 1)^{\mathrm{T}} \tag{3.125}$$

and

$$r_1^{u,v} = \begin{pmatrix} \cos\alpha & 0 & \sin\alpha & \cos\alpha \cdot d^{u,v} + \sin\alpha\frac{(d^{u,v})^2}{4F} + Xc_{u,v} - \frac{X_{P_r^{u,v}}+X_{P_s^{u,v}}}{2} \\ 0 & 1 & 0 & 0 \\ -\sin\alpha & 0 & \cos\alpha & -\sin\alpha \cdot d^{u,v} + \cos\alpha\frac{(d^{u,v})^2}{4F} + Zc_{u,v} - \frac{Z_{P_r^{u,v}}+Z_{P_s^{u,v}}}{2} \\ 0 & 0 & 0 & 1 \end{pmatrix} \tag{3.126}$$

where $\alpha = \arctan\frac{Z_{P_r^{u,v}}-Z_{P_s^{u,v}}}{X_{P_r^{u,v}}-X_{P_s^{u,v}}}$.

Next, the nodal coordinates of the uvth module in coordinate system $O_3^{u,v}X_3^{u,v}Y_3^{u,v}Z_3^{u,v}$ are transformed to the antenna parent paraboloid coordinate system $O_3X_3Y_3Z_3$

$$(X_{3u,v,i,j}, Y_{3u,v,i,j}, Z_{3u,v,i,j}, 1)^{\mathrm{T}} = r_2^{u,v}(X_{3i,j}^{u,v}, Y_{3i,j}^{u,v}, Z_{3i,j}^{u,v}, 1)^{\mathrm{T}} \tag{3.127}$$

and

$$r_2^{u,v} = \begin{pmatrix} \cos(\gamma - \frac{\pi}{2}) & -\sin(\gamma - \frac{\pi}{2}) & 0 & 0 \\ \sin(\gamma - \frac{\pi}{2}) & \cos(\gamma - \frac{\pi}{2}) & 0 & 0 \\ 0 & 0 & 1 & 0 \\ 0 & 0 & 0 & 1 \end{pmatrix} \tag{3.128}$$

Finally, the nodal coordinates in $O_3X_3Y_3Z_3$ are converted to the antenna coordinate system $O_4X_4Y_4Z_4$

$$(X_{4u,v,i,j}, Y_{4u,v,i,j}, Z_{4u,v,i,j}, 1)^{\mathrm{T}} = r_3(X_{3u,v,i,j}, Y_{3u,v,i,j}, Z_{3u,v,i,j}, 1)^{\mathrm{T}} \tag{3.129}$$

and

$$r_3 = \begin{pmatrix} \cos\varsigma & 0 & -\sin\varsigma & d \\ 0 & 0 & 0 & 0 \\ \sin\varsigma & 0 & \cos\varsigma & \frac{d^2}{4F} \\ 0 & 0 & 0 & 1 \end{pmatrix} \tag{3.130}$$

where $\varsigma = \arctan\frac{Z_{P_r} - Z_{P_s}}{X_{P_r} - X_{P_s}}$.

(3) Antenna height

(1) Deployed height

Figure 3.41 shows the section diagram of the modular deployable antenna in the plane $O_3X_3Z_3$. The front and rear support cable nets are fixed on the stand-offs, and the truss is located on the back of the support cable net, which together with the stand-offs, forms the support structure of the support cable net.

The nodes of the upper and lower structures are located on two concentric spheres, and the antenna reflector is a paraboloid. The spheres on which the upper and lower trusses located are concentric to the sphere fitted by the corresponding parent paraboloid. Now, we connect the two end points P_r and P_s and then move the line P_rP_s in parallel so that it is tangent to the parent paraboloid and the upper and lower frame structure. The distance between line P_rP_s and the tangent lines correspond to the height of the different parts of the antenna H_1, H_2, and H_3 (Fig. 3.41). The total height H equals the sum of the 3 parts as follows,

$$H = H_1 + H_2 + H_3 \tag{3.131}$$

On plane $O_3X_3Z_3$, for a point M on the parent paraboloid, the parent paraboloid equation should be satisfied, and the tangent slope of the parent paraboloid at point

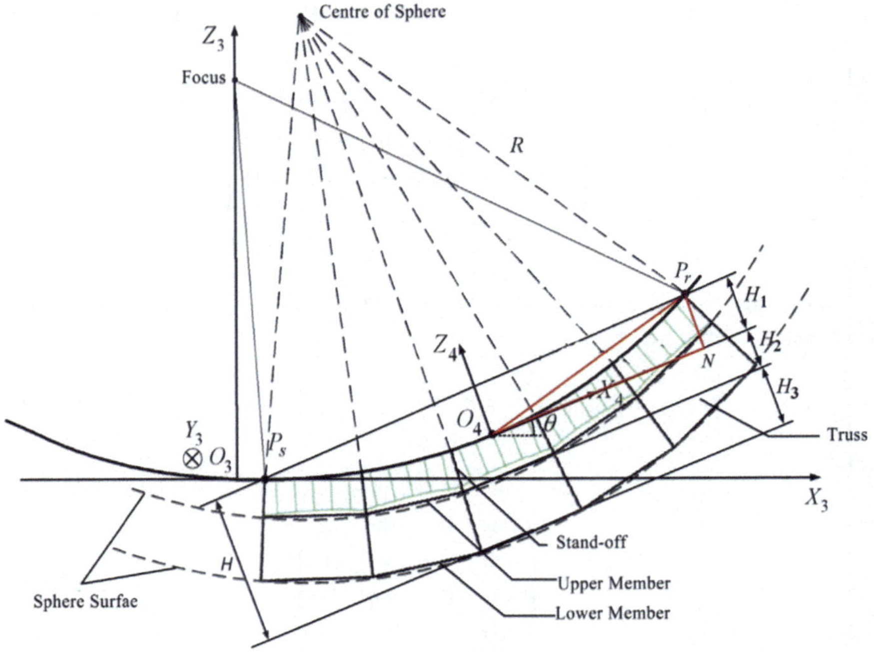

Fig. 3.41 Cross-sectional diagram of modular deployable antenna

M is equal to the slope of the line segment $P_r P_s$. The parent paraboloid equation is $Z_3 = \frac{X_3^2}{4F}$, and the coordinates (X_M, Z_M) of point M satisfy the equation

$$\begin{cases} Z_M = \frac{X_M^2}{4F} \\ \frac{2X_M}{4F} = \frac{Z_{P_r} - Z_{P_s}}{X_{P_r} - X_{P_s}} \end{cases} \tag{3.132}$$

By substituting the coordinates of P_r and P_s into Eq. (3.132), the coordinates obtained for point M are $(d, \frac{d^2}{4F})^{\mathrm{T}}$, that is, the origin of the local coordinate system O_4. Making a triangle $P_r Q_4 N$ yields the following angle:

$$\angle P_r O_4 N = \arctan \frac{Z_{P_r} - Z_{O_4}}{X_{P_r} - X_{O_4}} - \theta \tag{3.133}$$

where $\theta = \arctan \frac{d}{2F}$.

Then

$$\begin{aligned} H_1 &= \sqrt{(X_{P_r} - X_{O_4})^2 + (Z_{P_r} - Z_{O_4})^2} \cdot \cos \angle P_r O_4 N \\ &= \frac{D_e^2 (4F^2 + 1)}{16 F d^2} \cos \theta \end{aligned} \tag{3.134}$$

For the stability of the antenna structure, the stand-off should not be too long. Generally, $L : H_3 : H_2 = 20 : 5 : 1$ [9]; that is

$$H_3 = \frac{1}{4}L \tag{3.135}$$

$$H_2 = \frac{1}{20}L \tag{3.136}$$

The deployed height of the antenna can be obtained by Eqs. (3.134)–(3.136).

(2) Stowed height

In the stowed state, the modules are connected with bar IH, so IH length is equal. Since the curvature of the paraboloid increases outwards, the lower chord of the truss which is the sum of the lengths of DE and EF increases. The planar figure of the whole antenna in the stowed state is shown in Fig. 3.42. From left to right, the parts correspond to the first, second, third, …, $(N + 1)$th module in Fig. 3.41. The stowed height formula is given in the first part of this section.

(4) Truss design

First, the paraboloid is approximated with the spherical surface closest to it (Fig. 3.41). Then, according to the height requirements of the antenna parts, the radius of the spherical surface of the upper and lower trusses is determined; next, the spatial geometric positions of the key nodes on the support structure are calculated. Finally, the geometric parameters of each module are determined to ensure an accurate connection between the modules.

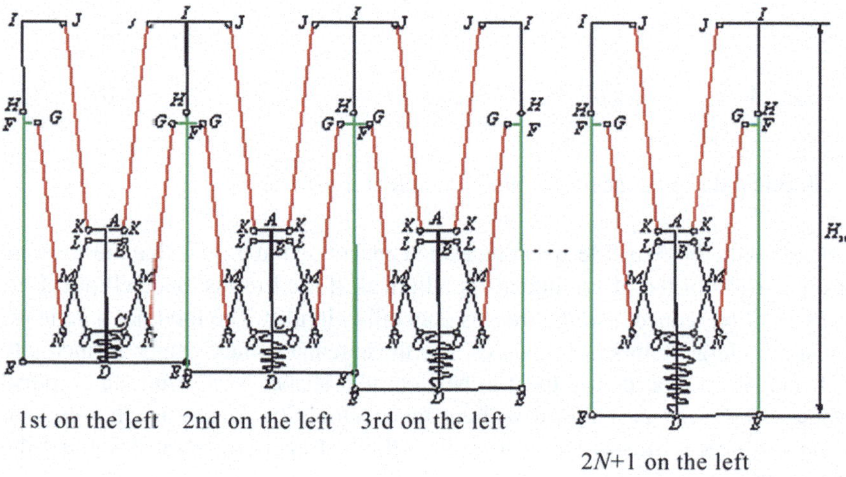

Fig. 3.42 Stowed state of modular deployable antenna

As defined in Sect. 3.1, the radius R of the sphere closest to the paraboloid is given by Eq. (3.1) where D is the length of $P_r P_s$.

The spherical radius of the upper truss is equal to the sum of R and the length of the stand-off H_2. Therefore, the equation of the spherical coordinates of the key nodes of the upper truss in coordinate system $O_4X_4Y_4Z_4$ is

$$X^2 + Y^2 + (Z + R)^2 = (R + H_2)^2 \tag{3.137}$$

The normal vector of point (X_i, Y_i, Z_i) on the sphere is

$$\overrightarrow{N_i} = [2X_i, 2Y_i, 2(Z_i + R)]^{\mathrm{T}} \tag{3.138}$$

and the corresponding unit vector is

$$\overrightarrow{N_i} = [\frac{X_i}{\sqrt{X_i^2 + Y_i^2 + (Z_i + R)^2}}, \frac{Y_i}{\sqrt{X_i^2 + Y_i^2 + (Z_i + R)^2}}, \frac{Z_i + R}{\sqrt{X_i^2 + Y_i^2 + (Z_i + R)^2}}]^{\mathrm{T}} \tag{3.139}$$

As mentioned above, the six vertices of the hexagonal modular support cable net can be expressed as $(X_{4u,v,n',j}, Y_{4u,v,n',j}, Z_{4u,v,n',j}, 1)^{\mathrm{T}}$. The key nodes of the upper truss are located at the vertex of the support cable net along the normal distance H_2 of the sphere, and the coordinates are

$$(X_{4u,v,U,j}, Y_{4u,v,U,j}, Z_{4u,v,U,j}, 1)^{\mathrm{T}} = (X_{4u,v,n',j}, Y_{4u,v,n',j}, Z_{4u,v,n',j}, 1)^{\mathrm{T}} + \overrightarrow{N_i} H_2 \tag{3.140}$$

Similarly, it can be inferred that the coordinates of the nodes of the lower truss are

$$(X_{4u,v,D,j}, Y_{4u,v,D,j}, Z_{4u,v,D,j}, 1)^{\mathrm{T}} = (X_{4u,v,n',j}, Y_{4u,v,n',j}, Z_{4u,v,n',j}, 1)^{\mathrm{T}} + \overrightarrow{N_i}(H_2 + H_3) \tag{3.141}$$

(5) Elliptical design concept for modular antenna structure

A problem with the AstroMesh antenna when designing the truss is that the working aperture surface of the offset antenna is elliptical. If the truss is made elliptical, its stiffness will be reduced, so the truss is generally circular. The modular antenna is made up of a large number of modules, and the antenna surface can be formed into any shape; it is not necessary for it to be a circular shape. As a result, an elliptical aperture surface can be realized without much difficulty. Therefore, the modular antenna can be designed directly from the elliptical aperture when designing the support cable net and the truss.

3.2.3 Rigid-Rib Antenna

Since the rigid-rib antenna is similar to the common umbrella, it is also called the umbrella antenna (Fig. 3.43). Its advantages are a simple deployment mechanism, high deployment reliability, high furled rate for a small aperture antenna, and high reflector precision. The disadvantage is that many ribs (especially for high precision requirements with large-diameter antennas) have difficulty achieving high consistency, which tends to cause uneven force, thus affecting the surface precision. Such antennas are commonly used in high precision reflector antennas with a diameter of less than 5 m.

The structural details of this type of antenna are shown in Fig. 3.9; Fig. 3.9a shows the support rib structure with all ribs connected to the ring support base of the central body by sliding hinges. There are many structural forms of support rigid-ribs. Commonly used forms are the truss (Fig. 3.10a) and the rib plate (Fig. 3.10b). The upper chord of the ribs is a parabola or shaped curve with focal length F.

The composition of the central body and the feed system shown in Fig. 3.9b is that the minor surface is fixed under the backing plate, and the backing plate, the brace, the support sleeve and the horn which are connected and fixed on the top surface of the central body in turn. The central body is a thin-walled barrel structure that supports the feed.

The deployment of the rigid-rib antenna is usually driven by a motor (at the base of the rib), as shown in Fig. 3.9c. When the antenna is deployed, the slider moves downward to rotate the rib hinged on the annular seat of the central body until it is fully deployed.

The support cable net is connected between adjacent ribs. When the antenna is fully deployed, the support cable net is in tension as shown in Fig. 3.9d.

For easy of discussion, the parabolic coordinate system $O_3X_3Y_3Z_3$ is established with the origin located at the center of the reflector of the rigid-rib antenna, the Z axis coincides with the axis of the paraboloid, and the X axis passes through a rib, as shown in Fig. 3.9d.

A three-dimensional support cable net between two adjacent ribs is projected into the plane $O_3X_3Y_3$ (Fig. 3.44), which includes the circumferential cables and other cables. The upper- and lower-chord nodes of the ribs are connected with the front

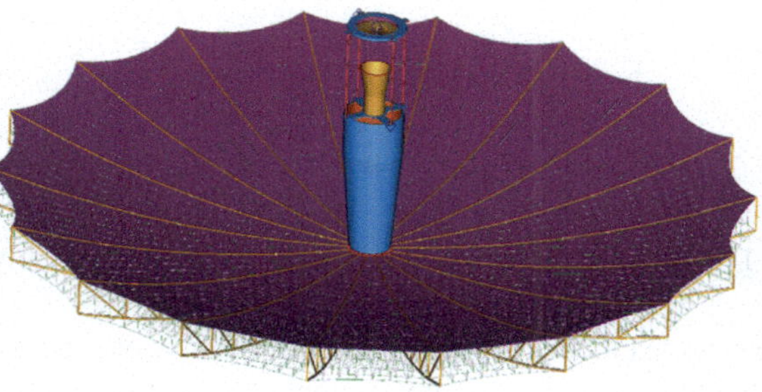

Fig. 3.43 Rigid-rib antena

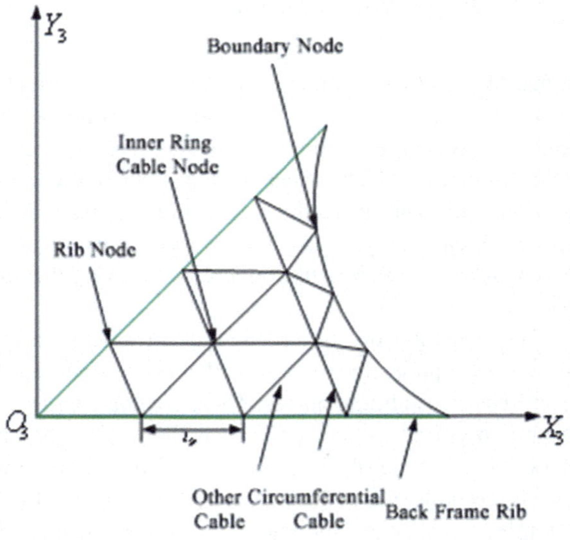

Fig. 3.44 Three-dimensional net between two supporting ribs

and rear support nets respectively. The inner ring cable nodes are the nodes of the support cable net other than the rib nodes, and the boundary nodes are the outer ring nodes of the support cable net other than the rib nodes. A mesh is laid on the rotational parabolic surface formed by the support rib and the front support cable net as a reflective surface.

1. Support cable net design

The design concept of the support cable net is to determine the edge length of the triangular net according to the physical aperture D by Eqs. (3.6) or (3.8). Then the total number of ribs n_{rb} and the total number of rings n_{rbi} divided among the ribs can be obtained from Eq. (3.13a). If the net generation method shown in Fig. 3.44 is used, the total number $nu1$ of triangular elements, cable number of the triangles $nu2$ of a single layer support cable net, and total number of nodes $nu3$ can be obtained

$$nu1 = n_{rb} \cdot n_{rbi}^2 \tag{3.142}$$

$$nu2 = n_{rb} \cdot \frac{n_{rbi}(1 + 3n_{rbi})}{2} \tag{3.143}$$

$$nu3 = 1 + n_{rb} \cdot \frac{n_{rbi}(1 + n_{rbi})}{2} \tag{3.144}$$

(1) Determination of internal nodes on the front support cable net

For convenience in the following discussion, the total number n_{rbij} of segments divided by the jth circumferential cable between adjacent ribs is introduced.

(1) Upper chord node coordinates on supporting rib

For the rigid rib shown in Fig. 3.9c, the coordinates of the upper chord node should satisfy

$$\begin{cases} 4Fz_{3,j} = x_{3,j}^2 \\ \left(z_{3,j} - z_{3,j-1}\right)^2 + \left(x_{3,j} - x_{3,j-1}\right)^2 = \left(\frac{S_l}{2n_{rbi}}\right)^2 \end{cases}, \ (j = 2, \ldots, n_{rbi} + 1) \quad (3.145)$$

Clearly, the coordinates of the center node 1 are zero.

To obtain the coordinates of the upper chord nodes on the corresponding ring of the other ribs, let $r_{ij} = x_{3,j} (i = 1, 2, \ldots, n_{rb}; j = 1, 2, \ldots, n_{rbi})$, $\bar{\theta} = 2\pi/n_{rb}$. Then the coordinates of the jth upper chord node of the ith rib are

$$\begin{cases} x_{3,ij} = r_{ij} \cos i\bar{\theta} \\ y_{3,ij} = r_{ij} \sin i\bar{\theta} \\ z_{3,ij} = z_{3,j} \end{cases}, \ (i = 2, \ldots, n_{rb}; \ j = 2, \ldots, n_{rbij} + 1) \quad (3.146)$$

(2) Internal nodes on the front support cable net

It can be seen from Fig. 3.9d that the cable length of the jth ($j = 1, 2, \ldots, n_{rbi}$) ring between the two ribs is $s_j = r_{ij}\bar{\theta}$, and the side length of the triangular net obtained according to the accuracy of the surface shape is l (from Eqs. (3.6) or (3.8)), the total number of segments divided by the ring is $n_{rbij} = s_j/l$. It is assumed that the coordinates of the starting and ending points of the jth ring of the ith and (i + 1)th radial ribs are $(x_{3,ijs}\ y_{3,ijs}\ z_{3,ijs})^T$ and $(x_{3,ije}\ y_{3,ije}\ z_{3,ije})^T$, respectively, and the coordinates of the nodes on the jth ring are

$$\begin{cases} x_{3,k} = \frac{n_{rbij}-k}{n_{rbij}} x_{3,ijs} + \frac{k}{n_{rbij}} x_{3,ije} \\ y_{3,k} = \frac{n_{rbij}-k}{n_{rbij}} y_{3,ijs} + \frac{k}{n_{rbij}} y_{3,ije} \\ z_{3,k} = \frac{x_{3,k}^2 + y_{3,k}^2}{4F} \end{cases}, \ (k = 1, 2, \ldots, n_{rbij}) \quad (3.147)$$

(2) Determination of internal nodes on the rear support cable net

Taking the back truss in the $O_3X_3Z_3$ plane (Fig. 3.10) as an example, the problem of determining the shape of the lower chord of the radial rib can be described as a mathematical programming problem when the parameters such as the material and cross-sectional dimensions are given. Its objective function is the root mean square (rms) error of the deviation of the upper chord from its ideal curve. The constraints include mass, natural frequencies, strength, etc. The design variables are the chord shapes (denoted as $z_3 = g(x_3)$). With this curve as the generatrix and the Z_3 axis as the axis of rotation, the rotational surface $z_3 = g(x_3, y_3)$ is obtained. The initial position of the lower support cable net nodes should be located on this surface.

In the initial design, the adjusting cables are set in the vertical direction, that is, the front and rear support net nodes have the same x_3, y_3 coordinates but different z_3 coordinates. Thus we have

$$(x'_{3,k}, y'_{3,k}, z'_{3,k}) = (x_{3,k}, y_{3,k}, z_{3,k} - \bar{h}_j),\ \ (j = 1, 2, \ldots, n_{rbi}; k = 1, 2, \ldots, n_{rbij}) \tag{3.148}$$

where $\bar{h}_j$ is the height of the rigid rib (Fig. 3.9d).

(3) Determination of node coordinates of the boundary support net

If the antenna is in a stowed state, the distance between the outermost support cable net nodes of two adjacent ribs will shrink inward. Figure 3.45 shows the corresponding rigid rib support cable net projected on the horizontal plane. R and R_0 are, respectively, half of the physical aperture D and the electronic aperture D_e projected on the horizontal plane, respectively. Theoretically, the boundary contraction δ_{sag} is related not only to the angle $\bar{\theta}$ between the two radial ribs but also to the cable tension. In the initial design stage, the geometric design of the support cable net is primarily used, so the cable tension can be temporarily ignored.

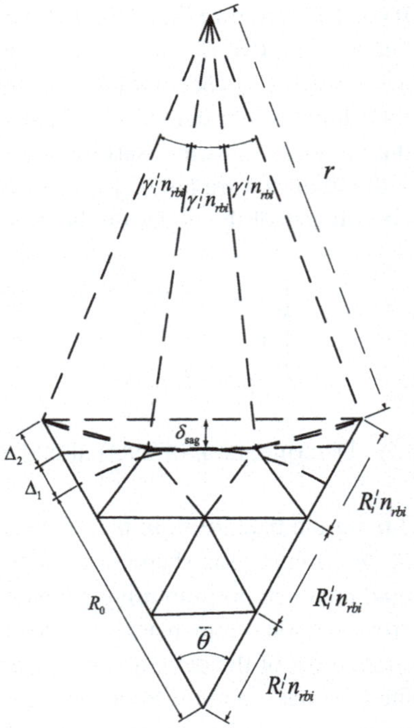

Fig. 3.45 Intercostal cable net division ($n_{rbi} = 3$) and sag-to-span ratio

The sag-to-span ratio can be defined as

$$\tau = \frac{\delta_{sag}}{2R_0 \tan(\bar{\theta}/2)} \tag{3.149}$$

Clearly, $\bar{\theta} = 2\pi/n_{rb}$.

Then

$$\Delta_1 = R_0 \frac{1 - \cos(\bar{\theta}/2)}{\cos(\bar{\theta}/2)} \tag{3.150}$$

$$\Delta_2 = \frac{\delta_{sag}}{\cos(\bar{\theta}/2)} \tag{3.151}$$

So

$$\frac{R}{R_0} = \frac{1 + 2\tau \tan(\bar{\theta}/2)}{\cos(\bar{\theta}/2)} \tag{3.152}$$

If R is divided equally into n_{rbi} (assumed to be 3 in the figure), and the angle γ corresponding to the outermost cable is also divided equally, we have

$$r = \frac{\delta_{sag}^2 + R^2 \sin^2(\bar{\theta}/2)}{2\delta_{sag}} \tag{3.153}$$

$$\gamma = 2 \arccos \frac{r - \delta_{sag}}{r} \tag{3.154}$$

Note that when n_{rbi} is an odd number, the actual value of the sag will be slightly smaller than the value in Fig. 3.45.

It is worth pointing out that the above method affects the sag δ_{sag} only for the outermost cable net nodes and not the internal nodes. This of course is not reasonable in reality, but it is acceptable in the initial design since the coordinates of the support cable net nodes will be changed in the detailed design later.

According to Fig. 3.45, when the parameters n_{rb} and n_{rbi} are given, the maximum sag will be

$$\delta_{sag}^{\max} = \frac{R\cos(\bar{\theta}/2)}{n_{rbi}} \tag{3.155}$$

and the maximum sag-to-span ratio is

$$\tau_{\max} = \frac{1}{2(n_{rbi} - 1)\tan(\bar{\theta}/2)} \tag{3.156}$$

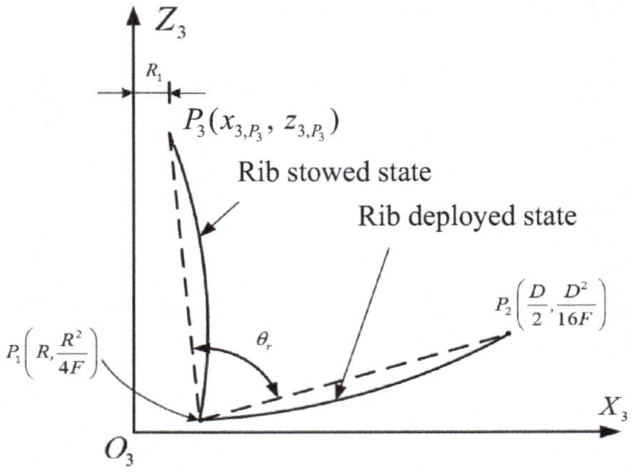

Fig. 3.46 Deployed state and stowed state of rib

2. Opening angle between the rigid rib in stowed and deployed states

As shown in Fig. 3.46, the upper chord of the antenna rotates in the counterclockwise direction with center P_1 and radius P_1P_2. When has rotated to point P_3, it is fixed on the cylinder with radius R_1.

Since the rigid rib must to satisfy the parabolic equation $z_3 = \frac{x_3^2}{4F}$, the coordinates of points P_1 and P_2 are $(R, \frac{R^2}{4F})$ and $(\frac{D}{2}, \frac{D^2}{16F})$, respectively. At the same time, the slope of the line passing through points P_1 and P_2 is

$$k = (\frac{\frac{D^2}{16F} - \frac{R^2}{4F}}{D/2 - R}) \tag{3.157}$$

The chord length when the rib is deployed and stowed are

$$|P_1P_2| = \sqrt{(R - D/2)^2 + (R^2/4F - D^2/16F)^2} \tag{3.158}$$

$$|P_1P_3| = \sqrt{(R - R_1)^2 + (R^2/4F - z_3)^2} \tag{3.159}$$

The value of z_3 can be found from the fact that the two chords are of equal length, and so the slope of the rib in the stowed state is

$$k' = \frac{|z_3 - R^2/4F|}{|R_1 - R|} \tag{3.160}$$

For the deployed state, the opening angle θ_r (that is, the angle between P_1P_2 and P_1P_3) is

$$\theta_r = \tan^{-1}\frac{k' - k}{1 + k'k} \tag{3.161}$$

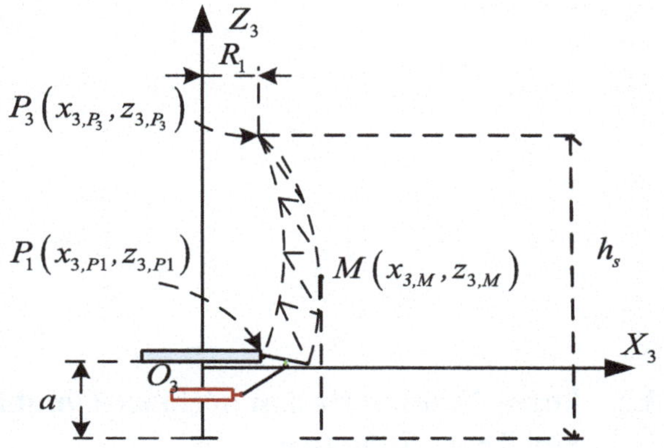

Fig. 3.47 Rigid rib stowed height computation

3. Deployed and stowed height of antenna

As shown in Fig. 3.9c, the height of the antenna in the deployed state is

$$h_d = \frac{D^2}{16F} + a \tag{3.162}$$

For the stowed state, if we want to obtain its corresponding height h_s, the coordinates (x_{3,P_3}, z_{3,P_3}) of point P_3 in Fig. 3.47 should be obtained first; these coordinates are exactly those of point P_2 after a rotation around point P_1 with angle θ_r and are determined by

$$\begin{bmatrix} x_{3,P_3} - R \\ z_{3,P_3} - \frac{R^2}{4F} \end{bmatrix} = \begin{bmatrix} \cos\theta_r \, \sin\theta_r \\ -\sin\theta_r \, \cos\theta_r \end{bmatrix} \begin{bmatrix} x_{3,P_2} - R \\ z_{3,P_2} - \frac{R^2}{4F} \end{bmatrix} \tag{3.163}$$

Therefore, the height in the stowed state is

$$h_s = \frac{R^2}{4F} - \left(\frac{D}{2} - R\right)\sin\theta_r + \left(\frac{D^2}{16F} - \frac{R^2}{4F}\right)\cos\theta_r + a \tag{3.164}$$

The radius of the stowed state is the coordinate value of point M in Fig. 3.47. According to the assumption that the lower chord of the rigid rib satisfies $z_3 = g(x_3)$, the chord curve function $z_3 = \tilde{g}(x_3)$ in the stowed state can be obtained by a rotational transformation.

4. Non-axisymmetric rigid-rib antenna

Due to the simplicity of the rigid-rib deployable antenna, it is sometimes used for the offset feed. For this case, the support ribs of the back truss will no longer be axially symmetrically distributed, but will be in the form shown in Fig. 3.48, which will not be described again here.

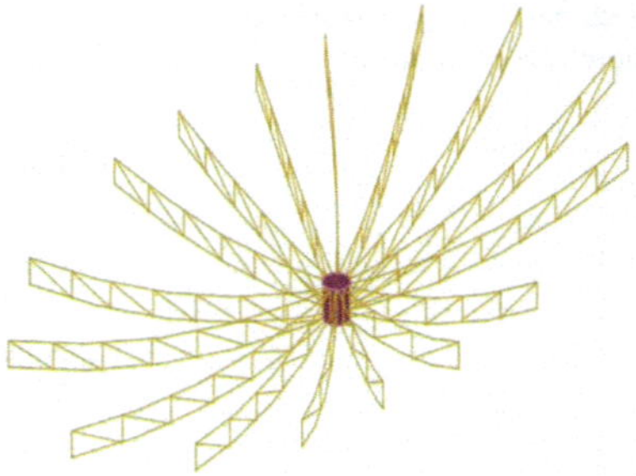

Fig. 3.48 Schematic diagram of non-axisymmetric rigid ribs

3.3 Form-Finding Design of Space Combined Cable and Truss Structure

3.3.1 Problem Presentation

As previously mentioned, the four developing tendencies of the spaceborne deployable antenna are a large aperture, a high precision, a light weight, and a high deploy/fold ratio. Large space mesh deployable antennas have been widely used. The mesh antenna is generally composed of a support truss, support cable net and mesh. The support cable net is tensioned to form a specific curved surface to meet the required electronic performance.

Since the support cable net structure itself is a highly geometrically nonlinear structure coupled with the truss structure, the interaction between them makes the problem more complicated. The foundation of such antenna design is analysis, and the difficulty of analysis is "shape" and "state" analysis. The "shape" means that the antenna needs to have high surface precision to ensure electronic performance. The "state" means that the structure has a reasonable tension distribution to ensure that the support cable net does not relax in the harsh thermal environment of space.

The methods of studying the form-finding of a support cable net structure are relatively mature [5, 9, 12–19] and mainly include the dynamic relaxation method, nonlinear finite element method, equilibrium matrix method and force density method. Unfortunately, the mesh needs a support structure composed of truss and cable net. The cable net and truss structures are coupled. When the truss is deformed, the boundary of the support cable net is changed, which may lead to a large change in the position and tension of the support cable net [21, 22]. This will reduce the precision and tension uniformity of the support cable net and may even result in slack cables. The existing methods usually first carry on the form-finding design of the support cable net and then take into account the influence of the truss deformation separately. In [20], the influence of the support truss structure on the support cable net is attributed to a change in the spatial position of the support cable net boundary points, and no tension design is carried out based on the cable truss structure. In [21], the support cable net is divided into internal and boundary cables connected with the truss, and the two parts are designed separately, which has certain limitations.

In [22], the inverse iterative method is used to improve the reflector precision of the cable truss net structure, but the tension of the support cable net cannot be effectively controlled.

None of the above methods [20–22] fundamentally solve the form-finding design problem of the cable-truss combined structures. Therefore, a form-finding analysis and design method for cable -truss combined structure is proposed below.

3.3.2 *Initial Pretension Design of Cable-Truss Structure*

For cable truss combined structures, the traditional initial pretension design method for the support cable net is not effective when the elastic deformation of the support truss structures is too large. For this reason, a new method is proposed, that is, to find a better “shape” and “state” by changing the initial length of the cable. The initial length of the cable is the length of the cable in the unstressed state. Suppose the initial length of the cable is l_0, the cable length and cable tension in the equilibrium state are l and T, respectively, and the elastic modulus and cross-sectional area of the cable are E and A respectively. Then

$$T = EA\frac{l - l_0}{l_0} \tag{3.165}$$

In the cable truss structure, assuming the total number of cable and truss elements is n, the total number of cable elements is m, and the initial length of the ith cable is l_{0i}, the initial length vector of the support cable net is

$$\mathbf{l}_0 = [l_{01}, l_{02}, \ldots, l_{0m}]^{\mathrm{T}} \tag{3.166}$$

The node with precision requirements in the cable truss structure is denoted as C, and the total number of nodes in C is nup. The sequence numbers are named as $q_1, \ldots, q_{\mathrm{nup}}$ respectively. When the initial length of the support cable net is $\mathbf{l}_0^{(0)}$, the coordinates of node q_i are $\mathbf{x}_{0q_i}$, and the node coordinates in C are $\mathbf{x}_{0q} = [\mathbf{x}_{0q_1}^{\mathrm{T}}, \mathbf{x}_{0q_2}^{\mathrm{T}}, \ldots, \mathbf{x}_{0q_{\mathrm{nup}}}^{\mathrm{T}}]^{\mathrm{T}}$. When the initial length vector becomes $\mathbf{l}_0 = \mathbf{l}_0^{(0)} + \Delta\mathbf{l}_0$, the coordinates of node q_i are assumed to be $\mathbf{x}_{q_i} = \mathbf{x}_{0q_i} + \Delta\mathbf{x}_{q_i}$, and the node coordinates in C are

$$\mathbf{x}_q = [\mathbf{x}_{q_1}^{\mathrm{T}}, \mathbf{x}_{q_2}^{\mathrm{T}}, \ldots, \mathbf{x}_{q_{nup}}^{\mathrm{T}}]^{\mathrm{T}} \tag{3.167}$$

Let q_i be located on the specified curve (surface) $z = L(x, y)$; then the z-direction error of node q_i can be expressed as

$$\delta_i(\mathbf{x}_{q_i}) = L(x_{q_i}, y_{q_i}) - z_{q_i} \tag{3.168}$$

Furthermore, the root mean square error of the nodes in the z-direction in set C is

$$f_1(\mathbf{l}_0) = \frac{1}{nup}\sum_{i=1}^{nup}\delta^2(\mathbf{x}_{q_i}) \tag{3.169}$$

In cable truss combined structures, the actual and allowable stresses of element i are σ_i and $[\sigma_i]$, respectively. The actual stress of the cable and truss element should be less than the allowable stress of the material; i.e.

$$h(\mathbf{l}_0) = \boldsymbol{\sigma} - [\boldsymbol{\sigma}] \le 0 \tag{3.170}$$

where $\boldsymbol{\sigma} = [\sigma_1, \sigma_2, \ldots, \sigma_n]^{\mathrm{T}}$ and $[\boldsymbol{\sigma}]_{n\times 1} = [[\sigma_1], [\sigma_2], \ldots, [\sigma_n]]^{\mathrm{T}}$.

If the initial length of the support cable net is $l_0^{(0)}$, the strain of the cable is denoted as $\varepsilon_0 = (\varepsilon_{01}, \varepsilon_{02}, \ldots, \varepsilon_{0m})^T$. If the initial length becomes $l_0 = l_0^{(0)} + \Delta l_0$, the strain of the cable is denoted as $\varepsilon = (\varepsilon_1, \varepsilon_2, \ldots, \varepsilon_m)^T$. Simply relying on the stress constraint (3.169) may not guarantee that the tension of the support cable net is greater than zero. To ensure that the support cable net is not slack, special strain constraints are introduced as

$$g(\mathbf{l}_0) = -\boldsymbol{\varepsilon} + \boldsymbol{\alpha} \le 0 \tag{3.171}$$

where $\boldsymbol{\alpha} = [\alpha, \alpha, \ldots, \alpha]^{\mathrm{T}}$ and α is a positive small value, and if the tension is greater than this value, the cable is considered not to be slack.

Up to this point, with the initial length of the cable as the design variable, the root mean square error of the nodes in the z-direction in set C as the objective function, and the structural strength and support cable net not being in the slack state as constraints, the following optimization model (PI) can be established:

$$\begin{aligned} &\text{find} \quad \mathbf{l}_0 = [l_{01}, l_{02}, \ldots, l_{0m}]^{\mathrm{T}} \\ &\min \quad f_1(\mathbf{l}_0) = \frac{1}{nup}\sum_{i=1}^{nup}\delta^2(\mathbf{x}_{q_i}) \\ &\text{s.t.} \quad h(\mathbf{l}_0) = \boldsymbol{\sigma} - [\boldsymbol{\sigma}] \le 0 \\ &\qquad\ \ \mathrm{g}(\mathbf{l}_0) = -\boldsymbol{\varepsilon} + \boldsymbol{\alpha} \le 0 \end{aligned} \tag{3.172}$$

In some cases, it is equally important to ensure the uniformity of the tension of the support cable net. In an actual antenna structure, the difference between cable tensions may be great. If the support cable net is divided into C_d types, and the mean value of strain of the 1st, 2nd, ..., C_d th cable is denoted, respectively, as $\bar{\varepsilon}_1, \bar{\varepsilon}_2, \ldots, \bar{\varepsilon}_{C_d}$, then the mean square value of the strain of the support cable net (corresponding to the cable tension) can be expressed as

$$f_2(\mathbf{l}_0) = \frac{1}{m}(\boldsymbol{\varepsilon} - \bar{\boldsymbol{\varepsilon}})^{\mathrm{T}}(\boldsymbol{\varepsilon} - \bar{\boldsymbol{\varepsilon}}) \tag{3.173}$$

where $\bar{\boldsymbol{\varepsilon}}_{m\times 1} = \left[\bar{\varepsilon}_1, \ldots, \bar{\varepsilon}_1, \bar{\varepsilon}_2, \ldots \bar{\varepsilon}_2, \ldots, \bar{\varepsilon}_{C_d}, \ldots \bar{\varepsilon}_{C_d}\right]^{\mathrm{T}}$ is the vector of the mean values of the cable strain.

Denoting the allowable displacement in the Z coordinate of the nodes in set C as $\bar{u}$, we have

$$p_i(\mathbf{l}_0) = \delta(\mathbf{x}_{q_i}) - \bar{u} \leq 0,\ i = 1, 2, \ldots, nup \tag{3.174}$$

Therefore, with the initial length of the support cable net as the design variable, the mean value of the support cable net strain is the objective function, and an optimization model (PII) is established with the constraints of structural strength, no slack cables in the support cable net and the displacement of the node coordinates in C being within the allowable range:

$$\begin{aligned}
&\text{find}\ \ \mathbf{l}_0 = [l_{01}, l_{02}, \ldots, l_{0m}]^{\mathrm{T}} \\
&\min\ \ f_2(\mathbf{l}_0) = \frac{1}{m}(\boldsymbol{\varepsilon} - \bar{\boldsymbol{\varepsilon}})^T(\boldsymbol{\varepsilon} - \bar{\boldsymbol{\varepsilon}}) \\
&\text{s.t.}\ \ h(\mathbf{l}_0) = \boldsymbol{\sigma} - [\boldsymbol{\sigma}] \leq 0 \\
&\qquad g(\mathbf{l}_0) = -\boldsymbol{\varepsilon} + \boldsymbol{\alpha} \leq 0 \\
&\qquad p_i(\mathbf{l}_0) = \delta(\mathbf{x}_{q_i}) - \bar{u} \leq 0,\ i = 1, 2, \ldots, nup
\end{aligned} \tag{3.175}$$

3.3.3 Solution and Convergence of Optimization Model

1. Solution of optimization model

Both the optimization model (PI) and (PII) are highly nonlinear programming problems because the cable strain and node coordinates in objective function and the constraints are all high-order nonlinear functions of the initial length of the cable. For this reason, they are addressed as follows.

For a nonlinear function $z = L(x, y)$, a linear expansion at the point $\left[x_{0q_i}, y_{0q_i}, z_{0q_i}\right]^{\mathrm{T}}$ with respect to the independent variable x, y is employed; then the Taylor expansion series at point $\left[x_{0q_i}, y_{0q_i}, z_{0q_i}\right]^{\mathrm{T}}$ of Eq. (3.168) is

$$\delta_i(\mathbf{x}_{q_i}) = L(x_{0q_i}, y_{0q_i}) + \frac{\partial L(x_{0q_i}, y_{0q_i})}{\partial x}\Delta x_{q_i} + \frac{\partial L(x_{0q_i}, y_{0q_i})}{\partial y}\Delta y_{q_i} - \left(z_{0q_i} + \Delta z_{q_i}\right) \tag{3.176}$$

Let $\mathbf{A}_i = \left(\frac{\partial L(x_{0q_i}, y_{0q_i})}{\partial x}, \frac{\partial L(x_{0q_i}, y_{0q_i})}{\partial y}, -1\right)$, $\Delta\mathbf{u}_{q_i} = \left(\Delta x_{q_i}, \Delta y_{q_i}, \Delta z_{q_i}\right)^{\mathrm{T}}$, and $b_i = L(x_{0q_i}, y_{0q_i}) - z_{0q_i}$; then

$$\delta_i(\mathbf{x}_{q_i}) = \mathbf{A}_i \Delta\mathbf{u}_{q_i} + b_i \tag{3.177}$$

Thus, Eq. (3.169) can be written as

$$f_3(\mathbf{l}_0) = \frac{1}{nup}\sum_{i=1}^{nup}\left(\mathbf{A}_i\Delta\mathbf{u}_{q_i} + b_i\right)^2 = \Delta\mathbf{u}_q^{\mathrm{T}}\mathbf{A}^{\mathrm{T}}\mathbf{A}\Delta\mathbf{u}_q + 2\mathbf{B}^{\mathrm{T}}\mathbf{A}\Delta\mathbf{u}_q + \mathbf{B}^{\mathrm{T}}\mathbf{B} \quad (3.178)$$

where $\mathbf{A} = \frac{1}{\sqrt{Nup}}\mathrm{diag}\left(\mathbf{A}_1, \mathbf{A}_2, \ldots, \mathbf{A}_{Nup}\right)$ and $\mathrm{B} = \frac{1}{\sqrt{Nup}}\left[\mathrm{b}_1, \mathrm{b}_2, \ldots, \mathrm{b}_{Nup}\right]^{\mathrm{T}}$.

Considering the geometric nonlinearity of the structure, the node coordinates are nonlinear functions of the initial length. These are difficult to work with, and the computation is time consuming. For this reason, the node coordinates $\mathbf{x}_q$ of the front cable net are expanded as a Taylor series with respect to $\mathbf{l}_0$ at $\mathbf{l}_0^{(0)}$; i.e.,

$$\mathbf{x}_q = \mathbf{x}_{0q} + \frac{\partial\mathbf{x}_q}{\partial\mathbf{l}_0}\Delta\mathbf{l}_0 \quad (3.179)$$

where

$$\frac{\partial\mathbf{x}_q}{\partial\mathbf{l}_0} = \begin{bmatrix} \frac{\partial\mathbf{x}_{q1}}{\partial l_{01}} & \frac{\partial\mathbf{x}_{q1}}{\partial l_{02}} & \cdots & \frac{\partial\mathbf{x}_{q1}}{\partial l_{0m}} \\ \frac{\partial\mathbf{x}_{q2}}{\partial l_{01}} & \frac{\partial\mathbf{x}_{q2}}{\partial l_{02}} & \cdots & \frac{\partial\mathbf{x}_{q2}}{\partial l_{0m}} \\ \vdots & \vdots & \ddots & \vdots \\ \frac{\partial\mathbf{x}_{qnup}}{\partial l_{01}} & \frac{\partial\mathbf{x}_{qnup}}{\partial l_{02}} & \cdots & \frac{\partial\mathbf{x}_{qnup}}{\partial l_{0m}} \end{bmatrix}_{3nup\times m} \quad (3.180)$$

is the sensitivity matrix of the node coordinates in C with respect to the initial length of the cable.

According to (3.179), we have

$$\Delta\mathbf{u}_q = \mathbf{x}_q - \mathbf{x}_{0q} = \frac{\partial\mathbf{x}_q}{\partial\mathbf{l}_0}\Delta\mathbf{l}_0 \quad (3.181)$$

Substituting Eq. (3.181) into Eq. (3.178) yields

$$f_3 = (\Delta\mathbf{l}_0)^{\mathrm{T}}\left(\mathbf{A}\frac{\partial\mathbf{x}_q}{\partial\mathbf{l}_0}\right)^{\mathrm{T}}\left(\mathbf{A}\frac{\partial\mathbf{x}_q}{\partial\mathbf{l}_0}\right)\Delta\mathbf{l}_0 + 2\mathbf{B}^{\mathrm{T}}\mathbf{A}\frac{\partial\mathbf{x}_q}{\partial\mathbf{l}_0}\Delta\mathbf{l}_0 + \mathbf{B}^{\mathrm{T}}\mathbf{B} \quad (3.182)$$

Similarly, the Taylor expansion of the strain $\boldsymbol{\varepsilon}$ with respect to $\mathbf{l}_0$ at $\mathbf{l}_0^{(0)}$ can be expressed as

$$\boldsymbol{\varepsilon} = \boldsymbol{\varepsilon}_0 + \Delta\boldsymbol{\varepsilon} = \boldsymbol{\varepsilon}_0 + \frac{\partial\boldsymbol{\varepsilon}}{\partial\mathbf{l}_0}\Delta\mathbf{l}_0 \quad (3.183)$$

where

$$\frac{\partial \boldsymbol{\varepsilon}}{\partial \mathbf{l}_0} = \begin{bmatrix} \frac{\partial \varepsilon_1}{\partial l_{01}} & \frac{\partial \varepsilon_1}{\partial l_{02}} & \cdots & \frac{\partial \varepsilon_1}{\partial l_{0m}} \\ \frac{\partial \varepsilon_2}{\partial l_{01}} & \frac{\partial \varepsilon_2}{\partial l_{02}} & \cdots & \frac{\partial \varepsilon_2}{\partial l_{0m}} \\ \vdots & \vdots & \ddots & \vdots \\ \frac{\partial \varepsilon_m}{\partial l_{01}} & \frac{\partial \varepsilon_m}{\partial l_{02}} & \cdots & \frac{\partial \varepsilon_m}{\partial l_{0m}} \end{bmatrix}_{m \times m} \tag{3.184}$$

is the sensitivity matrix of the cable strain with respect to the initial length of the cable.

Substituting Eq. (3.183) into Eq. (3.171) yields

$$g(\mathbf{l}_0) = -\boldsymbol{\varepsilon}_0 - \frac{\partial \boldsymbol{\varepsilon}}{\partial \mathbf{l}_0}\Delta \mathbf{l}_0 + \boldsymbol{\alpha} \le 0 \tag{3.185}$$

In actual mesh antenna shape design, the strength constraint is generally a passive constraint, which can be ignored for the moment, and will be considered in the material strength check at the end. Therefore, the nonlinear programming problem (PI) can be simplified to the following sequential quadratic programming problem (PIII)

$$\begin{aligned} &\text{find} \quad \Delta \mathbf{l}_0 = [\Delta l_{01}, \Delta l_{02}, \ldots, \Delta l_{0m}]^{\mathrm{T}} \\ &\min \quad f_3 = (\Delta \mathbf{l}_0)^{\mathrm{T}}\left(\mathbf{A}\frac{\partial \mathbf{x}_q}{\partial \mathbf{l}_0}\right)^{\mathrm{T}}\left(\mathbf{A}\frac{\partial \mathbf{x}_q}{\partial \mathbf{l}_0}\right)\Delta \mathbf{l}_0 + 2\mathbf{B}^{\mathrm{T}}\mathbf{A}\frac{\partial \mathbf{x}_q}{\partial \mathbf{l}_0}\Delta \mathbf{l}_0 + \mathbf{B}^{\mathrm{T}}\mathbf{B} \\ &\text{s.t.} \quad g = -\boldsymbol{\varepsilon}_0 - \frac{\partial \boldsymbol{\varepsilon}}{\partial \mathbf{l}_0}\Delta \mathbf{l}_0 + \boldsymbol{\alpha} \le 0 \end{aligned} \tag{3.186}$$

Suppose that the cable length and increment in the kth iteration are $\mathbf{l}_0^{(k)}$ and $\Delta \mathbf{l}_0^{(k)}$, respectively; it can be seen that the cable length $\mathbf{l}_0^{(k+1)}$ in the (k + 1)th iteration is

$$\mathbf{l}_0^{(k+1)} = \mathbf{l}_0^{(k)} + \gamma \Delta \mathbf{l}_0^{(k)} \tag{3.187}$$

The step size factor γ can be determined according to the degree of nonlinearity of the function.

Substituting Eq. (3.183) into Eq. (3.173) and expanding it results in

$$f_4(\mathbf{l}_0) = \mathbf{D}^{\mathrm{T}}\mathbf{D} + 2e\mathbf{D}^{\mathrm{T}}\Delta\boldsymbol{\varepsilon} + e^2\Delta\boldsymbol{\varepsilon}^{\mathrm{T}}\Delta\boldsymbol{\varepsilon} \tag{3.188}$$

where $\mathbf{D} = \sqrt{\frac{1}{m}}(\boldsymbol{\varepsilon}_0 - \bar{\boldsymbol{\varepsilon}})$ and $e = \sqrt{\frac{1}{m}}$. According to Eq. (3.19)

$$\Delta\boldsymbol{\varepsilon} = \boldsymbol{\varepsilon} - \boldsymbol{\varepsilon}_0 = \frac{\partial \boldsymbol{\varepsilon}}{\partial \mathbf{l}_0}\Delta \mathbf{l}_0 \tag{3.189}$$

Substituting into Eq. (3.188), this becomes

$$f_4 = \mathbf{D}^{\mathrm{T}}\mathbf{D} + 2e\mathbf{D}^{\mathrm{T}}\frac{\partial \boldsymbol{\varepsilon}}{\partial \mathbf{l}_0}\Delta \mathbf{l}_0 + e^2 \Delta \mathbf{l}_0^{\mathrm{T}} \frac{\partial \boldsymbol{\varepsilon}}{\partial \mathbf{l}_0}^{\mathrm{T}} \frac{\partial \boldsymbol{\varepsilon}}{\partial \mathbf{l}_0}\Delta \mathbf{l}_0 \tag{3.190}$$

According to Eqs. (3.177) and (3.181), we have

$$\delta_i(\mathbf{x}_{q_i}) = \mathbf{A}_i \frac{\partial \mathbf{x}_{q_i}}{\partial \mathbf{l}}\Delta \mathbf{l} + b_i \tag{3.191}$$

Thus, the high-order nonlinear programming problem (PII) can be transformed into the following sequential quadratic programming problem (PIV):

$$\begin{aligned}
&\text{find} \quad \Delta \mathbf{l}_0 = [\Delta l_{01}, \Delta l_{02}, \ldots, \Delta l_{0m}]^{\mathrm{T}} \\
&\min \quad f_4 = \mathbf{D}^{\mathrm{T}}\mathbf{D} + 2e\mathbf{D}^{\mathrm{T}}\frac{\partial \boldsymbol{\varepsilon}}{\partial \mathbf{l}_0}\Delta \mathbf{l}_0 + e^2 \Delta \mathbf{l}_0^{\mathrm{T}} \frac{\partial \boldsymbol{\varepsilon}}{\partial \mathbf{l}_0}^{\mathrm{T}} \frac{\partial \boldsymbol{\varepsilon}}{\partial \mathbf{l}_0}\Delta \mathbf{l}_0 \\
&\text{s.t.} \quad g = -\boldsymbol{\varepsilon}_0 - \frac{\partial \boldsymbol{\varepsilon}}{\partial \mathbf{l}_0}\Delta \mathbf{l}_0 + \boldsymbol{\alpha} \le 0 \\
&\qquad p = \mathbf{A}_i \frac{\partial \mathbf{x}_{q_i}}{\partial \mathbf{l}}\Delta \mathbf{l} + b_i - \bar{u} \le 0, i = 1, 2, \ldots, nup
\end{aligned} \tag{3.192}$$

2. Convergence of optimal models

After transforming the nonlinear problems (PI) and (PII) into the sequential quadratic programming problems (PIII) and (PIV), respectively, the cable length increment in the kth iteration can be obtained. In the above transformation, the objective and constraint functions are expanded in terms of the first and second Taylor series respectively, which may lead to iterative oscillation or even non-convergence. For this reason, the following method is proposed:

(1) To avoid the non-convergence of the form-finding analysis after the increment of cable length in the kth iteration, the increment of the cable length should be limited. Generally, the cable length increment $\Delta \mathbf{l}_0^{(k)}$ is restricted to the range of $\pm 5\% \mathbf{l}_0^{(k)}$.
(2) We introduce the step size factor γ. For simplicity, a simplified method is adopted here; that is, the step factor is planned in a single-variable nonlinear way according to the objective function and constraints of the actual model. Taking the optimization model (PIII) as an example, the optimization model of the step size factor (PV) is:

$$\begin{aligned}
&\text{find} \quad \gamma \\
&\min \quad \mathrm{f}(\gamma) \\
&\mathrm{s.}t. \quad \mathrm{g}(\gamma) \le 0 \\
&\qquad \gamma \in [0, 1]
\end{aligned} \tag{3.193}$$

In the formula, f $= \frac{1}{nup}\sum_{i=1}^{nup}\delta^2(\mathbf{x}_{q_i})$; g($\gamma$) $= -\boldsymbol{\varepsilon} + \boldsymbol{\alpha}$. Since the step factor is determined according to the actual model, the error caused by linear approximation can be minimized to avoid oscillation during iteration.

3.3.4 Initial Shape Design of Mesh Deployable Antenna

The initial shape design of the cable truss structure derived above is aimed at obtaining an ordinary cable truss structure, which is also applicable to a cable truss deployable antenna. Of course, the following specifics of the mesh antenna should be taken into account:

(1) Considering the requirements of electronic performance, the coordinates $\boldsymbol{x}_{q_i} = \left[x_{q_i}, y_{q_i}, z_{q_i}\right]^{\mathrm{T}}$ of the front cable net node $\boldsymbol{q}_i$ of the mesh antenna should be located on the paraboloid, that is, satisfy the paraboloid function

$$x^2 + y^2 = 4F_{up}(z - h) \tag{3.194}$$

where F_{up} is the focal length of the front cable net and h is the z coordinate of the paraboloid vertex. Thus, the z-direction error of node $\boldsymbol{q}_i$ is

$$\delta_i\left(\boldsymbol{x}_{q_i}\right) = S\left(x_{q_i}, y_{q_i}\right) - z_{q_i} = \frac{x_{q_i}^2 + y_{q_i}^2}{4F_{up}} + h - z_{q_i} \tag{3.195}$$

Then, in the optimization model (PIII),

$$\mathbf{A} = \frac{1}{\sqrt{N_{up}}}\mathrm{diag}\left(\boldsymbol{A}_1, \boldsymbol{A}_2, \ldots, \boldsymbol{A}_{N_{up}}\right),\ \mathbf{B} = \frac{1}{\sqrt{N_{up}}}\left[b_1, b_2, \ldots, b_{N_{up}}\right]^{\mathrm{T}}$$
$$\boldsymbol{A}_i = \frac{1}{2F_{up}}\left[2x_{0q_i}, 2y_{0q_i}, -4F_{up}\right],\ \Delta\boldsymbol{u}_{q_i} = \left[\Delta x_{q_i}, \Delta y_{q_i}, \Delta z_{q_i}\right]^{\mathrm{T}}$$
$$b_i = x_{0q_i}^2 + y_{0q_i}^2 - 4F_{up}\left(z_{0q_i} - h_0\right).$$

(2) When considering the tension uniformity of the support cable net, the support cable net can be divided into three types: front, rear and vertical cables. In the optimization model (PIV), the mean value of the support cable net strain is $\bar{\boldsymbol{\varepsilon}} = [\bar{\varepsilon}_1, \ldots, \bar{\varepsilon}_1, \bar{\varepsilon}_2, \ldots, \bar{\varepsilon}_2, \bar{\varepsilon}_3, \ldots, \bar{\varepsilon}_3]^{\mathrm{T}}$ where $\bar{\varepsilon}_1, \bar{\varepsilon}_2$ and $\bar{\varepsilon}_3$ are the mean value strain of the front, rear and vertical cables, respectively.

3.3.5 Solution Method and Process

In summary, the solution method of problem (PIII), which is largely the same as for problem (PIV), can be described with the following algorithm and the flow chart in Fig. 3.49.

1^0 The finite element analysis model of the support cable truss structure is established, and the accuracy requirements β_u of surface error convergence, the lower limit strain of the support cable net α and the allowable stress $[\sigma]$ are given; let k = 0;

2^0 Given the initial support cable net unstressed length $\mathbf{l}_0^{(0)}$, the finite element analysis of the structure is carried out, and the mean square value $f_3^{(k)}$ of the support cable net surface error, the support cable net strain vector $\boldsymbol{\varepsilon}^{(k)}$ and the structure strength $\boldsymbol{\sigma}^{(k)}$ are calculated;

3^0 The sensitivity matrix of the node displacement $\left[\frac{\partial \mathbf{x}_q}{\partial \mathbf{l}_0}\right]^{(k)}$ and cable strain $\left[\frac{\partial \boldsymbol{\varepsilon}}{\partial \mathbf{l}_0}\right]^{(k)}$ for the initial length of the cable are solved by Eqs. (3.16) and (3.20), respectively;

4^0 According to the quadratic programming problem (PIII), the initial length increment $\Delta\mathbf{l}_0^{(k)}$ of the support cable net is obtained, and the new initial length $\mathbf{l}_0^{(k+1)}$ of the support cable net is obtained by Eq. (3.23). The structural model is updated, and finite element analysis of the structure is carried out to determine the mean square value error $f_3^{(k+1)}$ of the reflective surface, the strain $\boldsymbol{\varepsilon}^{(k+1)}$ of the support cable net and the structural strength $\boldsymbol{\sigma}^{(k+1)}$.

5^0 If the shape error satisfies the convergence condition and the strain and strength constraints of the support cable net, i.e.,

$$\begin{aligned} &\left| f_3^{(k+1)} - f_3^{(k)} \right| \le \beta_u \\ &-\boldsymbol{\varepsilon}^{(k+1)} + \boldsymbol{\alpha} \le 0 \\ &\boldsymbol{\sigma}^{(k+1)} - [\boldsymbol{\sigma}] \le 0 \end{aligned} \tag{3.196}$$

then iteration stops; otherwise, set k = k+1, and go to 3^0.

It should be noted that the initial length in 2^0 can be obtained in two ways. One is that the group of cable tensions $\mathbf{T} = [T_1, T_2, \ldots, T_m,]^{\mathrm{T}}$ are obtained by the equilibrium matrix, and then $\mathbf{l}_0^{(0)}$ is calculated according to Eq. (3.33). The other is to provide a set of tensions arbitrarily, such as by letting all cable forces be uniform or specifying the tension interval. Then, the initial length of the group of cables is obtained by

$$l_{0i}^{(0)} = \frac{EA}{T_i + EA} l_i^{(0)}, \quad i = 1, 2, \ldots, m \tag{3.197}$$

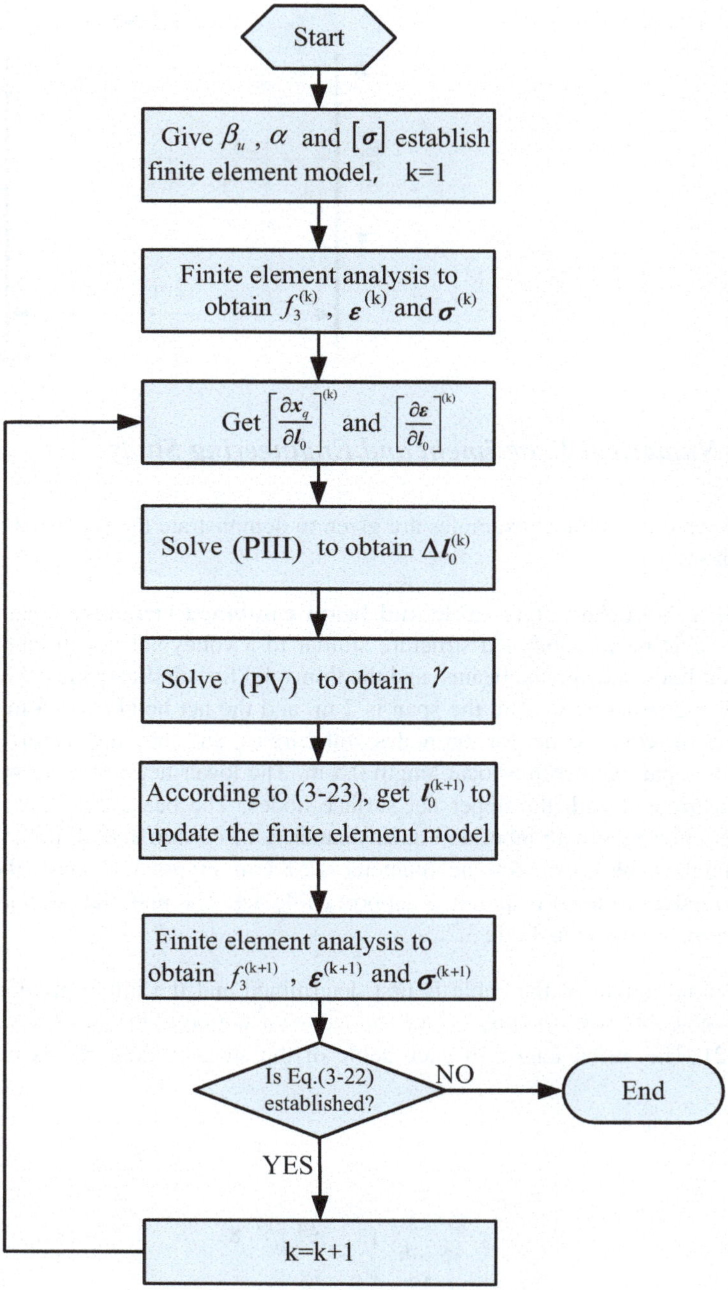

Fig. 3.49 Form-finding flow chart of cable truss combined structure

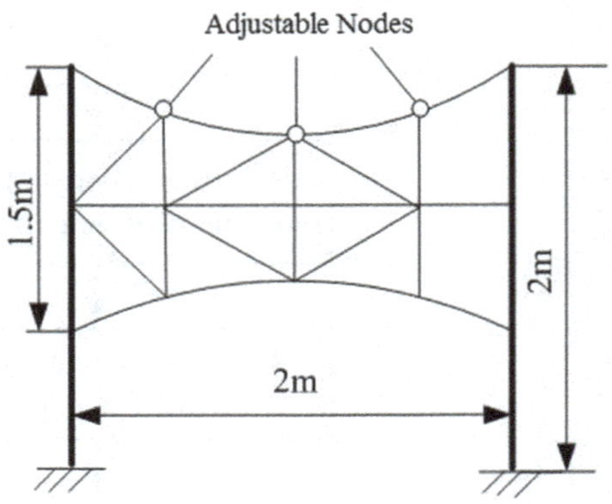

Fig. 3.50 Volleyball net beam structure

3.3.6 *Numerical Experiment and Engineering Study*

The following two typical examples are given to demonstrate the proposed models and methods.

Example 1 Volleyball style cable and beam combined structure Figure 3.50 shows a cable beam combined structure similar to a volleyball net structure. The thick solid lines indicate the beams, and the thin solid lines indicate the cables. The height of the structure is 2 m, the span is 2 m, and the net height is 1.5 m. There are precision requirements for the nodes with circles, and they are required to be located on a parabola with a focal length 0.5 m. The lower net nodes are symmetrically distributed with the upper net surface nodes. The beam is a carbon fiber hollow circular tube with inner and outer diameters of 40 mm and 48 mm, respectively, and the cable cross-sectional diameter is 2.8 mm. Figure 3.51 shows the finite element model of the corresponding support cable net. The material parameters of the structure are given in Table 3.2.

The initial length of the cable is first determined and the initial pretension of the support cable net structure is determined with the equilibrium equation (see Fig. 3.52). The initial length of each cable of the support cable net is obtained

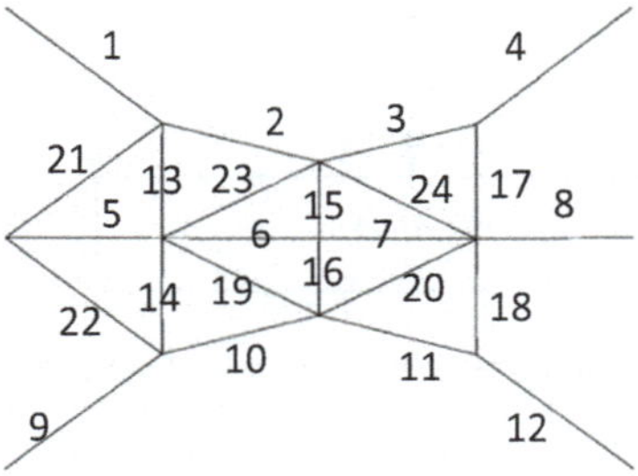

Fig. 3.51 Finite element model of support cable net

Table 3.2 The material properties of structure

	Material	Elastic modulus/GPa	Density (g/cm^3)	Poisson's ratio	Allowable stress/MPa	TEC/(1/°C)
Truss	Carbon fiber	150	1.6	0.3	600	5e−7
Cable-net	Aramid	20	1.7	0.3	500	−2e-6

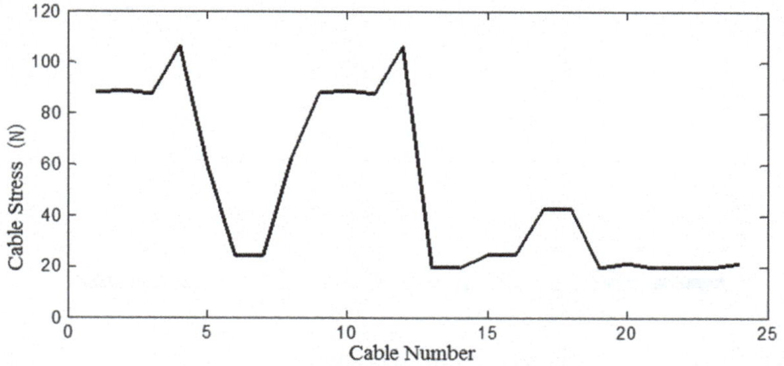

Fig. 3.52 The initial pretention of cable

according to Eq. (3.33). Finite element static analysis is performed on the cable-beam structure, and the stress distribution of the deformed support cable net is shown in Fig. 3.53. It can be seen that 6 cable elements—the 6th, 7th, 8th, 21st, 23nd and 24th elements—are slack, and the reflector precision error is 4.82 mm, which is obviously not necessary to meet the requirement. There are two purposes to achieve in the initial shape design: one is to meet the requirement of the shape accuracy, and the other is to guarantee that no cables are slack. For this reason, with the same initial cable tensions, the following optimizations (PIII) and (PIV) are performed separately, and the results are observed.

(1) We apply the optimization model (PIII) to the form-finding design, and set the convergence accuracy of the root mean square error of the shape as $\left|\sqrt{f_3^{(k+1)}} - \sqrt{f_3^{(k)}}\right| < 10^{-3}$mm, and $\alpha = 1.63 \times 10^{-3}$ corresponding to the cable pretension 2 N. It converges after 11 iterations. Figure 3.54 shows the iteration history of the shape error, and the shape error decreases to 0.003 mm. Figure 3.55 shows the iteration history for the maximum and minimum cable tensions. Figure 3.56 shows the stress distribution of the support cable net. There are no slack cables, and the minimum tension also reaches the lower limit of 2 N.
(2) We apply the optimization model (PIV) to the form-finding design, and the allowable nodal displacement is 0.1 mm and $\alpha = 1.63 \times 10^{-3}$ corresponding to the cable pretension 2 N. The convergence condition of the root mean square

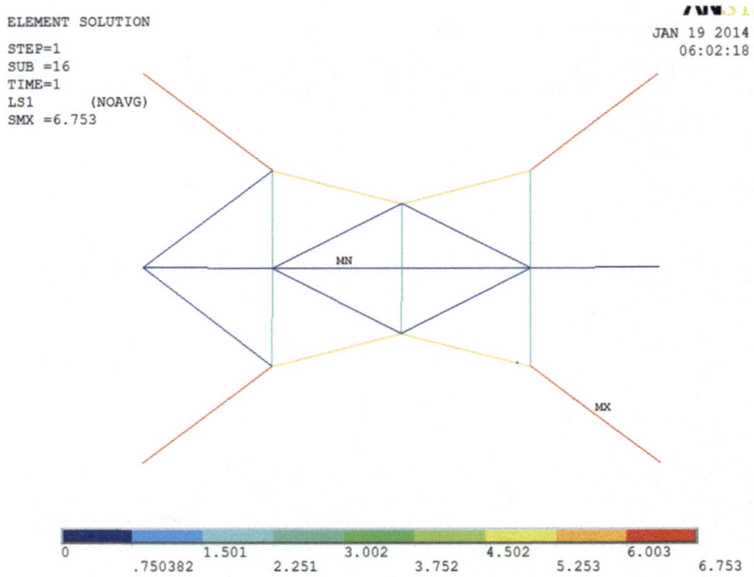

Fig. 3.53 Initial stress distribution of support cable net

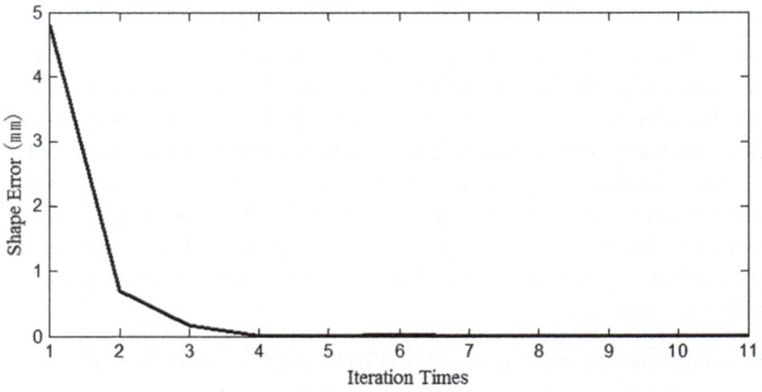

Fig. 3.54 Iteration history of shape error

error of the cable stress is $\left|f_4^{(k+1)} - f_4^{(k)}\right| < 10^{-3}$, and it converges after 13 iterations. Figure 3.57 shows the iteration history of the maximum and minimum tension, and Fig. 3.58 shows the iterative history of the shape error. The optimized surface accuracy is 0.07 mm. Figure 3.59 shows the stress distribution of the support cable net after form-finding. It can be seen from the figure that there are no slack cables, and the minimum tension reaches the lower limit of 2 N.

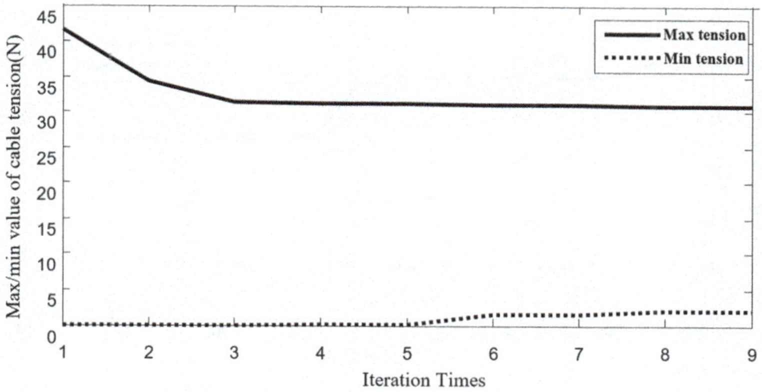

Fig. 3.55 Iteration history of the minimum and maximum pretention

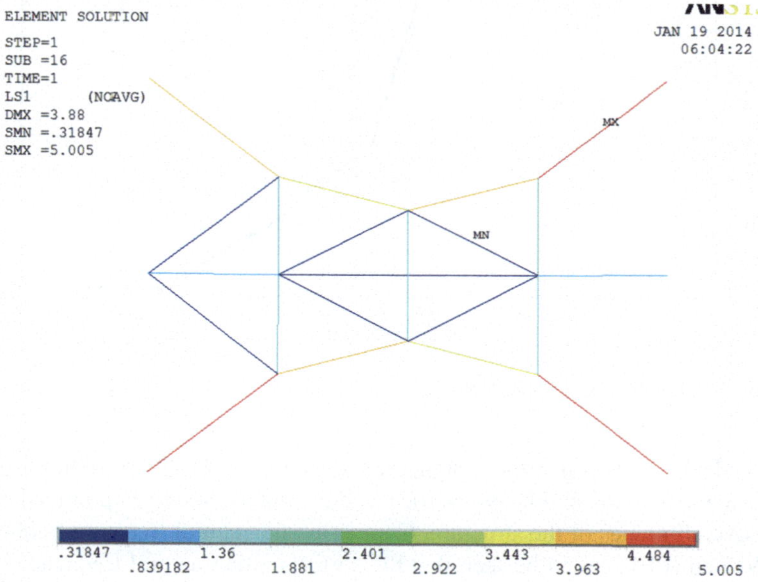

Fig. 3.56 Stress distribution of the support cable net

The structural characteristics of the cable truss structure before and after the form-finding design are shown in Table 3.3. Both the application of the optimization model (PIII) and (PIV) can obtain an equilibrium state that satisfies the constraint. Since the optimization model (PIII) targets the minimum of the surface error, the surface error of the resulting design is 0.003 mm, which is obviously better than that of the optimization model (PIV), 0.06 mm.

Table 3.4 lists the changes in the natural frequencies before and after the form-finding design. The structural natural frequency is significantly improved. Because

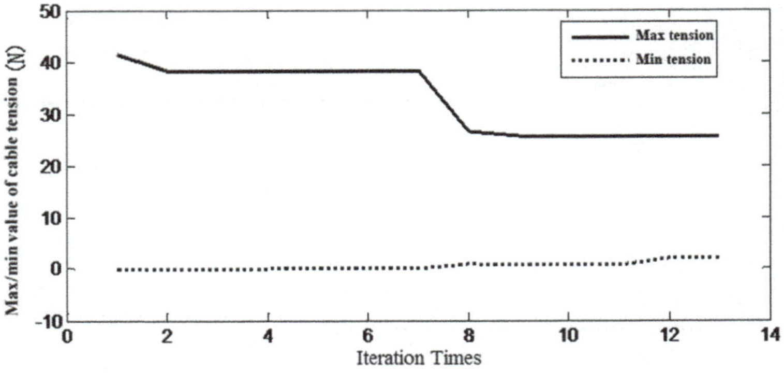

Fig. 3.57 Iteration history of the minimum and maximum pretention

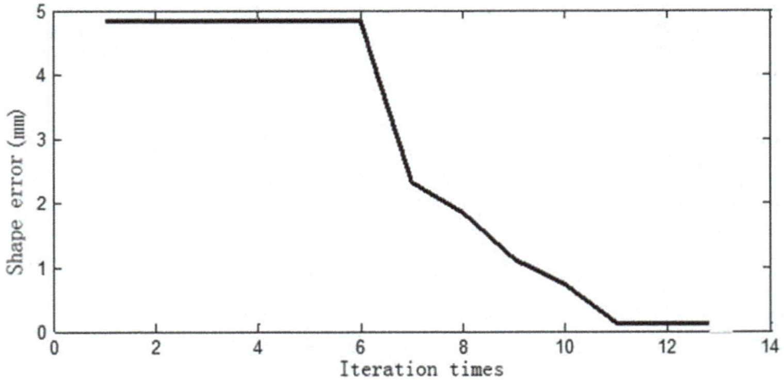

Fig. 3.58 Iteration history of shape error

there are slack cables before the design, the second mode (25.29 Hz) of the structure is the local mode of the middle two vertical cables, and the first five optimized modes are the overall modes of the structure. Comparing the natural frequencies obtained by (PIV) and (PIII), it can be seen that the corresponding natural frequency of the former is slightly higher because the uniformity of the tension of the support cable net contributes to the fundamental frequency.

After calculation, it is seen that if the given initial tension is arbitrary, the convergence progress is slower but has little effect on the final optimization result, which indicates that the method is not sensitive to the initial value of the cable length.

Example 2 AstroMesh antenna with offset paraboloid The antenna is shown in Fig. 3.60, in which a three-dimensional support cable net is employed. The diameter of the antenna is 10 m, and the focal lengths of the front and rear net are 5 m and 30 m, respectively, and the offset distance is 6 m. The main cable is divided into 10 segments; the total number of cable elements of the front and rear net is 288 and the total number of adjustable cables is 85. The cross-sectional diameter of the cable is

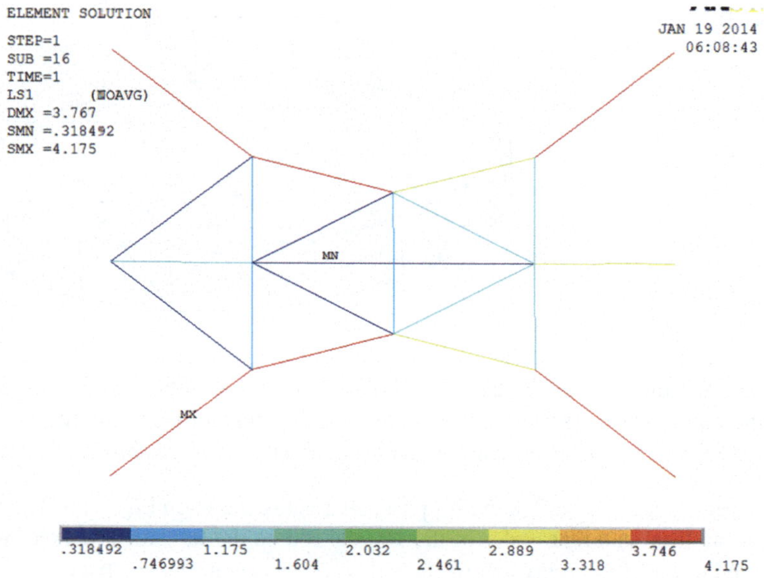

Fig. 3.59 Cable stress distribution

Table 3.3 Structural characteristics before and after optimization

	Original value	Optimization results (PIII)	Optimization results (PIV)
Reflector precision/mm	4.82	0.003	0.07
Maximum stress Cable/beam/MPa	5/10	11.7/6	11.4/4.18
Support cable net tension Max/Min/N	41.56/0	30.8/1.95	25.69/1.96
Max deflection of beam/mm	4.5	3.9	3.8

Table 3.4 Nature frequency comparison before and after optimization

		First order/Hz	Second order/Hz	Third order/Hz	Fourth order/Hz	Fifth order/Hz
Before form-finding design		20.80	25.29	31.62	57.07	99.81
After form-finding design	model(PIII)	26.88	40.61	94.98	165.66	166.95
	model(PIV)	26.88	40.65	95.15	165.66	166.97

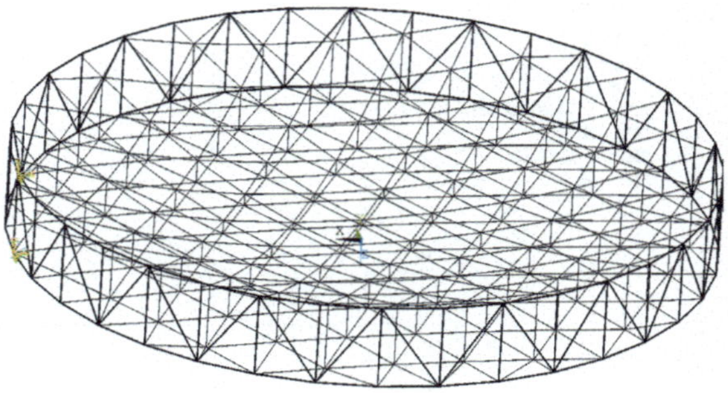

Fig. 3.60 FEM model of offset AstroMesh antenna

1.4 mm. Each member of the truss is a hollow, circular carbon fiber tube, and the inner and outer diameters of the cross section are 14 mm and 15.2 mm, respectively. The material parameters of the support cable net and the truss are shown in Table 3.2.

The finite element model of the support cable net is shown in Fig. 3.60. The black points at the front and rear nodes of the vertical rod on the left side are fixed, which is intended to faithfully represent the actual working conditions in space.

The lower strain limit is taken as 1.2992e-005 when it is designed using the model (PIV), which is equivalent to a tension of 0.4 N. The convergence condition of the mean square value of the strain serving as the objective function is $\left|f_3^{(k+1)} - f_3^{(k)}\right| \leq \beta_\varepsilon$, where $\beta_\varepsilon = 1.0550 \times 10^9$. The allowable displacement $\bar{u} = 0.1$ mm

It converges after 10 iterations, and Figs. 3.61 and 3.62 show the iteration history of the mean square value of the cable strain and the error of the reflection surface, respectively. The symbol X 11 in the lower right corner of the figures indicates the number of iterations, and the values next to Y represent the root mean square of the strain and shape error respectively. Table 3.5 gives the optimization results of the reflector surface error, stress of the strut element, cable tension, maximum truss deformation and natural frequency.

The mean square strain of the support cable net before form-finding is 5×10^{-8}, and the corresponding root mean square tension is 6.88 N. In addition, the tension uniformity is poor, there are slack cables in the front cable net, and the root mean square error of the reflector is 2.42 mm. The mean square strain decreases to 9.2×10^{-10} after the design, and the corresponding root mean square value of the cable tension is 0.927 N. The tension uniformity is also clearly improved. The ratios of the maximum to minimum tension of the front and rear support cable net and the vertical cable are 1.4, 1.03, and 4.02, respectively. The root mean square error of the reflector is 0.09 mm. This shows that the method has obvious effects on improving the tension uniformity of the support cable net and the precision of the reflector surface.

A strength check of the optimized design results shows that the stresses of the support cable net and truss are within the allowable value ranges. The improvement in the natural frequency before and after optimization is not obvious, and we found that the first five modes are overall vibrations, indicating that the tension of the support cable net has little effect on the low-order natural frequencies. The optimized

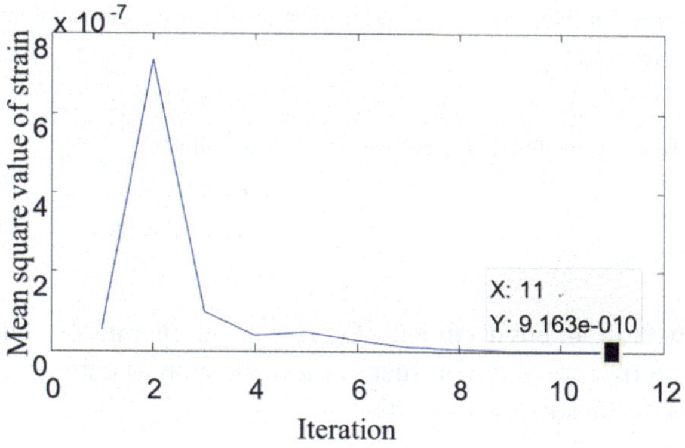

Fig. 3.61 Iteration history of strain mean

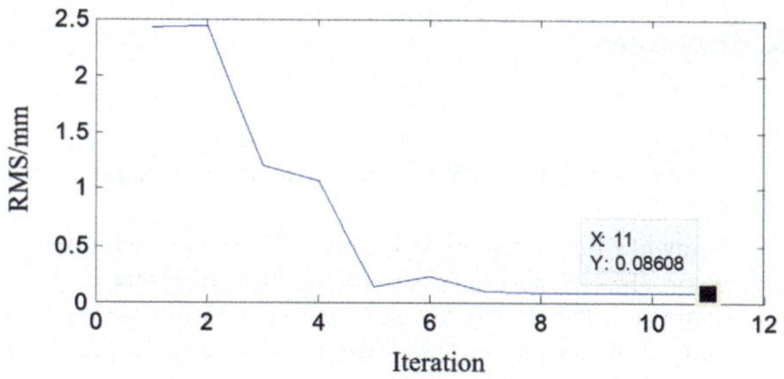

Fig. 3.62 Iteration history of shape precision

Table 3.5 Optimization results of structure performance

Reflector precision/mm		Initial values	Optimization results
		2.42	0.09
Max stress of cable net/truss/MPa		56.7/23.35	52.87/105.6
Tension of cable net max/min/N	Front net	48.92/0	23.72/16.97
	Rear net	87.23/43.93	82.31/79.78
	Vertical cables	4.70/1.24	5.84/1.45
Max truss deformation/mm		12.0	85
First three natural frequencies/Hz		0.33/1.22/5.74	0.34/1.22/5.81

maximum deformation is 85 mm, which occurs on the support cable net, and the maximum deformation before optimization is 12 mm, which occurs on the truss. This is caused by the self-balancing of the cable truss structure after the initial length is optimized. The large deformation of the overall support cable net is adapted to the uniformity of the cable tension.

To investigate the influence of high and low temperatures in space on the performance of the antenna, the optimized cable truss structure is applied at temperature of ±200 °C. The tension distribution in the equilibrium state and the reflector surface error are shown in Table 3.6. It can be seen that the support cable net is not in slack

Table 3.6 Reflector error and tension distribution when applying temperature load

RMS/mm		+200 °C	−200 °C
		0.4901	0.4586
Tension of cable net max/min (N)	Front net	20.39/31.46	21.39/7.10
	Rear net	104.3/86.97	71.91/53.19
	Vertical cables	6.94/1.87	4.47/0.66

even in an environment of ±200 °C temperatures. The designed antenna structure can overcome the problem of a slack support cable net when the antenna enters and exits the shadow of Earth.

References

1. Tibert AG (2003) Optimal design of tension truss antennas. The 44th AIAA/ASME/ASCE/AHS/ASC structures, structural dynamics and materials conference, Norfolk, USA, 2003
2. Agrawal PK, Anderson MS, Card MF (1981) Preliminary design of large reflectors with flat facets. IEEE Trans Antenna Propag 29(4):688–694
3. Hedgepeth JM (1980) Accuracy potentials for large space antenna structures. The 39th annual conf. of the society of allied weight engineers, St. Louis, MO, USA
4. Miura K, Tanizawa K (1998) Tension truss antenna-concept, reality and future. IUTAM-IASS symposium on deployable structures: theory and applications. Cambridge, UK
5. Tibert G (2002) Deployable tensegrity structures for space applications [PhD Dissertation]. Stockholm, The Royal-Institute of Technology, Sweden
6. Li T, Zhou M, Duan B (2008) A method of form-finding analysis for flexible cable net structures of deployable antennas. J Astron 29(3):795–798
7. Shi H, Yang B, Thomson M et al (2012) Automatic surface mesh generation for design of space deployable mesh reflectors. The 53nd AIAA/ ASME/ ASCE/AHS/ASC structures, structural dynamics and materials conference, Honolulu of USA
8. Shi H, Yang B, Fang H. Offset-feed surface mesh generation for design of space deployable mesh reflectors. Proceedings of the 54th AIAA/ASME/ASCE/AHS/ASC Structures, Structural Dynamics and Materials, Boston ofUSA,2013
9. Shi-kun Z, You-zhi J, Zhao-yun C et al (2014) Design and analysis of HJ-1-C satellite SAR antenna. J Radars 3(3):266–273 (in Chinese)
10. Zongquan D (2013) Design of space folding mechanism. Harbin Institute of Technology Press, Harbin (in Chinese)
11. Meguroa A, Harada S, Watanabe M (2003) Key technologies for high-accuracy large mesh antenna reflectors. Acta Astronaut 53(11):899–908
12. Santiago J, Baier H (2013) Advances in deployable structures and surfaces for large apertures in space. Ceas Space J 5(3):89–115
13. Thomson MW (2000) The AstroMesh deployable reflector. IUTAM-IASS symposium on deployable structures: theory and applications. Springer Netherlands, 435–446
14. Veenendaal D, Block P (2012) An overview and comparison of structural form finding methods for general networks. Int J Solids Struct 49(37):41–53
15. Dongwu Y, Yuanying Q, Hong B (2012) Least-norm method for pretension optimization of mesh reflector. J Mech Eng 48(21):22–27
16. Gang L, Fu-ling G (2005) Pretension optimization in parabolic cable net of AstroMesh deployable reflector. J Zhejiang Univ 39(1):1557–1560 (in Chinese)

17. Yang B, Shi H, Thomson M (2009) Optimal design of initial surface profile of deployable mesh reflectors via static modeling and quadratic programming. The 50th AIAA/ASME/ASCE/AHS/ASC structures, structural dynamic and material conference. Palm Springs, California, USA
18. Liu W, Li DX (2013) Simple technique for form-finding and tension determining of cable-network antenna reflectors. J Spacecraft Rockets 50(2):479–481
19. Morterolle S, Maurin B, Quirant J (2012) Numerical form-finding of geotensoid tension truss for mesh reflector. Acta Astronaut 76:154–163
20. Tanaka H, Shimozono N, Natori MC (2008) A design method for cable network structures considering the flexibility of supporting structures. Trans Jpn Soc Aeronaut Space Sci 50(170):267–273
21. Guo-qiang Y, Jie Z (2012) Structural Analysis and design for cable net of deployable antenna. Eng Mech 29(11):306–312
22. Ma XF, Song YP, Li ZJ et al (2013) Mesh reflector antennas: form-finding analysis review. The 54th AIAA/ASME/ASCE/AHS/ASC conference on Structures, Structural Dynamic and Material, Boston, USA

Chapter 4
Analysis and Control of Flexible Multibody Deployment

4.1 Introduction

The law of universal gravitation, proposed by Newton (Fig. 4.1), provides the theoretical basis for the development of artificial satellites, which consist of the proprietary system and support system. The deployable antenna is an important part of the proprietary system.

There are two states of deployable antennas: the folded and deployed states. During satellite launching, the antenna is stored in the rocket fairing, which means it is in the folded state. After entering orbit, the antenna is released and deployed in the predesigned frame shape, which means it is in the deployed state. The whole movement of the antenna from the folded state to the deployed state under the driving force (moment) is called the deployment and it is as shown in Fig. 4.2. The deployment of deployable antennas is a complex nonlinear mechanical process that performs the conversion from an unsteady state to a steady state and from mechanism to structure. Smooth and precise deployment is the basis for antenna application; additionally, it is a process in which malfunction is apt to occur. The deployment of the mesh antenna includes two drives. One is the torsion spring driven at the joint. Once the spring is unlocked, the strain energy stored in the torsion spring will drive the antenna to deploy to a certain angle. The other is the motor that drives to overcome the resistance caused by gradually tensioned cable net with the increasing deployment angle. The requirements for the deployment are not limited to ensuring that the antenna is accurately deployed according to the predesigned trajectory. At the same time, the initial and final stages of the deployment require avoiding intense impacts and damage; the vibration caused by the flexibility factor can be quickly suppressed during deployment, which ensures the stability of antenna deployment. Hence, there is significant theoretical and engineering value in researchers accurately predicting the structural (mass, volume, natural frequency) and dynamic properties of the deployment and implementing reasonable deployment trajectory and control algorithms.

B. Duan et al., *Large Deployable Satellite Antennas*,
Springer Tracts in Mechanical Engineering,
https://doi.org/10.1007/978-981-15-6033-0_4

Fig. 4.1 Isaac Newton, 1643–1728, physicist, mathematician, and astronomer, who is honored as the father of physics. He proposed the law of universal gravitation and Newton's three laws, was the founder of classical mechanics, invented the reflective optical telescope, and developed calculus with Leibnitz

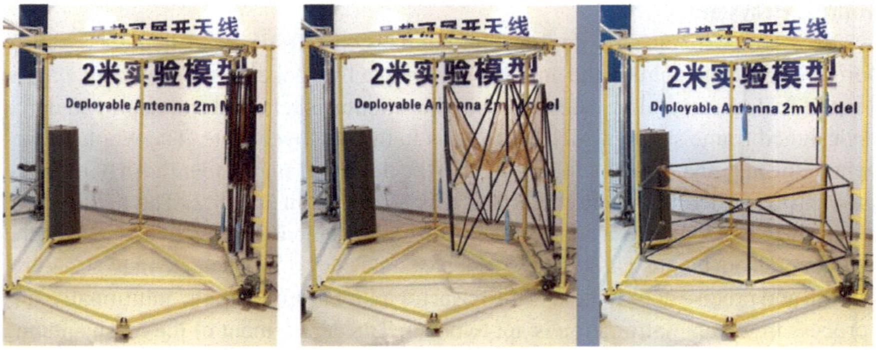

Fig. 4.2 The deployment of a 2 m-diameter mesh antenna model

4.2 Flexible Multibody Mechanics (Kinematic and Dynamic) Model of Trusses

The deployable antenna is a typical flexible multibody system. During the deployment, the flexibility of this component cannot be neglected; it is involved with large-scale motion of the overall components and leads to strong nonlinearity of the system motion. Its dynamic characteristics are quite complicated. To reflect the dynamic characteristics completely and exactly, it is necessary to utilize the technology of multibody flexible dynamic analysis. The main characteristic of a multibody flexible dynamic problem is that the flexible part will experience both large rigid body motion and flexible deformation motion at the same time. These two motions are tightly coupled, which increases the complexity.

4.2.1 Structural Components and Simplified Model

The deployable truss, which is assembled with several geometrically consistent parallelogram truss cells, is the antenna motion part of folding and deploying and the fundamental support structure of the deployable antenna. Each truss cell consists of a three-dimensional hinge (points B, D, E), a five-dimensional hinge (points A, C, F), horizontal links (BC and AD), perpendicular links (AB and CD) and diagonal sleeves (AC and CF), as shown in Fig. 4.3. Assuming that the perpendicular link AB is fixed, the truss cell can be deployed into a rectangle with the driving force of the torsion springs inside the five-dimensional hinges and by pulling the cable through the diagonal sleeves AC and CF.

Based on the synchronous hinge, the parallelogram truss cells can simultaneously deploy, and then the entire deployable truss can achieve integral deployment. Based on the symmetry of the system and the synchronicity of the cell deployment, an arbitrary truss cell (the quadrilateral cell ABCD in Fig. 4.3) can be regarded as a simplified model for dynamic modeling and analysis. Finally, a unified assembly is performed to obtain the deployment dynamics equation of the entire deployable antenna system.

4.2.2 Kinematics Analysis of Deployment

According to the discussion above, the truss of the deployable antenna consists of n identical parallelogram truss cells. As shown in Fig. 4.4, $Ox_iy_iz_i$ is the body coordinate system of the ith truss cell, where points A_i, B_i, C_i and D_i are the end points of the truss cell, θ is the deployment angle between the horizontal link A_iD_i and

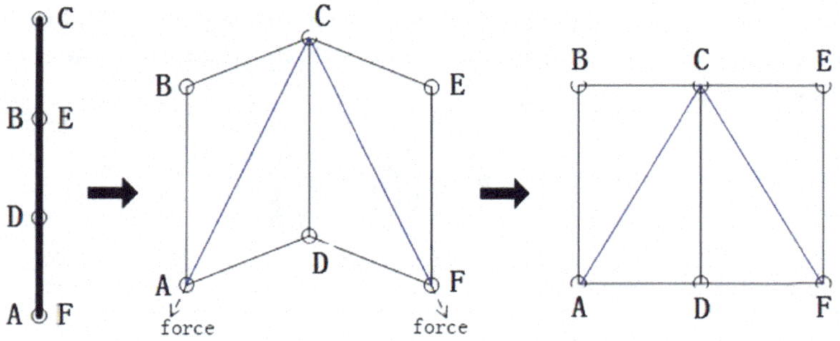

Fig. 4.3 Deployment of the deployable truss

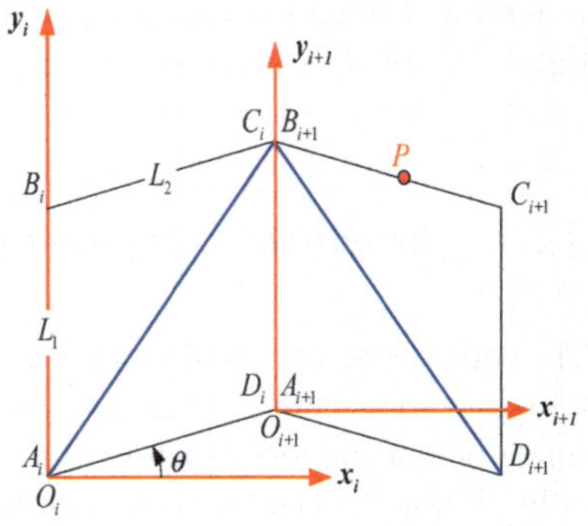

Fig. 4.4 Coordinate system of truss cell

the x_i axis. The body coordinate system $Ox_iy_iz_i$ coincides with the inertial coordinate system. In each truss cell, the y_i axis coincides with the link A_iB_i, and all points in the rectangular cell are in the plane $Ox_iy_iz_i$. Since the angle between the body coordinate system $Ox_iy_iz_i$ and $Ox_{i+1}y_{i+1}z_{i+1}$ is $\varphi = 360°/n$, the displacement, velocity, and acceleration for each point P_i on the truss cell can be deduced mathematically.

During deployment, the coordinate system transformation from system $Ox_iy_iz_i$ to $Ox_{i+1}y_{i+1}z_{i+1}$ can be seen as equivalent to translating the body coordinate system from point O_i to point O_{i+1} (point D_i) first and then rotating φ degrees counterclockwise around the y_i axis. The transformation matrix can be expressed as

$$
{}^{i}\boldsymbol{T}_{i+1} = \begin{bmatrix} \cos\varphi & 0 & \sin\varphi & X_{Di} \\ 0 & 1 & 0 & Y_{Di} \\ -\sin\varphi & 0 & \cos\varphi & 0 \\ 0 & 0 & 0 & 1 \end{bmatrix} \tag{4.1}
$$

In the coordinate systems $Ox_iy_iz_i$ and $Ox_{i+1}y_{i+1}z_{i+1}$, points $\boldsymbol{P}_i = [Px_i, Py_i, Pz_i, 1]^{\mathrm{T}}$ and $\boldsymbol{P}_{i+1} = [Px_{i+1}, Py_{i+1}, Pz_{i+1}, 1]^{\mathrm{T}}$ obey

$$
\boldsymbol{P}_i = {}^{i}\boldsymbol{T}_{i+1} \cdot \boldsymbol{P}_{i+1} \tag{4.2}
$$

L_1 and L_2 denote the lengths of the perpendiculars AB and BC, respectively. By the recurrence equation, the displacement of an arbitrary point P in the inertial coordinate system $Ox_1y_1z_1$ transformed from the body coordinate system $Ox_iy_iz_i$ can be described as

$$
\boldsymbol{P}_1 = {}^{1}\boldsymbol{T}_i \cdot \boldsymbol{P}_i = \left(\prod_{k=2}^{i} {}^{k-1}\boldsymbol{T}_k \right) \cdot \boldsymbol{P}_i \tag{4.3}
$$

Then we have

$$\boldsymbol{P}_1 = \begin{bmatrix} \cos[(i-1)\varphi] & 0 & \sin[(i-1)\varphi] & \left\{\sum_{k=2}^{i} \cos[(k-2)\varphi]\right\} L_2 \cos\theta \\ 0 & 1 & 0 & 0.5[1-(-1)^{i-1}] L_2 \sin\theta \\ -\sin[(i-1)\varphi] & 0 & \cos[(i-1)\varphi] & -\left\{\sum_{k=2}^{i} \sin[(k-2)\varphi]\right\} L_2 \cos\theta \\ 0 & 0 & 0 & 1 \end{bmatrix} \cdot \boldsymbol{P}_i \tag{4.4}$$

Differentiating Eq. (4.3), we obtain

$$\dot{\boldsymbol{P}}_1 = {}^1\dot{\boldsymbol{T}}_i \cdot \boldsymbol{P}_i + {}^1\boldsymbol{T}_i \cdot \dot{\boldsymbol{P}}_i \tag{4.5}$$

For the point $\boldsymbol{P}_i$ in the body coordinate system $Ox_iy_iz_i$, the velocity components v_{x1}, v_{y1} and v_{z1} in the inertial coordinate system $Ox_1y_1z_1$ can be expressed mathematically as follows

$$\begin{cases} v_{x1} = -\left\{\sum_{k=2}^{i} \cos[(k-2)\varphi]\right\} L_2\dot{\theta}\sin\theta \\ v_{y1} = 0.5[1-(-1)^{i-1}] L_2\dot{\theta}\cos\theta \\ v_{z1} = \left\{\sum_{k=2}^{i} \sin[(k-2)\varphi]\right\} L_2\dot{\theta}\sin\theta \end{cases} \tag{4.6}$$

The second derivative of Eq. (4.3) is

$$\ddot{\boldsymbol{P}}_1 = {}^1\ddot{\boldsymbol{T}}_i \cdot \boldsymbol{P}_i + 2{}^1\dot{\boldsymbol{T}}_i \cdot \dot{\boldsymbol{P}}_i + {}^1\boldsymbol{T}_i \cdot \ddot{\boldsymbol{P}}_i \tag{4.7}$$

For the point $\boldsymbol{P}_i$ in the body coordinate system $Ox_iy_iz_i$, the acceleration components a_{x1}, a_{y1} and a_{z1} in the inertial coordinate system $Ox_1y_1z_1$ can be expressed mathematically as follows

$$\begin{cases} a_{x1} = -\left\{\sum_{k=2}^{i} \cos[(k-2)\varphi]\right\} L_2(\dot{\theta}^2\cos\theta + \ddot{\theta}\sin\theta) \\ a_{y1} = 0.5[1-(-1)^{i-1}] L_2(-\dot{\theta}^2\sin\theta + \ddot{\theta}\cos\theta) \\ a_{z1} = \left\{\sum_{k=2}^{i} \sin[(k-2)\varphi]\right\} L_2(\dot{\theta}^2\cos\theta + \ddot{\theta}\sin\theta) \end{cases} \tag{4.8}$$

The driving moment is provided by the driving cable through diagonal sleeves. Assuming that the driving cable length l is equal to the diagonal sleeve length in the truss cell, the relationship between the deployment angle θ and driving cable l can be described as

$$\theta = \arccos\left(\frac{L_1^2 + L_2^2 + l^2}{2L_1L_2}\right) - \frac{\pi}{2} \tag{4.9}$$

$$\dot{\theta} = \frac{l\dot{l}}{L_1L_2\cos\theta} \tag{4.10}$$

$$\ddot{\theta} = \frac{\dot{l}^2 + l\ddot{l} + L_1L_2\dot{\theta}^2\sin\theta}{L_1L_2\cos\theta} \tag{4.11}$$

4.2.3 Dynamic Analysis of Deployment

Kinematics analysis is the description of the geometrical configuration of mechanisms. In practical applications, the dynamic characteristics of mechanisms should be considered. The system dynamic equation can be expressed as

$$\frac{\mathrm{d}}{\mathrm{d}t}\frac{\partial E_k}{\partial \dot{\boldsymbol{q}}} - \frac{\partial E_k}{\partial \boldsymbol{q}} + \frac{\partial (E_p + E_v)}{\partial \boldsymbol{q}} = \boldsymbol{Q}_F \tag{4.12}$$

where $\boldsymbol{q}$ and $\dot{\boldsymbol{q}}$ are the generalized coordinate and velocity, respectively; E_k, E_p and E_v are the kinetic, elastic and geopotential energy, respectively; and $\boldsymbol{Q}_F$ is the generalized moment.

If the mass of a truss is distributed on the four nodes, and the equivalent mass is m, the kinetic energy of the trusses of the antenna can be obtained from Eq. (4.6) as

$$\begin{cases} E_k = m\sum\limits_{i=2}^{n}[(\xi\dot{\theta}\sin\theta)^2 + (\psi\dot{\theta}\cos\theta)^2 + (\zeta\dot{\theta}\sin\theta)^2] \\ \xi = \left\{\sum\limits_{k=2}^{i}\cos[(k-2)\varphi]\right\}L_2 \\ \psi = 0.5[1-(-1)^{i-1}]L_2 \\ \zeta = \left\{\sum\limits_{k=2}^{i}\sin[(k-2)\varphi]\right\}L_2 \end{cases} \tag{4.13}$$

Hypothetically, the configuration of the deployed antenna is in the zero-potential position, and the overall geopotential energy can be described as

$$E_v = -\sum_{i=1}^{n} mgL_2\sin\theta = -nmgL_2\sin\theta \tag{4.14}$$

where g is the gravitational acceleration with $g = 9.8$ m/s^2 for ground tests and $g = 0$ m/s^2 for space simulations.

A moving coordinate system is established as shown in Fig. 4.5. The x axis is chosen as the line connecting nodes A and B of the non-deformed link and the y axis

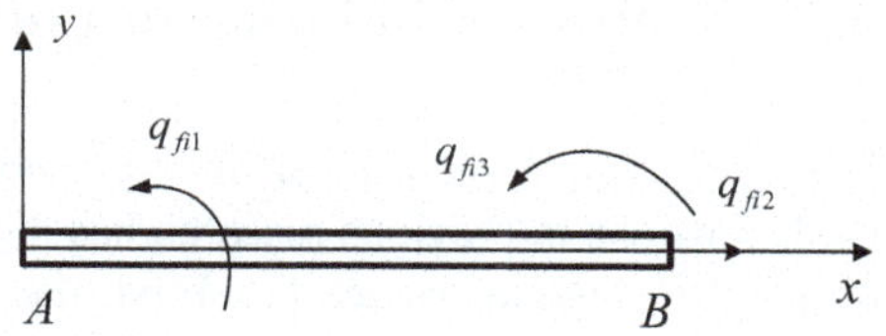

Fig. 4.5 Description of elasticity link

is determined in terms of the right coordinate system. The parameters $q_{f\,i1}$ and $q_{f\,i2}$ are the rotational angles of nodes A and B, respectively, and q_{fi3} is the displacement along the x axis. Therefore, the flexible deformation movement of the link relative to the moving coordinate system is equivalent to an elastic vibration of a simply supported beam. The shape function is

$$\boldsymbol{\Phi} = \begin{bmatrix} 0 & \varepsilon & 0 \\ l_b(\varepsilon - 2\varepsilon^2 + \varepsilon^3) & 0 & l_b(\varepsilon^3 - \varepsilon^2) \end{bmatrix} \tag{4.15}$$

where l_b is the length of the beam element and $\varepsilon = x/l_b$.

The generalized coordinate for describing flexible deformations is defined as $\boldsymbol{q}_{fi} = [q_{f\,i1}\ q_{f\,i2}\ q_{f\,i3}]^{\mathrm{T}}$.

The elastic potential energy of an arbitrary component can be described as:

$$E_{p\,i} = \frac{1}{2}\left[\frac{4EI}{l_b}\left(q_{f\,i1}^2 + q_{f\,i1}\cdot q_{f\,i3} + q_{f\,i3}^2\right) + \frac{EA_s}{l_b}q_{f\,i2}^2\right] \tag{4.16}$$

where I is the processional moment, A_s is the cross-sectional area, and E is the Young's modulus.

Therefore, the overall potential energy of elasticity is

$$E_p = \sum_{i=1}^{3n} E_{p\,i} \tag{4.17}$$

Combining Eqs. (4.13)–(4.17) with Eq. (4.12) gives the flexible multibody dynamic model as

$$\boldsymbol{M}\ddot{\boldsymbol{q}} + \boldsymbol{K}\boldsymbol{q} = \boldsymbol{Q}_F + \boldsymbol{Q}_v \tag{4.18}$$

where $\boldsymbol{M}$ and $\boldsymbol{K}$ are the mass and stiffness matrices, respectively, and $\boldsymbol{Q}_F$ is the generalized moment. The generalized force $\boldsymbol{Q}_v$ related to the quadratic of velocity, can be expressed as

$$\boldsymbol{Q}_v = \frac{\partial}{\partial \boldsymbol{q}}\left(\frac{1}{2}\dot{\boldsymbol{q}}^T\boldsymbol{M}\dot{\boldsymbol{q}}\right) - \dot{\boldsymbol{M}}\dot{\boldsymbol{q}} \tag{4.19}$$

4.3 Analysis of Influence of Cable Net on Deployment

Because of the lack of clear and systematic mechanics analysis, the influence of cable tensions in previous research has mostly depended on two assumptions based on engineering experience [1]. First, the forces of the cable net start to affect the deployable truss at a particular time, which is usually thought to be at 90% of the total deployment time. Second, the forces of the cable net are equivalent to spring forces, which increase from zero at that particular moment to the predesigned tension of the deployed state. In this treatment, the assumptions are derived from engineering experience and lack theoretical support, which may lead to a large deviation between the analysis and test results. Clearly and accurately determining the shape of slack cables is the foundation of the analysis of the influence of cable net tension during deployment. When only the boundary conditions and the initial length (unstressed) of each cable segment are known, a key problem is to determine the shape of the cable net at an arbitrary point during deployment.

4.3.1 Cable Net Modeling Based on ECE (Elastic Catenary Element)

Precise modeling of the cable net structure is a precondition of obtaining the variation of the cable configuration and the tension distribution. Dynamic modeling of general cable-net structures gas mainly employed link or beam elements in the past. In these cases, the approximate error depends on the quantity of elements, and reducing the error by increasing the number of elements will result in a huge computational cost. Since the ECE is an analytical element, one ECE alone can exactly model a cable (whether in a gravity or microgravity environment), which makes it suitable for modeling the cable net structure.

For the ECE MN in the plane $O'x'z'$, as shown in Fig. 4.6, L_0 is the original length. h and l are the sag and span between the two nodes, respectively; F_1', F_3', F_4' and F_6'

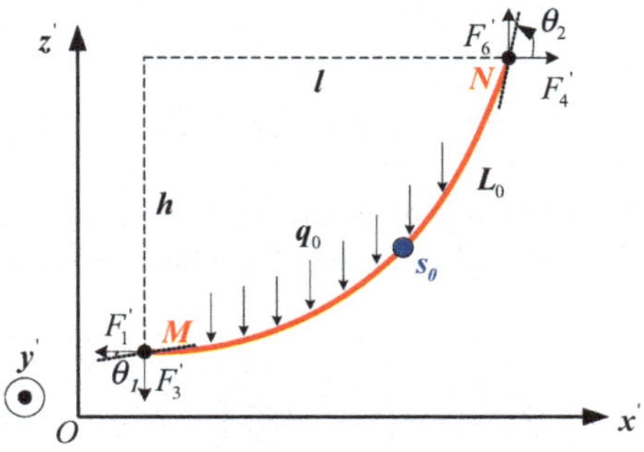

Fig. 4.6 Description of elastic catenary element in the $O'x'z'$ plane

are the nodal forces in x' and z', the load q_0 is the gravitational load distributed along the horizontal span of the cable; and θ_1 and θ_2 are the angles between the horizontal and the tangent to the cable at the nodes. Kashani [2] derived this model in detail, and a brief explanation is given here.

The tangent stiffness matrix for an ECE changes with the variation of node positions, unlike the matrices for frequently used finite elements (e.g., links, beams, planes and shells).

$$\boldsymbol{K}' = \begin{bmatrix} k' & -k' \\ -k' & k' \end{bmatrix} \tag{4.20}$$

where

$$k' = \alpha \begin{bmatrix} \frac{q_0 L_0}{AE} + \sin\theta_2 - \sin\theta_1 & 0 & \cos\theta_1 - \cos\theta_2 \\ 0 & \frac{H}{\alpha l} & 0 \\ \cos\theta_1 - \cos\theta_2 & 0 & \frac{q_0 l}{H} + \sin\theta_1 - \sin\theta_2 \end{bmatrix} \tag{4.21}$$

$$\alpha = \frac{q_0}{4\sin\frac{\theta_2-\theta_1}{2}\left[\frac{q_0}{2}\left(\frac{l}{H} - \frac{L_0}{AE}\right)\cos\frac{\theta_2+\theta_1}{2} - \sin\frac{\theta_2-\theta_1}{2}\right] + \frac{q_0^2 L_0 l}{AEH}} \tag{4.22}$$

$$\theta_1 = \arctan\left\{\sinh\left[\cosh^{-1}\left(\frac{q_0 L_0}{2H\sinh\left(\frac{q_0}{2}\left(\frac{l}{H} - \frac{L_0}{AE}\right)\right)}\right) - \frac{q_0}{2}\left(\frac{l}{H} - \frac{L_0}{AE}\right)\right]\right\} \tag{4.23}$$

$$\theta_2 = \arctan\left\{\sinh\left[\cosh^{-1}\left(\frac{q_0 L_0}{2H\sinh\left(\frac{q_0}{2}\left(\frac{l}{H} - \frac{L_0}{AE}\right)\right)}\right) + \frac{q_0}{2}\left(\frac{l}{H} - \frac{L_0}{AE}\right)\right]\right\} \tag{4.24}$$

The horizontal force H, which should be calculated in advance to obtain the tangent stiffness matrix $\boldsymbol{K}'$, satisfies

$$\begin{aligned} f(H) = {} & \frac{4H^2}{q_0^2}\sinh^2\left(\frac{q_0}{2}\left(\frac{l}{H} - \frac{L_0}{AE}\right)\right) \\ & + \frac{h^2}{\left[1 + \left(\frac{q_0 L_0}{2AE}\right)\coth\left(\frac{q_0}{2}\left(\frac{l}{H} - \frac{L_0}{AE}\right)\right)\right]^2} - L_0^2 \equiv 0 \end{aligned} \tag{4.25}$$

H can be obtained by solving the following equation based on an iterative algorithm. Then, the cable tension can be obtained as

$$\boldsymbol{F}' = \begin{bmatrix} F_1' \\ F_2' \\ F_3' \\ F_4' \\ F_5' \\ F_6' \end{bmatrix} = \begin{bmatrix} -H \\ 0 \\ -H\sinh\left[\cosh^{-1}\left(\frac{q_0 L_0}{2H\sinh\left(\frac{q_0}{2}\left(\frac{l}{H}-\frac{L_0}{AE}\right)\right)}\right) - \frac{q_0}{2}\left(\frac{l}{H}-\frac{L_0}{AE}\right)\right] \\ H \\ 0 \\ H\sinh\left[\cosh^{-1}\left(\frac{q_0 L_0}{2H\sinh\left(\frac{q_0}{2}\left(\frac{l}{H}-\frac{L_0}{AE}\right)\right)}\right) + \frac{q_0}{2}\left(\frac{l}{H}-\frac{L_0}{AE}\right)\right] \end{bmatrix} \tag{4.26}$$

Furthermore, the tension and position coordinates for an arbitrary point s_0 on an ECE can be expressed as

$$T(s_0) = [H^2 + (q_0 s_0 - F_3')^2]^{1/2} \tag{4.27}$$

$$\begin{cases} x'(s_0) = \frac{Hs_0}{AE} + \frac{H}{q_0}\left[\sinh^{-1}\left(\frac{F_3'}{H}\right) + \sinh^{-1}\left(\frac{q_0 s_0 - F_3'}{H}\right)\right] \\ y'(s_0) = 0 \\ z'(s_0) = \frac{F_3' s_0}{AE}\left(\frac{q_0 s_0}{2F_3'} - 1\right) + \frac{H}{q_0}\left[\left(1 + \left(\frac{q_0 s_0 - F_3'}{H}\right)^2\right)^{1/2} - \left(1 + \left(\frac{F_3'}{H}\right)^2\right)^{1/2}\right] \end{cases} \tag{4.28}$$

4.3.2 Form-Finding Method for Slack Cable Net

The form-finding process for a cable-net structure aims to determine the equilibrium state after the boundary conditions, loads and original lengths are established. In other words, it is a procedure to obtain the coordinates of the free nodes P_1 and P_2 in the equilibrium state and the corresponding tension distribution of the cable net when the coordinates of the fixed nodes P_3, P_4, P_5 and P_6 are known as shown in Fig. 4.7. For the arbitrary free node c at equilibrium, the resultant force $\boldsymbol{F}_c$ should be zero:

$$\begin{cases} F_{cx} = 0 \\ F_{cy} = 0 \\ F_{cz} = 0 \end{cases} \tag{4.29}$$

For the slack cable net system modeled by ECEs, in view of Eq. (4.26), the nodal force $\boldsymbol{F}_{ci}$ of an arbitrary element i is an implicit function of the sag h_i and span l_i between two nodes. Furthermore, since the sag and span of a catenary are determined by two node coordinates, the nodal force is also an implicit function of the coordinate

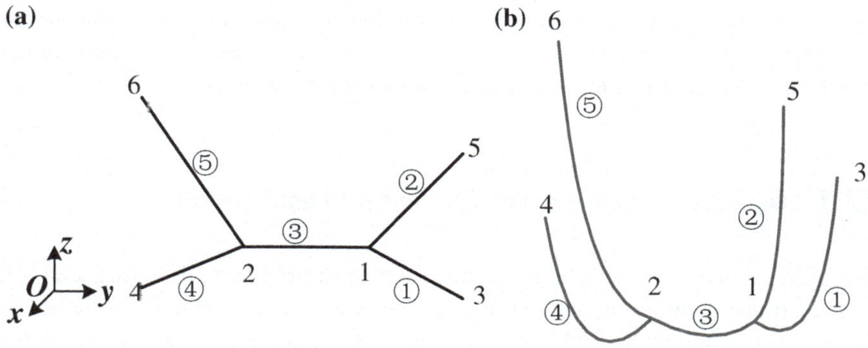

Fig. 4.7 Cable net system. **a** Tensioned cable net. **b** Slack cable net

$\boldsymbol{P_c} = [x_c\ y_c\ z_c\ ]^{\mathrm{T}}$ of the free node c and the coordinate $\boldsymbol{P_{ki}} = [x_{ki}\ y_{ki}\ z_{ki}\ ]^{\mathrm{T}}$ of node k_i, as $\boldsymbol{F_{ci}} = \boldsymbol{F_{ci}}(\boldsymbol{P_{ki}}, \boldsymbol{P}_c)$. Therefore, the equilibrium equations for the node c of a slack cable net can be written as

$$\begin{cases} \sum_{i=1}^{c_N} F_{cix} = \sum_{i=1}^{c_N} \pm H_i \cos\phi_i = 0 \\ \sum_{i=1}^{c_N} F_{ciy} = \sum_{i=1}^{c_N} \pm H_i \sin\phi_i = 0 \\ \sum_{i=1}^{c_N} F_{ciz} = \sum_{i=1}^{c_N} \pm H \sinh\left[\cosh^{-1}\left(\frac{q_0 L_0}{2H\sinh(\frac{q_0}{2}(\frac{l}{H}-\frac{L_0}{AE}))}\right) \pm \frac{q_0}{2}\left(\frac{l}{H}-\frac{L_0}{AE}\right)\right] = 0 \end{cases} \tag{4.30}$$

For a generalized slack cable net, the optimization model for form-finding can be expressed as

$$\begin{aligned} &\text{find} \quad \boldsymbol{P}_f = \{\boldsymbol{P}_{fi}\} = \{x_{fi}\ y_{fi}\ z_{fi}\}^{\mathrm{T}}\ (i = 1, 2, \ldots n_f) \\ &\min \quad \boldsymbol{F}_{plus} = \sum_{i=1}^{n_f} \left|\boldsymbol{F}_i(\boldsymbol{P}_r, \boldsymbol{P}_f)\right| \\ &\text{s.t.} \quad \boldsymbol{P}_r = \{\boldsymbol{P}_{ri}\} = \{x_{ri}\ y_{ri}\ z_{ri}\}^{\mathrm{T}}\ (i = 1, 2, \ldots n_r) \\ &\qquad x_L \le x_{fi} \le x_U \\ &\qquad y_L \le y_{fi} \le y_U \\ &\qquad z_{fi} \le z_U \end{aligned} \tag{4.31}$$

where $\boldsymbol{P}_f$ and $\boldsymbol{P}_r$ are the coordinate tensors of the free and fixed nodes, respectively, n_f and n_r are the numbers of free and fixed nodes, respectively, $\boldsymbol{F}_i$ is the resultant force of the *i*th free node; x_L and y_L are the lower bounds of the coordinates of free nodes in the *x* and *y* directions; and x_U, y_U and z_U are the upper bounds of the coordinates of free nodes in the *x, y* and *z* directions, respectively.

4.3.3 Influence of Cable Net Tension on Deployment

Through kinematics equations, the motion of an arbitrary point of the truss cell is a function of the deployment angle θ. Because the cable net is generally connected to three-dimensional and five-dimensional hinges, which are located at the end points of the perpendicular links, the coordinates of the fixed nodes of the cable net are also functions of the deployment angle θ. Figure 4.8 is a bird's-eye view of a deployable antenna. When the perpendicular link P_1P_7 is a fixed boundary, the fixed nodes of the cable net will move along the trajectories shown by the dashed lines in Fig. 4.8.

Then the deployment can be simply discretized to several instantaneous the angle vector $\boldsymbol{\theta}$ related structures $\boldsymbol{S(\theta)}$,

$$\boldsymbol{\theta} = \{\theta_i\}^{\mathrm{T}} \ (i = 1, 2, \ldots n_b) \tag{4.32}$$

where n_b is the total number of discrete conditions.

For an arbitrary condition $S(\theta_i)$, the coordinates of the fixed nodes $\boldsymbol{P}_{\boldsymbol{r},\theta i}$ can be calculated by Eq. (4.4). Based on the form-finding optimization method mentioned in Sect. 4.3.2, the coordinates $\boldsymbol{P}_{f,\theta i}$ of the free nodes can be found. As a result, the nodal forces of the fixed nodes can be obtained as

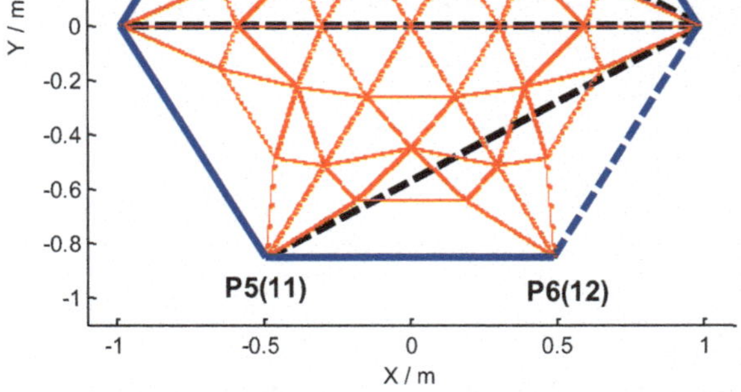

Fig. 4.8 Trajectory of fixed nodes of cable-net

$$\boldsymbol{F}_{\boldsymbol{P}_{r,\theta i}} = \left\{\boldsymbol{F}_{t,\theta i}\right\}\ (t = 1, 2, \cdots n_r) \tag{4.33}$$

where $\boldsymbol{F}_{t,\theta i}$ is the force vector of the tth fixed node.

It should be noted that there are several cables connected with one fixed node and, therefore

$$\boldsymbol{F}_{t,\theta i} = \sum_{w=1}^{n_w} \boldsymbol{F}_{t,\theta i,w} \tag{4.34}$$

where n_w is the total number of cables connected with the tth fixed node and $\boldsymbol{F}_{t,\theta i,w}$ is the force of the wth cable connected with the tth fixed node in the θ_i condition.

For each condition $S(\theta_i)$, the nodal force of each fixed node can be found with Eqs. (4.33) and (4.34). By synthetically considering all instantaneous structures, the variation of the forces on the fixed nodes (and likewise on the end points of the perpendicular links) during deployment can be obtained as $\boldsymbol{F}_{\boldsymbol{cable}} = \left\{\boldsymbol{F}_{\boldsymbol{cable},t}\right\}$ $(t = 1, 2, \cdots n_r)$, where

$$\boldsymbol{F}_{\boldsymbol{cable},t} = \begin{cases} 0 & 0^{\mathrm{o}} \le \theta \le \theta_1 \\ \boldsymbol{F}_{\boldsymbol{P}_{r,\theta 1}} & \theta_1 \le \theta \le \theta_2 \\ \vdots & \\ \boldsymbol{F}_{\boldsymbol{P}_{r,\theta b}} & \theta_b \le \theta \le 90^{\mathrm{o}} \end{cases} \tag{4.35}$$

The forces caused by the cable net tension can be added in the deployment dynamic model (Eq. 4.18) as the generalized force

$$\boldsymbol{M}\ddot{\boldsymbol{q}} + \boldsymbol{K}\boldsymbol{q} = \boldsymbol{Q}_F + \boldsymbol{Q}_v + \boldsymbol{F}_{\boldsymbol{cable}} \tag{4.36}$$

The dynamic model can be solved based on the Newmark method, and then the influence of the cable net tension on the deployment can be determined.

4.3.4 Verification Experiment

1. Experimental model design of cable net
   In accordance with the accuracy of the required cable net, a positive-feeding cable net structure with three rings in the radial direction and a focal length ratio of 0.6 is designed based on the 2 m-diameter prototype model.
   The front and rear cable nets are symmetrical, and the cable net design is obtained based on the force-density method to ensure the uniform force of each joint in the deployed state. The cross-sectional area of the cables is circular with a 1 mm diameter, the material is Kevlar, the modulus of elasticity is 20 GPa and the density is. The total numbers of cables for the front and rear cable net and the

vertical cables are 90, 90, and 31 respectively. In total, 211 ECEs are used to construct the finite element model of the cable net. The element numbers are shown in Fig. 4.9, and the original lengths of the cable elements are given in Table 4.1.

2. Experimental scheme

   (1) 3D photogrammetry is adopted to measure the node positions of the cable net in several typical conditions. The root mean square (RMS) errors of all node positions between the simulation and experimental results are calculated.
   (2) A tensometer is used to measure the tensions of typical cables in all typical conditions, and these results were then compared with the simulation results. The measurement range of the tensometer is from 0 to 200 N with a measuring error less than ±2N. The cable tension value can be read when it is clamped on the dynamometer pulley. Then, typical cables are selected, the tension values of which are measured under all typical working conditions.

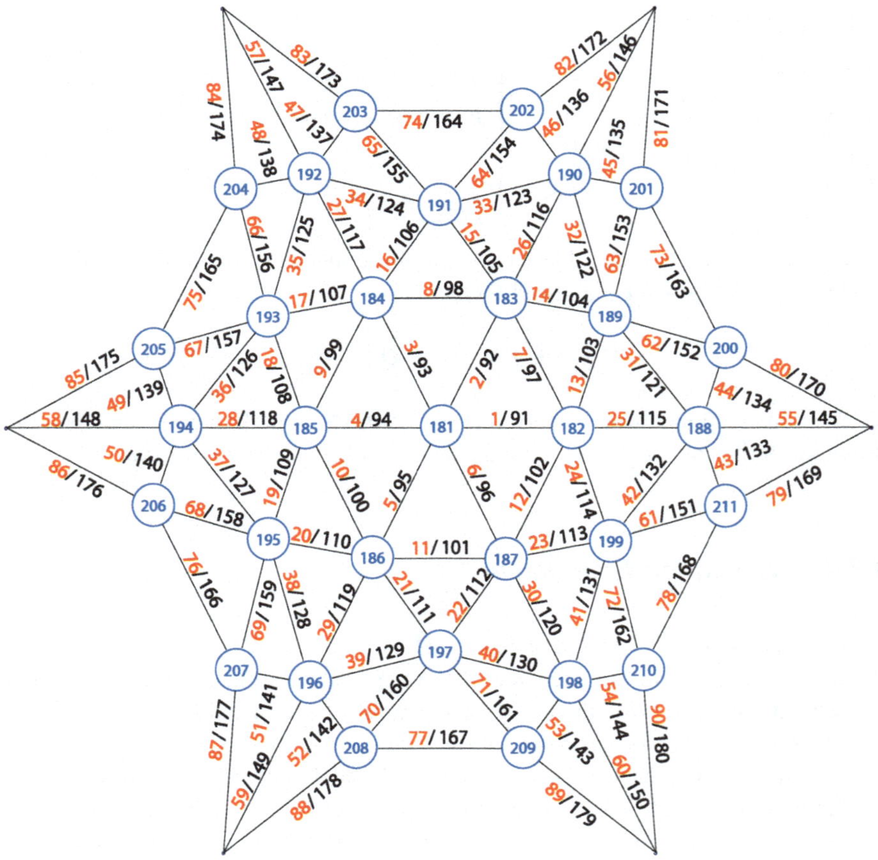

Fig. 4.9 Element number of cable-net

Table 4.1 Original length of cable elements

Part	Original length			
Front net	l_1–l_6 = 295.12	l_7–l_{12} = 294.78	l_{13}–l_{24} = 233.68	l_{25}–l_{30} = 284.60
	l_{31}–l_{42} = 295.89	l_{43}–l_{54} = 161.82	l_{55}–l_{60} = 396.94	l_{61}–l_{72} = 266.04
		l_{73}–l_{78} = 370.57	l_{79}–l_{90} = 369.88	
Rear net	l_{91}–l_{96} = 295.12	l_{97}–l_{102} = 294.78	l_{103}–l_{114} = 233.68	l_{115}–l_{120} = 284.60
	l_{121}–l_{132} = 295.89	l_{133}–l_{144} = 161.82	l_{145}–l_{150} = 396.94	l_{151}–l_{162} = 266.04
		l_{163}–l_{168} = 370.57	l_{169}–l_{180} = 369.88	
Vertical cables	l_{181} = 290.02	l_{182}–l_{187} = 318.56	l_{188}–l_{194} = 399.31	l_{194}–l_{200} = 352.96
	l_{200}–l_{211} = 430.22			

(3) The tension of the driving cable is measured in real time and compared with the simulated value.
A tensometer is used to measure the tension of the driving cable. As the motor operates, the antenna is slowly deployed, and the tensometer records the tension of the driving cables in real time, and finally outputs the numerical result.

3. Discussion

(1) Comparison of node positions of cable net between simulation and experiment
Figure 4.10 shows the configuration variation of the cable net structure during deployment in several typical conditions. The blue lines are the slack cables and the red lines indicate the tensioned cables. There are three phases of the cable net deployment: slack, combined slack/tensional and complete tensional states.
The RMS errors of the positions of all nodes in the simulation and experimental results are calculated, as shown in Table 4.2. It can be seen that the RMS errors are all controlled within 3 mm. Considering the fabrication, assembly errors during the model manufacturing process and the systematic errors of photogrammetry, and the RMS value of the cable net can be less than 1 mm.

(2) Comparison of cable tension of cable net between simulation and experiment
The tension of typical cables is measured with tensometers and the experimental data are compared to the simulation results (Table 4.3). Figure 4.11 shows that the variation trend of the cable tension remains basically the same, and the maximum difference value of the tension is less than 15%.
Since the elasticity modulus of the cable material (Kevlar) is fairly high (20 GPa), a small change in the cable length would cause a large tension deviation.

(3) Comparison of driving cable tension of cable net between simulation and experiment

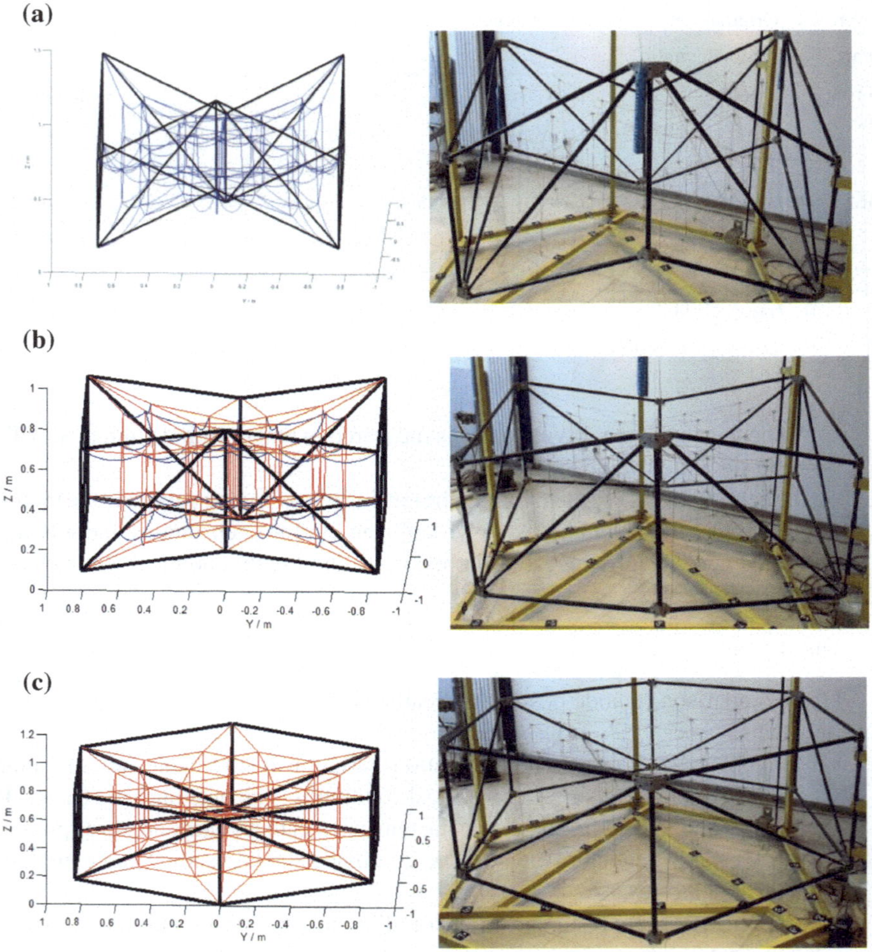

Fig. 4.10 Configuration variation of cable net. **a** Deployment angle 65° (completely slack). **b** Deployment angle 82° (partial slack/tension). **c** Deployment angle 90° (completely tensional)

Table 4.2 RMS in typical conditions

Deployment angle/°	65	79	82	87	90
RMS/mm	2.57	2.78	2.48	1.91	.94

Figure 4.12 shows the variation curve of the forces on hinges caused by cable net tension during deployment. As a positive-feeding antenna model, the cable net structure has axial symmetry. Therefore, four kinds of hinges are chosen for analysis: three-dimensional hinges, and five-dimensional hinges (which are connected to the front cable net) as well as three-dimensional hinges and five-dimensional hinges (which are connected to the rear cable net).

Table 4.3 Tension measurement results of typical cables (N)

Element number	65°		79°		82°		87°		90°	
	Sim.	Exp.	Sim.	Exp.	Sim.	Exp.	Sim.	Exp.	Sim.	Exp.
7	0	0	30.08	29.36	49.9	57.37	22.57	25.03	15.01	17.11
80	0	0	11.75	13.25	26.31	24.52	20.36	23.36	33.07	37.96
81	0	0	89.94	100.3	146.8	171.2	78.21	86.36	33.07	38.62
32	0	0	0	0	0	0	9.70	8.29	42.11	45.25
55	0	0	0	0	0	0	5.31	3.52	15.01	18.01

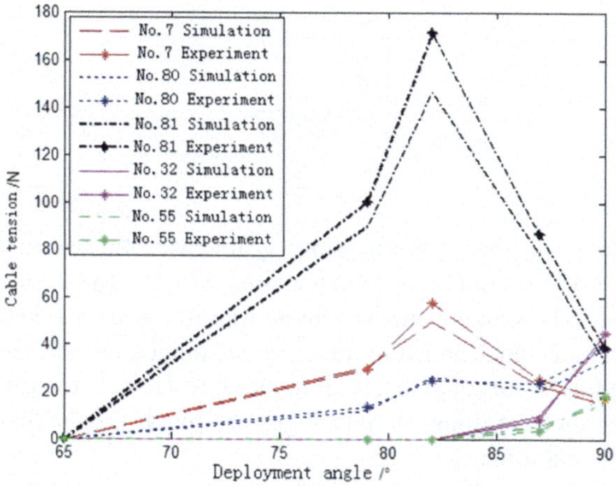

Fig. 4.11 Comparison of simulation and experiment of typical cable tension

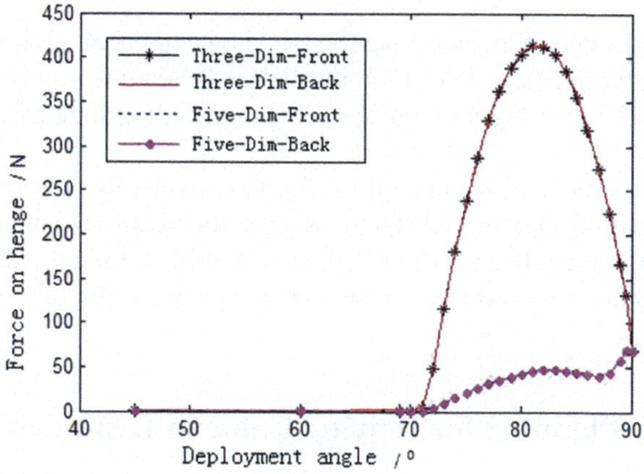

Fig. 4.12 Forces on hinges during deployment

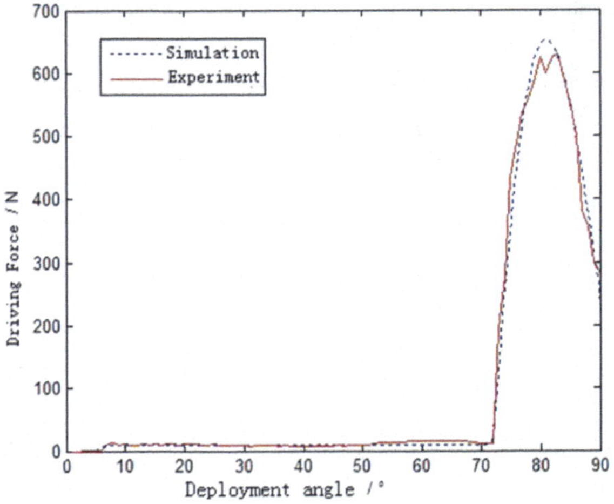

Fig. 4.13 Driving force curve in deployment

① Since the front and rear cable nets are symmetric, the variation trend of the forces on five-dimensional hinges connecting with the front and rear cable nets are coincident. The same results are found for three-dimensional hinges.
② Both three-dimensional and five-dimensional hinges are stressed at the same time, and the forces are equal in the deployed state. However, during deployment, the forces on the five-dimensional hinges are much higher than those on the three-dimensional hinges.
③ The maximum forces do not appear at the ending state ($\theta = 90°$, 68.1 N), but at the state of before the ending point ($\theta = 82°$, 414.2 N). The maximum force is as great as six times the force at the ending state.

The curve of the deployment driving force is shown in Fig. 4.13. It is obvious that the maximum force appears in the point just before the ending point ($\theta = 82°$). The peak value of 652.3 N is higher than 288.8 N in the ending state by approximately 126%.

The above experimental results tell us that in correctively selecting the driving motor, we need to consider multiple factors systematically instead of only the ending state force, such as the influence of cable nets with different topologies, the original cable length and the designed pre-tension, and the peak tension of the cable net.

4.4 Velocity Planning for Driving Cable in Deployment

To ensure the stability of deployment, the peak value of acceleration should be kept as small as possible. The Bernstein function is selected as the basis function; thus by optimizing the position of several control points, a Bezier curve is used to synthesize the length variation curve of the driving cable so that the deployment of the antenna can meet engineering requirements.

Here the nth-order Bezier curve is used to define the displacement of the driving cable as

$$l(t) = \sum_{i=0}^{n} P_i \times B_{i,n}(t)\ 0 \le t \le 1 \tag{4.37}$$

where t is the generalized normalization time range from 0 to 1.

The winding velocity of the driving cable is obtained by differentiating Eq. (4.37)

$$\dot{l}(t) = \frac{\mathrm{d}l(t)}{\mathrm{d}t} = \sum_{i=0}^{n} P_i \times \frac{\mathrm{d}B_{i,n}(t)}{\mathrm{d}t} \tag{4.38}$$

where

$$\frac{\mathrm{d}B_{i,n}(t)}{\mathrm{d}t} = \frac{n!}{(i-1)!(n-i)!}t^{i-1}(1-t)^{n-i} - \frac{n!}{i!(n-i-1)!}t^{i}(1-t)^{n-i-1} \tag{4.39}$$

The winding acceleration of the driving cable is obtained by differentiating Eq. (4.37) twice.

$$\ddot{l}(t) = \frac{\mathrm{d}^2 l(t)}{\mathrm{d}t^2} = \sum_{i=0}^{n} P_i \times \frac{\mathrm{d}^2 B_{i,n}(t)}{\mathrm{d}t^2} \tag{4.40}$$

where

$$\frac{\mathrm{d}^2 B_{i,n}(t)}{\mathrm{d}t^2} = \frac{n!}{(i-2)!(n-i)!}t^{i-2}(1-t)^{n-i} - 2\times\frac{n!}{(i-1)!(n-i-1)!}t^{i-1}(1-t)^{n-i-1} - \frac{n!}{i!(n-i-2)!}t^{i}(1-t)^{n-i-2}$$

The relation between the kinematics states of the deployment angle θ and the collecting length of driving cable l can be described by Eqs. (4.9)–(4.11). The connections among the deployment angular velocity, deployment angular acceleration and l, $\dot{l}$ and $\ddot{l}$ can be established directly, and they can be transformed to

$$\theta = \arcsin\left(\frac{l^2 - L_1^2 - L_2^2}{2L_1L_2}\right) \tag{4.41}$$

$$\dot{\theta} = 2\cdot Q^{-\frac{1}{2}}\cdot l\cdot\dot{l} \tag{4.42}$$

$$\ddot{\theta} = 2\cdot Q^{-\frac{1}{2}}\cdot\dot{l}^2 + 2\cdot Q^{-\frac{1}{2}}\cdot l\cdot\ddot{l} + 4\cdot Q^{-\frac{3}{2}}\cdot l^2\cdot(l^2 - L_1^2 - L_2^2)\cdot\dot{l}^2 \tag{4.43}$$

where

$$Q = -l^4 + 2(L_1^2 + L_2^2) \cdot l^2 - (L_1^2 - L_2^2)^2$$

Since the winding length, velocity and acceleration of the driving cable can be obtained, the corresponding deployment angle, angular velocity and acceleration can be known according to Eqs. (4.41)–(4.43).

In other words, the winding length variation of the driving cable fit by a Bezier curve can be obtained by changing the control point parameters P_i. Then, combined with kinematic relationship between the driving cable length and the deployment angle, the corresponding antenna deployment law can be obtained and the deployment trajectory of the antenna can be determined. As a result, a functional relationship can be established between the Bezier curve control point P_i and the antenna deployment law. Among the control points P_i ($i = 0, 1, 2, \ldots, n-1$), P_0 and P_{n-1} are the boundary conditions at the starting point and ending point, respectively. Therefore, based on the optimization method, the remaining $n-1$ control points $(P_1, P_2, \ldots, P_{n-1})$ are chosen as design variables for the deployment drive system. The objective is to minimize the maximum angular acceleration while ensuring the kinematic performance of deployment.

The optimization model [3] can be described as

$$\begin{aligned} &\text{find} \quad \mathbf{P} = (P_1 \cdots P_{n-1})^{\mathrm{T}} \\ &\text{min} \quad f = \max(abs(\ddot{\theta})) \\ &\text{s.t} \quad h_j(P_1 \cdots P_{n-1}) = 0 \; j = 1 \cdots n_h \\ &\qquad\; g_k(P_1 \cdots P_{n-1}) < 0 \; k = 1 \cdots n_g \end{aligned} \tag{4.44}$$

where n_h and n_g are the total numbers of equality and inequality constraints, which are decided according to the kinematics performance.

The beginning and ending states of the deployment are both static, which means the velocity and acceleration are zero, and the equality constraints can be represented as

$$\begin{cases} h_1(P_1 \cdots P_{n-1}) = \dot{l}(0) = 0 \\ h_2(P_1 \cdots P_{n-1}) = \dot{l}(1) = 0 \\ h_3(P_1 \cdots P_{n-1}) = \ddot{l}(0) = 0 \\ h_4(P_1 \cdots P_{n-1}) = \ddot{l}(1) = 0 \end{cases} \tag{4.45}$$

If there is requirement that the deployment angular velocity remains constant during a certain period of time, then the inequality constraint can be described as

$$g(P_1 \cdots P_{n-1}) = \int_{t=t_1}^{t=t_2} \left|\dot{\theta}(t) - \dot{\theta}(t_1)\right| dt - \chi < 0 \tag{4.46}$$

where $[t_1, t_2]$ is the time interval of uniform motion and χ is a positive small number representing the range of error.

Thus far, the objective function, design variables and constraints have been developed. The optimal control point parameters $(P_1, P_2, \cdots, P_{n-1})$ can be obtained by sequential quadratic programming (SQP). Then, the collecting progress of the driving cables can be obtained by combining it with Eqs. (4.37), (4.38), and (4.40). Finally, the kinematics state of antenna deployment can be obtained by solving Eqs. (4.41)–(4.43).

4.5 Deployment Progress Control

Deployment stability is one of the key points of deployment control for deployable antennas. The key is to ensure deployment accuracy without large shocks at the beginning and ending states of deployment. At the same time, in space applications, it is difficult to control deployment through a large number of sensors. Therefore, in general, the driving moment required by the motor is obtained by pre-designing the driving cable velocity (the flexible multi-body dynamic model described in Sect. 4.2 is subjected to inverse dynamic analysis), and then open loop control is utilized to control the deployment of the deployable antenna.

Considering the requirements of low power and mass for the motor in space application, the motor power is chosen as the objective function, and the shock in deployment is chosen as the constraint. Then, the optimization model can be mathematically stated as the following nonlinear programming problem:

$$\begin{aligned} &\text{find} \quad \mathbf{P} = (P_1 \cdots P_{n-1})^{\mathrm{T}} \\ &\min \quad f = \max(abs(P)) \\ &\text{s.t.} \quad g(P_1 \cdots P_{n-1}) = \max\big(abs\big(\ddot{\theta}\big)\big) \le \alpha_0 \end{aligned} \tag{4.47}$$

The inequality constraint represents that the maximum angular acceleration in deployment is equal to or less than the allowable value.

Theoretical simulation and experimental verification are conducted based on a 2 m-diameter prototype model. The motor power for antenna deployment is obtained by the product of the collecting velocity of the driving cable and the driving force recorded in the experiment. To compare both the data (Figs. 4.14 and 4.15) of the simulation and experiment, the actual deployment time is normalized.

Comparing the driving force curves (Figs. 4.14 and 4.15), it can be seen that the driving cable force is mainly determined by the nonlinear factors such as hinge friction and clearance in the early stage of deployment and by the cable net structure in the later stage.

The theoretical peak value of the driving force (17.15 N) and the experimental value (18.12 N) are substantially equal (the deviation is approximately 5.35%).

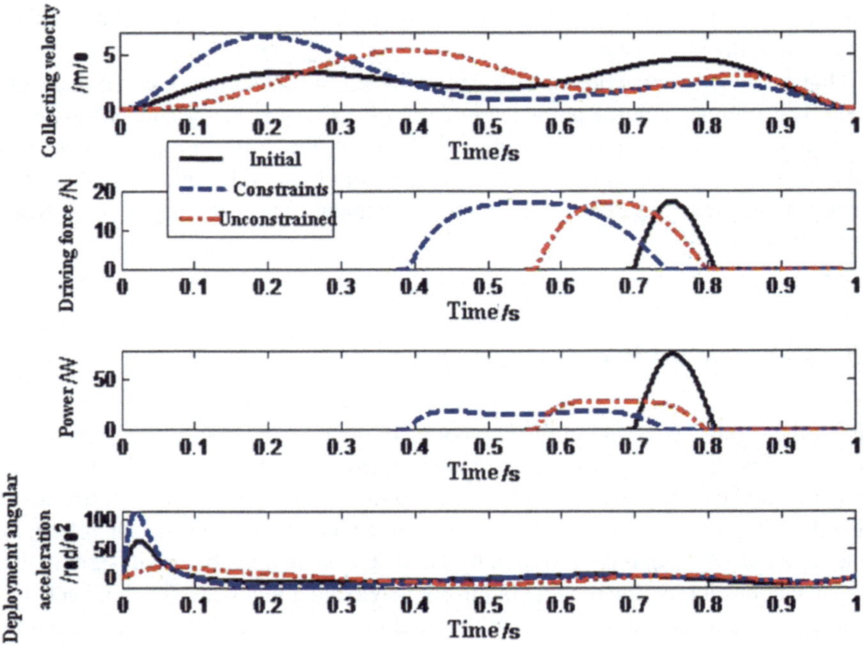

Fig. 4.14 Comparison of simulation results

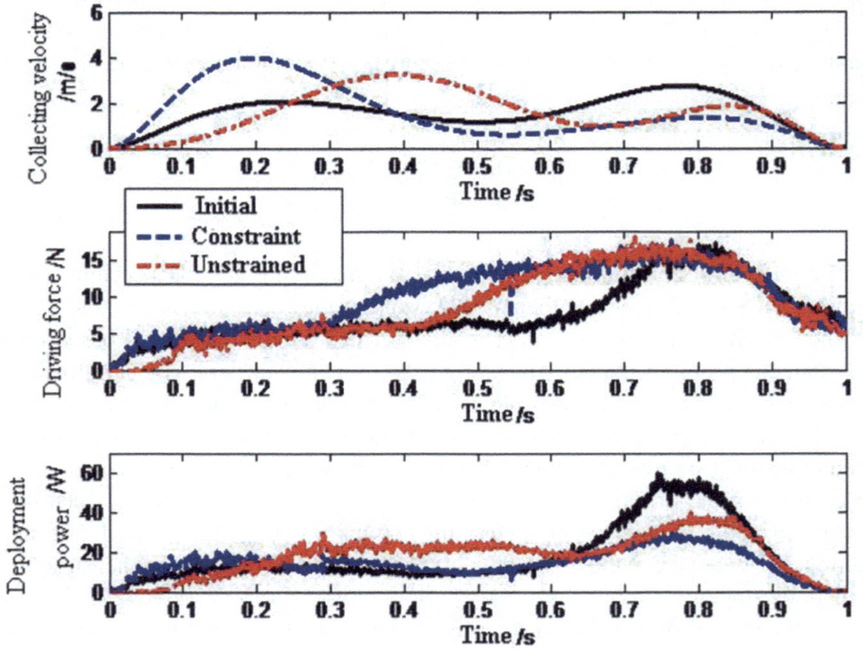

Fig. 4.15 Comparison of experiment results

In terms of motor power, the peak value of power before optimization is 60.14 W. After optimization without the angular acceleration constraint, the value is 30.65 W, which is 49.03% lower than that in the initial case. After optimization with the angular acceleration constraint, the value is 40.17 W, which is 33.21% lower than the initial value and 23.69% higher than that of optimization without constraint.

The results show that the optimal collecting velocity of the driving cable increases rapidly in the initial stage and then maintains a low value near the peak of angular acceleration, thereby reducing the peak value of the motor power. When the angular acceleration constraint is added, the initial driving cable velocity is smoother, thus the peak power is larger, which is consistent with the conclusion of the theoretical results.

References

1. Li TJ (2012) Deployment analysis and control of deployable antenna. Aerosp Sci Technol 18(1):42–47
2. Kashni K (1988) Representation of cables in space subjected to uniformly distributed loads. Int J Space Struct 3(4):221–230
3. Zhang YQ, Yang DW, Li SL (2016) An integrated control and structural design approach for mesh reflector deployable space antennas. Mechatronics 35:71–81

Chapter 5
Electronic Performance Analysis and Equivalent of Mesh Reflector Antenna

5.1 Introduction

Spaceborne antennas are designed for the tasks of space-to-ground and space-to-space communication, navigation, precision observation, space-based warning systems, etc. The electronic performance is the most important and final target that all other factors serve. On April 10, 2019, astronomers from China, the United States, Europe, Japan, Australia and Chile announced the observation of a black hole 55 million light-years away from the Earth using very long baseline interferometry (VLBI) technology composed of eight antennas distributed across the Earth, thus verifying Einstein's theory of relativity (Fig. 5.1).

The calculation of the electronic performance of large spaceborne mesh antennas is far more complicated and difficult than for general reflector antennas. The mesh reflector is not a solid surface reflector and cannot be meshed in a conventional manner. Theoretically, since each wire segment is a line current element, all the wire segments need to be regarded as current elements, and the total far-field pattern of the antenna is then formed by the superposition of far-field patterns generated by all the current elements [1].

5.2 Reflection Matrix and Transmission Matrix of Space Mesh Antenna

5.2.1 Incident, Reflection, and Transmission Fields of Reflecting Mesh

When an electromagnetic wave radiated from transmitting antenna feed or received by a receiving antenna is incident on the mesh of a reflector, it will be divided into two parts, namely reflection and transmission parts leaking through the mesh openings.

B. Duan et al., *Large Deployable Satellite Antennas*,
Springer Tracts in Mechanical Engineering,
https://doi.org/10.1007/978-981-15-6033-0_5

Fig. 5.1 Albert Einstein, 1879–1955, physicist, was honored as the father of modern physics because he established special relativity and general relativity. He discovered the photoelectric effect

According to electromagnetic field theory, there are three electromagnetic fields in space, incident, reflection and transmission fields as shown in Fig. 5.2. In each field, arbitrarily polarized electromagnetic waves can be decomposed into transverse electric (TE) waves and transverse magnetic (TM) waves. A TE wave is a linearly polarized wave whose electric field vector is perpendicular to the incident plane (the plane determined by the incident vector and the z axis), which is also called vertical polarization. The TM wave is a linearly polarized wave whose magnetic field vector is parallel to the incident plane, which is also called parallel polarization.

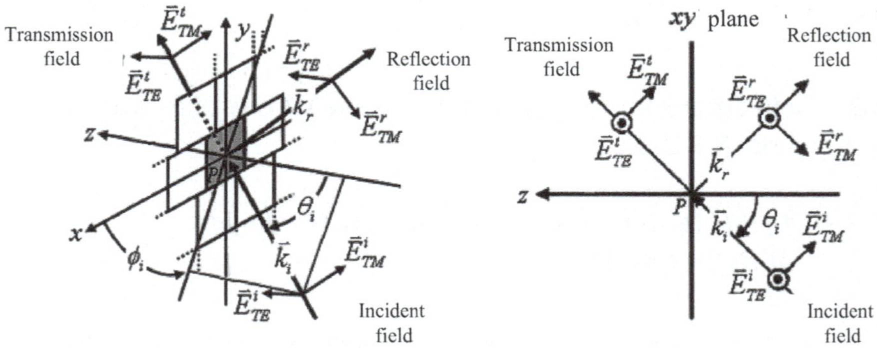

Fig. 5.2 Relationship among incident field, reflection field and transmission field

E_{TE}^{i} the amplitudes of TE waves in the incident field
E_{TM}^{i} the amplitudes of TM waves in the incident field
E_{TE}^{r} the amplitudes of TE waves in the reflection field
E_{TM}^{r} the amplitudes of TM waves in the reflection field
E_{TE}^{t} the amplitudes of TE waves in transmission field
E_{TM}^{t} the amplitudes of TM waves in transmission field, superscripts i, r and t represent the incident, reflection and transmission fields, respectively.

Because of the very small thickness of the mesh, the assumption is that the incident angle is equal to the radiation angle, and the electric field on both sides of the mesh is continuous, which can be described as

$$\begin{cases} E_{TE}^{t} = E_{TE}^{i} + E_{TE}^{r} \\ E_{TM}^{t} = E_{TM}^{i} - E_{TM}^{r} \end{cases} \tag{5.1}$$

Using the reflection matrix $\boldsymbol{R}$, the relationship between the amplitude of the reflection field $\left(E_{TE}^{r}, E_{TM}^{r}\right)$ and that of the incident field $\left(E_{TE}^{i}, E_{TM}^{i}\right)$ can be described as

$$\begin{bmatrix} E_{TE}^{r} \\ E_{TM}^{r} \end{bmatrix} = R\begin{bmatrix} E_{TE}^{i} \\ E_{TM}^{i} \end{bmatrix} = \begin{pmatrix} R_{TEco} & R_{TMcr} \\ R_{TEcr} & R_{TMco} \end{pmatrix}\begin{bmatrix} E_{TE}^{i} \\ E_{TM}^{i} \end{bmatrix} \tag{5.2}$$

The relationship between the amplitude of the transmission field $\left(E_{TE}^{t}, E_{TM}^{t}\right)$ and that of the incident field $\left(E_{TE}^{i}, E_{TM}^{i}\right)$ using the transmission matrix $\boldsymbol{T}$ can be described as

$$\begin{bmatrix} E_{TE}^{t} \\ E_{TM}^{t} \end{bmatrix} = T\begin{bmatrix} E_{TE}^{i} \\ E_{TM}^{i} \end{bmatrix} = \begin{pmatrix} T_{TEco} & T_{TMcr} \\ T_{TEcr} & T_{TMco} \end{pmatrix}\begin{bmatrix} E_{TE}^{i} \\ E_{TM}^{i} \end{bmatrix} \tag{5.3}$$

5.2.2 Surface Induced Current of Flat Mesh

Figure 5.3 describes the situation of TE and TM waves incident on the flat mesh. According to the electromagnetic boundary condition on the interface derived from the Maxwell equations, the induced current at point (0, 0, 0) of the mesh can be expressed as [1].

$$\vec{J} = \hat{n} \times \left(\vec{H}_i + \vec{H}_r\right) + (-\hat{n}) \times \vec{H}_t = \vec{J}_+ + \vec{J}_- \tag{5.4}$$

where $\hat{n}$ is the normal vector of the mesh plane, and $\hat{n} = -\hat{z}$.

Synthesizing the incident, reflection and transmission fields yields the induced currents $\vec{J}_+$ and $\vec{J}_+$ on the front and back surfaces of the mesh, as follows:

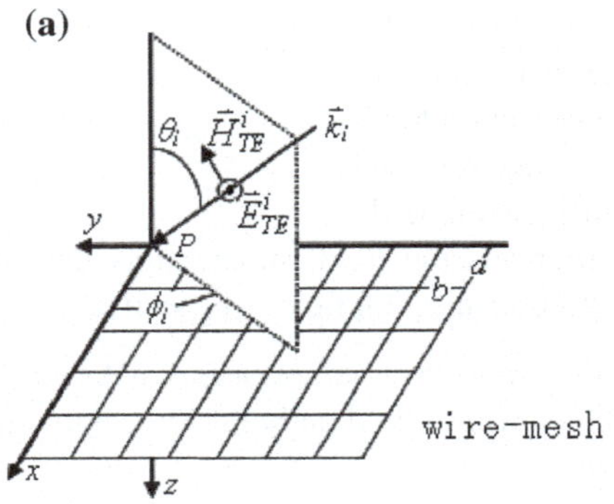

Fig. 5.3 Plane wire-mesh model of incident TE wave and TM wave. **a** TE incident wave (vertical polarization). **b** TM incident wave (horizontal polarization)

$$\vec{J}_+ = \hat{n} \times \left(\vec{H}_i + \vec{H}_r\right) = \hat{n} \times \frac{1}{\eta_0}\left[\hat{\theta}_i\left(E^i_{TE} - E^r_{TE}\right) + \hat{\phi}_i\left(E^i_{TM} + E^r_{TE}\right)\right] \quad (5.5)$$

$$\vec{J}_- = \left(-\hat{n}\right) \times \vec{H}_t = -\hat{n} \times \frac{1}{\eta_0}\left[\hat{\theta}_i E^t_{TE} + \hat{\phi}_i E^t_{TM}\right] \quad (5.6)$$

The surface induced current $\vec{J}$ of the mesh can be derived as [1]

$$\vec{J} = \vec{J}_+ + \vec{J}_- = 2\hat{n} \times \vec{H}_i - \hat{n} \times \frac{2}{\eta_0}\left(\hat{\theta}_i E^t_{TE} + \hat{\phi}_i E^t_{TM}\right) = \vec{J}_{PO} + \vec{J}_m \quad (5.7)$$

where $\vec{J}_{PO} = 2\hat{n} \times \vec{H}_i$ is the induced current of physical optics on the solid surface and $\vec{J}_m$ is the residual induced current due to energy leakage of the mesh, which weakens the current $\vec{J}_{PO}$.

The surface induced current of the curved mesh can be obtained by calculating the surface current of an infinitely large tangent plane, and coordinate conversion is required during the calculation.

Once the reflection and transmission matrices of the mesh are determined, the induced current can be calculated using the methods for general reflectors.

5.2.3 Passive Inter-modulation

Passive inter-modulation (PIM) occurs when a linear combination of carrier signals generated by two or more transmitting carrier waves meeting in a passive device falls into the receiving pass band, creating an interference phenomenon [2]. For example, two carrier waves with frequencies f_1 and f_2, when passing through a weakly nonlinear passive device, will produce a series of new waves with frequencies $mf_1 + nf_2$ $(m, n = 0, \pm 1, \pm 2, \ldots)$, where $|m| + |n|$ is the order of the passive inter-modulation results. Since the new frequencies falling into the receiving channel of the communication system when passing through the passive device cannot be filtered by the filter, the system will receive spurious signals.

There are several reasons for the passive inter-modulation of passive devices, mainly material and contact nonlinearities.

Material nonlinearity refers to the nonlinear constitutive parameters of materials used in microwave communication systems, such as the dielectric constant, magnetic permeability and metal conductivity. The reasons why materials have nonlinear constitutive parameters may include hysteresis and oxidation corrosion of metal surfaces.

Contact nonlinearity is the non-ideal perfect contact that may occur when two metals contact each other, in which the actual contact area is smaller than the ideal one. If the metal surface itself is of certain roughness, or if there is an air film or oxide film on metal contact surface, this causes a nonlinear relationship that does not conform to Ohm's law when metals contact. And it also attributes to material nonlinearity. In general, material and contact nonlinearities are essentially caused by nonlinear constitutive parameters.

5.3 Equivalence Rules for Typical Mesh

5.3.1 Equivalence Rules for Mesh

To acquire the equivalent electromagnetic parameters (dielectric constant, magnetic permeability, and electrical conductivity), the mesh is taken as equivalent to a solid medium whose reflection coefficient is the same as that of the mesh so that it is convenient to calculate the electronic performance of the mesh antenna with the traditional method.

1. Reflection coefficient of solid medium

Suppose the complex dielectric constant and complex permeability of the equivalent solid medium are denoted by $\tilde{\varepsilon}_r = \varepsilon_r' - j\varepsilon_r''$ and $\tilde{\mu}_r = \mu_r' - j\mu_r''$, respectively, and the incident angle of the electromagnetic wave is denoted by θ_m^i. The dielectric constant and magnetic permeability in vacuum are denoted by ε_0 and μ_0, respectively and the free space wave number is denoted by k_0; then, the wave impedance in vacuum is $Z_0 = \sqrt{\mu_0 / \varepsilon_0}$. The corresponding wave impedance of the solid medium under the irradiation of the vertical and parallel polarized electromagnetic waves can be obtained as [3].

$$\text{Vertical polarized} \quad \bar{Z}_{\perp} = \frac{k_0(\mu_r' - j\mu_r'')}{\gamma' - j\gamma''} \tag{5.8}$$

$$\text{Parallel polarized} \quad \bar{Z}_{//} = \frac{\gamma' - j\gamma''}{k_0(\varepsilon_r' - j\varepsilon_r'')} \tag{5.9}$$

where, $\bar{Z}_{\perp}$ and $\bar{Z}_{//}$ are the corresponding wave impedances of the solid medium under the irradiation of the vertical and parallel polarized electromagnetic waves, respectively; γ' is the phase constant and γ'' is the attenuation constant [3]:

$$\gamma' = \frac{k_0}{\sqrt{2}}\left[\sqrt{(\mu_r'\varepsilon_r'' + \mu_r''\varepsilon_r')^2 + (\mu_r'\varepsilon_r' - \mu_r''\varepsilon_r'' - \sin^2\theta_m^i)^2} + (\mu_r'\varepsilon_r' - \mu_r''\varepsilon_r'' - \sin^2\theta_m^i)\right]^{1/2} \tag{5.10}$$

$$\gamma'' = \frac{k_0}{\sqrt{2}}\left[\sqrt{(\mu_r'\varepsilon_r'' + \mu_r''\varepsilon_r')^2 + (\mu_r'\varepsilon_r' - \mu_r''\varepsilon_r'' - \sin^2\theta_m^i)^2} - (\mu_r'\varepsilon_r' - \mu_r''\varepsilon_r'' - \sin^2\theta_m^i)\right]^{1/2} \tag{5.11}$$

Under the irradiation of vertical and parallel polarized electromagnetic waves, the reflection coefficients of the interface between the solid medium and vacuum are (where both are infinite media)

$$\text{Vertical polarized} \quad \Gamma_{\perp} = \frac{\bar{Z}_{\perp}\cos\theta_m^i - 1}{\bar{Z}_{\perp}\cos\theta_m^i + 1} \tag{5.12}$$

$$\text{Parallel polarized} \quad \Gamma_{//} = \frac{\cos\theta_m^i - \bar{Z}_{//}}{\cos\theta_m^i + \bar{Z}_{//}} \tag{5.13}$$

2. Reflection coefficient of mesh

The reflection coefficient of the mesh, in different weaving forms at different incidence angles, can be obtained with experiments or numerical simulations. Under the irradiation of vertical and parallel polarized electromagnetic waves, the reflection coefficient of the mesh is

$$\text{Vertical polarized} \quad \Gamma_{\perp}' = R_{\phi\phi} \tag{5.14}$$

$$\text{Parallel polarized} \quad \Gamma_{//}' = R_{\theta\theta} \tag{5.15}$$

where $R_{\phi\phi}$ and $R_{\theta\theta}$ are obtained from experiments or numerical simulations.

3. Equivalent electromagnetic parameters

To obtain the equivalent electromagnetic parameters of the solid medium corresponding to the mesh, a nonlinear system of equations is established by combining Eqs. (5.8), (5.9), (5.14), and (5.15). The equivalent electromagnetic parameters of the solid medium can be obtained by solving the system of equations following the rule that the real part Re[] and imaginary part Im[] are equal [4, 5]:

$$\begin{cases} \text{Re}[\Gamma_{\perp}] = \text{Re}[\Gamma'_{\perp}] \\ \text{Im}[\Gamma_{\perp}] = \text{Im}[\Gamma'_{\perp}] \\ \text{Re}[\Gamma_{//}] = \text{Re}[\Gamma'_{//}] \\ \text{Im}[\Gamma_{//}] = \text{Im}[\Gamma'_{//}] \end{cases} \tag{5.16}$$

Equation (5.16) is a system of nonlinear equations in terms of the equivalent parameters $\tilde{\varepsilon}_r = \varepsilon'_r - j\varepsilon''_r$ and $\tilde{\mu}_r = \mu'_r - j\mu''_r$, which can be solved with the minimization method of the modulus function. The basic idea of the minimization method of the modulus function is to transform the nonlinear system of equations into a minimization problem of a multi-variable nonlinear function by constructing a modulus function.

For example, a binary nonlinear equation is expressed as

$$\begin{cases} f_1(x, y) = 0 \\ f_2(x, y) = 0 \end{cases} \tag{5.17}$$

where x and y are the variables to be solved, and f_1 and f_2 are nonlinear equations.

For the above nonlinear equations, the modulus equation is

$$\Phi(x, y) = [f_1(x, y)]^2 + [f_2(x, y)]^2 \tag{5.18}$$

where Φ is the modulus equation constructed from the nonlinear system of equations, and the squaring operation prevents negative value.

The solution (x^*, y^*) of the nonlinear Eqs. (5.16) is the minimum of the modulus function of Eq. (5.18). Therefore, the solution of the original nonlinear system of Eq. (5.16) can be obtained by finding the minimum of the modulus function of Eq. (5.18).

Geometrically, the modulus function (5.18) denotes a spatially curved surface whose tangential point (x^*, y^*) to the x–y plane is the minimum point. For each given value, the modulus function Φ corresponds to a closed curve (i.e. contour line) parallel to the x–y plane, and each contour line also corresponds to a modulus function of the value Φ. In the vicinity of the minimum point (x^*, y^*), the contour lines form a family of closed curves centered on this point, and the corresponding function value gradually decreases from outside to inside; when it reaches the point (x^*, y^*), the value drops to zero.

Generally, an n-variable nonlinear system of equations can be denoted by

$$\begin{cases} f_1(x_1, x_2, \ldots, x_n) = 0 \\ f_2(x_1, x_2, \ldots, x_n) = 0 \\ \ldots\ldots \\ f_n(x_1, x_2, \ldots, x_n) = 0 \end{cases} \tag{5.19}$$

where x_1, x_2,..., and x_n are the variables to be solved, and f_1, f_2..., and f_n are nonlinear functions.

The vector form of Eq. (5.19) can be expressed as

$$\boldsymbol{F}(\boldsymbol{X}) = \mathbf{0}, \quad \mathbf{X} = [x_1, x_2, \ldots, x_n]^{\mathrm{T}} \tag{5.20}$$

where **X** denotes the column vector of the n variables, and $\boldsymbol{F}$ is the system of nonlinear equations.

The modulus function is constructed as

$$\Phi(\boldsymbol{X}) = f_1(\boldsymbol{X})^2 + f_2(\boldsymbol{X})^2 + \cdots + f_n(\boldsymbol{X})^2 \tag{5.21}$$

and the iterative solution is found by the steepest descent method with the iteration form.

$$\boldsymbol{X}_{k+1} = \boldsymbol{X}_k - \lambda_k \boldsymbol{g}_k, \quad k = 0, 1, 2, \ldots \tag{5.22}$$

where $\boldsymbol{X}_{k+1}$ and $\boldsymbol{X}_k$ are the n-variable column vectors in the $(k + 1)$th and kth iterations, respectively, $\boldsymbol{g}_k$ is the gradient vector of the modulus function in the kth iteration, and λ_k is the optimal step factor in the kth iteration. The ith component of the gradient vector can be expressed as

$$g_{i,k} = \left(\frac{\partial \Phi}{\partial x_i}\right)_k = 2\left(\left(\frac{\partial f_1}{\partial x_i}\right)_k (f_1)_k + \left(\frac{\partial f_2}{\partial x_i}\right)_k (f_2)_k + \cdots + \left(\frac{\partial f_n}{\partial x_i}\right)_k (f_n)_k\right) \tag{5.23}$$

According to the steepest descent method, the optimal step factor in the kth iteration is obtained as

$$\lambda_k = \frac{(\boldsymbol{g}_k)^T \boldsymbol{g}_k}{2(\boldsymbol{J}_k \cdot \boldsymbol{g}_k)^T (\boldsymbol{J}_k \cdot \boldsymbol{g}_k)}, \quad k = 0, 1, 2, \ldots \tag{5.24}$$

where $\boldsymbol{J}_k$ denotes the Jacobi matrix of the modulus function Φ in the kth iteration.

$$\boldsymbol{J}_k = \begin{bmatrix} \frac{\partial f_1}{\partial x_1} & \frac{\partial f_1}{\partial x_2} & \cdots & \frac{\partial f_1}{\partial x_n} \\ \frac{\partial f_2}{\partial x_1} & \frac{\partial f_2}{\partial x_2} & \cdots & \frac{\partial f_2}{\partial x_n} \\ \cdots & \cdots & \cdots & \cdots \\ \frac{\partial f_n}{\partial x_1} & \frac{\partial f_n}{\partial x_2} & \cdots & \frac{\partial f_n}{\partial x_n} \end{bmatrix}_{\boldsymbol{X}=\boldsymbol{X}_k} \tag{5.25}$$

The termination condition of the iteration is

$$\|\boldsymbol{X}_{k+1} - \boldsymbol{X}_k\| \le \varepsilon \tag{5.26}$$

where ε is the convergence precision.

The iteration procedure is described in Fig. 5.4.

4. Equivalent precision analysis

Regarding the equivalent precision of the mesh equivalence criterion, there are three points to be clarified:

(1) In the equivalence process, due to the unique property of the mesh material, the reflection coefficient matrix can be expressed as

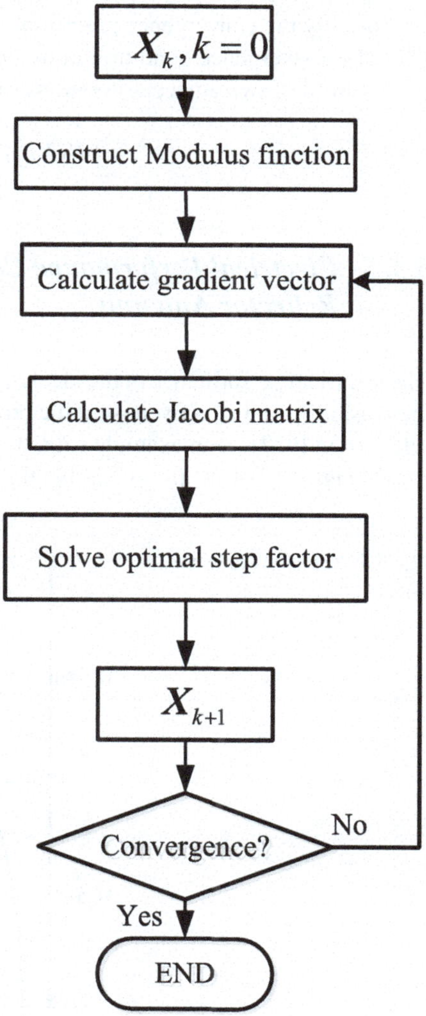

Fig. 5.4 Flow chart of iteration

$$R = \begin{bmatrix} R_{\theta\theta} & R_{\theta\phi} \\ R_{\phi\theta} & R_{\phi\phi} \end{bmatrix} \tag{5.27}$$

That is, there are cross terms between the two polarizations (vertical and parallel polarizations). According to [1], the contribution of these two cross terms is very small, and the difference with respect to the main diagonal terms is almost 8–9 dB, so it can be ignored.

(2) In the process of solving the nonlinear system of equations, the minimization method of the modulus function is used to solve iteratively. the convergence condition is given in Eq. (5.26) where the convergence precision ε is the calculation precision of the equivalent process and can be artificially set as needed. usually the convergence precision is set to $\varepsilon = 10^{-6}$ in the calculation.

(3) The convergence condition in the iteration process can also be changed to the 2-norm of two adjacent iterations satisfying the convergence precision.

$$\| \boldsymbol{F}(\boldsymbol{X}_{k+1}) - \boldsymbol{F}(\boldsymbol{X}_k) \| \le \varepsilon \tag{5.28}$$

5.3.2 Electrical Performance Calculation for Typical Mesh Reflector Antenna

The electrical performance of the ideal mesh reflector antenna is calculated to confirm the feasibility and effectiveness of the proposed method. Taking the offset paraboloid reflector in [6, 7] as an example (Fig. 5.5), the aperture of the reflector is 0.6 m, the focal length is 0.4 m, the offset height is 0.4 m, the feed inclination angle ψ_0 is

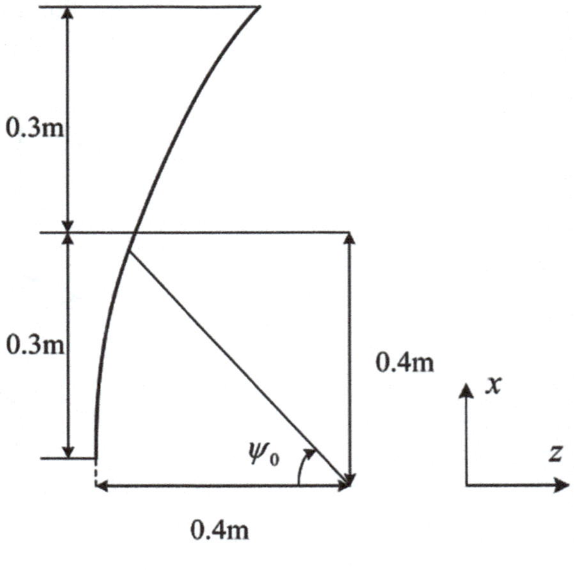

Fig. 5.5 The offset reflector sketch

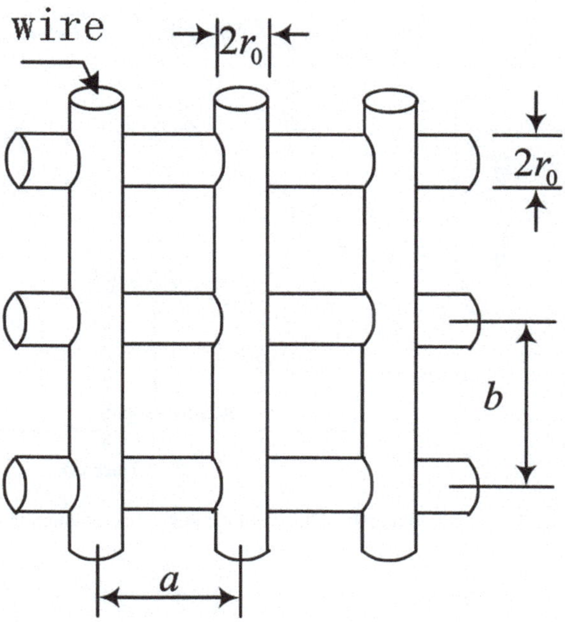

Fig. 5.6 The wire-grid scheme

48.31°, and the operating frequency is 20 GHz. A Cosine-Q feed is employed in the right-hand circular polarization excitation mode with an edge taper −18 dB and feed parameter $Q_x = Q_y = 11.01$. A simple woven grid [6] element is used in the mesh (Fig. 5.6) with the element parameters of crosswise spacing a, longitudinal spacing b and cross-sectional radius r_0. Two sets of mesh structure parameters are used in the calculation, namely $a = 0.02\lambda$, $b = 0.04\lambda$, $r_0 = 0.001\lambda$ and $a = 0.02\lambda$, $b = 0.04\lambda$, $r0 = 0.001\lambda$ (λ is the operating wavelength).

In this simulation, only the far-field pattern of the ideal reflector antenna with mesh transmission is calculated and compared with the results in [6]. Figure 5.7 shows the simulated far-field pattern of the x–z plane where the dashed line represents the result of the mesh reflector with $a = 0.02\lambda$ and $b = 0.04\lambda$, the dotted line represents that of the mesh reflector with $a = 0.02\lambda$ and $b = 0.04\lambda$, and the solid line represents that of the solid reflector. Figure 5.7 shows that the mesh structure has a great influence on the far-field cross-polarization pattern, which results in an increase in the cross-polarization level and has almost no influence on the main polarization pattern. Figure 5.8 shows the far-field pattern in [6]. By comparing the simulation results with those in [6], the results of this paper are seen to have little difference from those in [6], which demonstrates the effectiveness of the proposed method.

An inspection of the calculation cost and number of elements of the model using line elements and of the equivalent model shows that the number is line elements reaches approximately 138,000 and that 9.2 h are needed for calculation when line elements are used to model and analyze the mesh reflector; in contrast, only approximately 10,000 plane elements are used and 0.4 h are needed for the equivalent model.

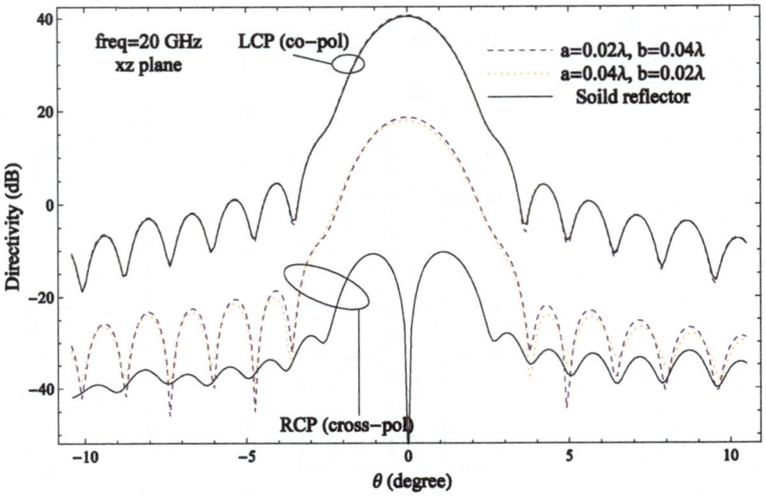

Fig. 5.7 Far-field pattern of ideal wire-mesh reflector antenna using proposed method

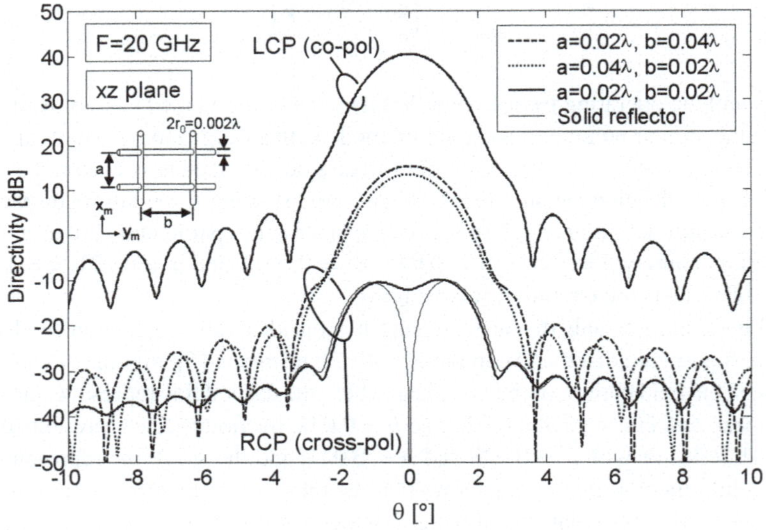

Fig. 5.8 Far-field pattern of ideal wire-mesh reflector antenna in [6, 7]

Clearly, the equivalent model for the electrical performance analysis of the wire-mesh reflector antenna can greatly reduce the computation cost while maintaining the same precision.

5.4 Electromechanical Integrated Design of Space Mesh Reflector Antenna

A spaceborne mesh reflector antenna is a common type of electronic equipment, and its structural and electromagnetic characteristics interact with and restrict each other. To design a high-performance spaceborne mesh reflector antenna, it is necessary to propose an electromechanical integrated design model based on an electromechanical coupling model from the perspective of electromechanical synthesis and interdisciplinary intersections. This section conducts a preliminary exploration of the electromechanical integrated design of a space mesh reflector antenna.

The space mesh reflector antenna is generally composed of a front support cable net, rear support cable net, vertical cables, a mesh and a deployable truss. The cross-sectional area of the elements in the mesh reflector antenna is denoted by A, and the pre-tension is denoted by F. The cable net consists of main cables, assistant cables and secondary cables, and the deployable truss consists of horizontal bars, vertical bars and diagonal bars. Therefore, the design variables of the cross-sectional area and pre-tension can be merged according to engineering requirements.

The combined design variables of the cross-sectional area are denoted as $A_1, A_2, \ldots, A_n$, and the combined design variables of the pre-tension are denoted as $F_1, F_2, \ldots, F_m$ where n and m are the total number of the combined design variables of the cross-sectional area and of the pre-tension.

Taking the combined cross-sectional area and pre-tension as design variables and the performance of the antenna weight, antenna gain and side-lobe level as objective functions, an electromechanical comprehensive optimization model of the space mesh reflector antenna can be mathematically described as follows:

$$\text{find} \quad X = (A_1, \ldots, A_n)^{\mathrm{T}} \quad \text{and} \quad (F_1, F_2, \ldots, F_m)^{\mathrm{T}} \tag{5.29}$$

$$\min \quad W = \sum_{i=1}^{n} \rho A_i \sum_{j=1} L_{ij} \quad \text{or} \quad -Gain \quad \text{and} \quad E_{\text{side}} \tag{5.30}$$

$$\text{s.t.} \quad Gain(X) \geq G^0 \quad \text{and} \quad E_{side} \leq E_{side}^0 \quad \text{or} \quad W \leq W_{\max} \tag{5.31}$$

$$D_k^{deploy} - D^L \leq 0 \quad (k = 1, 2, \ldots, PNUM) \tag{5.32}$$

$$\sigma_{r,k}^{deploy} - [\sigma_{cable}^{deploy}] \leq 0 \quad (r = 1, 2, \cdots CNUM) \tag{5.33}$$

$$\sigma_{e,k}^{deploy} - [\sigma_{truss}^{deploy}] \leq 0 \quad (e = 1, 2, \ldots TNUM) \tag{5.34}$$

$$-f_k^{deploy} + [f^{deploy}] \le 0 \quad (k = 1, 2, \ldots, PNUM) \tag{5.35}$$

$$-f^{folded} + [f^{folded}] \le 0 \tag{5.36}$$

$$V_{folded} - V^U \le 0 \tag{5.37}$$

$$A_i \in [A^L, A^U] \quad (i = 1, 2, \ldots, n) \tag{5.38}$$

$$F_i \in [F^L, F^U] \quad (i = 1, 2, \ldots, m) \tag{5.39}$$

where

(1) Design variables

A_i is the set of design variables of an element cross-sectional area, and n is the total number of variables of the cross-sectional areas.

F_i is the set of pre-tension design variables, and m is the total number of pre-tension design variables.

(2) Objective functions

W is the structure mass, ρ is the material density, L_{ij} is the element length corresponding to design variable A_i, *Gain* is the antenna gain, and E_{side} is the first side-lobe level.

(3) Constraint conditions

Equation (5.31) describes the antenna gain constraint, the first side-lobe level constraint and the mass constraint where G^0 is the minimum allowable value of the antenna gain, E_{side}^0 is the allowable value of the antenna's first side-lobe level, and $W_{\max}$ is the maximum allowable value of the antenna mass.

Equation (5.32) is the precision constraint of the deployed antenna where $PNUM$ is the total number of loading conditions of the deployed antenna, D_k^{deploy} is the actual surface precision of the deployed antenna in the kth loading condition and D^L is the allowable precision designed for the mesh antenna.

Equation (5.33) is the strength constraint of the cable elements of the deployed antenna where $\sigma_{r,k}^{deploy}$ is the actual stress of the cable elements in the kth loading condition in the deployed state, $[\sigma_{cable}^{deploy}]$ is the allowable stress of the cable elements, and $CNUM$ is the total number of cable elements.

Equation (5.34) is the strength constraint of the link elements of the ring truss of the deployed antenna where $\sigma_{e,k}^{deploy}$ is the actual stress of the link elements under the kth loading condition in the deployed state, $[\sigma_{truss}^{deploy}]$ is the maximum allowable stress of the link elements, and $TNUM$ is the total number of link elements.

Equation (5.35) is the fundamental frequency constraint in the deployed state where f_k^{deploy} is the fundamental natural frequency of the antenna in the deployed state in the kth working condition, and $[f^{deploy}]$ is the allowable fundamental frequency in the deployed state.

Equation (5.36) is the fundamental frequency constraint in the stowed state where f^{folded} is the fundamental natural frequency of the antenna in the stowed state, and $[f^{folded}]$ is the allowable fundamental natural frequency in the stowed state.

Equation (5.37) is the volume constraint in the stowed state, where V_{folded} is the actual volume of the antenna in the stowed state and V^U is the allowable volume in the stowed state.

Equation (5.38) is the boundary constraint of the cross-sectional area design variable where A^U and A^L are the upper and lower limits of the element cross-sectional area, respectively.

Equation (5.39) is the boundary constraint of the pre-tension design variable where F^L and F^U are the upper and lower limits of the pre-tension, respectively.

Thus far, the overall framework of the electromechanical integrated optimization design of spaceborne mesh reflector antennas has been established. Of course, it is also necessary to consider the solution method of the optimization model, coupling model, and derivation method of the sensitivity equation. This remains to be studied in depth and systematically.

5.5 Numerical Experiment and Analysis of Results

5.5.1 Electrical Performance Analysis of Array Feed and Deployable Truss Mesh Reflector Antenna

Here the structural model of a mesh reflector antenna is established in the integrated design platform SDADS by Xidian University of China, which will be introduced in detail in Chap. 9. The dual-polarized log-period antenna array is employed as the antenna array feed. The electrical performance analysis is performed on the pre-designed array feed mesh reflector antenna inside the SDADS. The antenna operates at 300 MHz and the geometric parameters are shown in Fig. 5.9. The geometric model of the reflector is shown in Fig. 5.10 and the CAD model of the array feed is shown in Fig. 5.11.

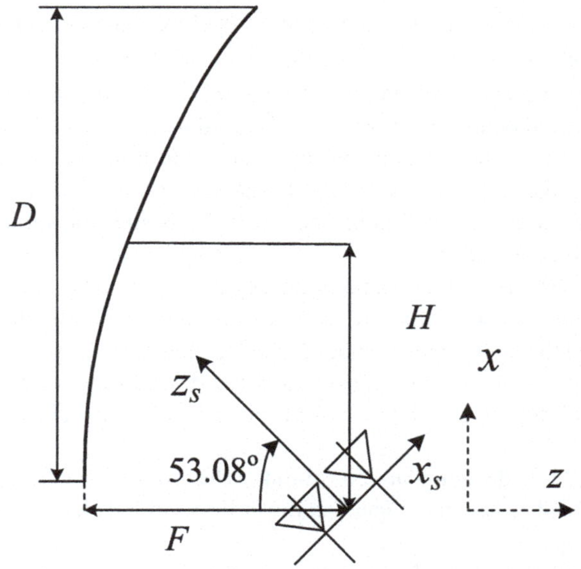

Fig. 5.9 Geometric diagram of mesh reflector antenna with offset array feed

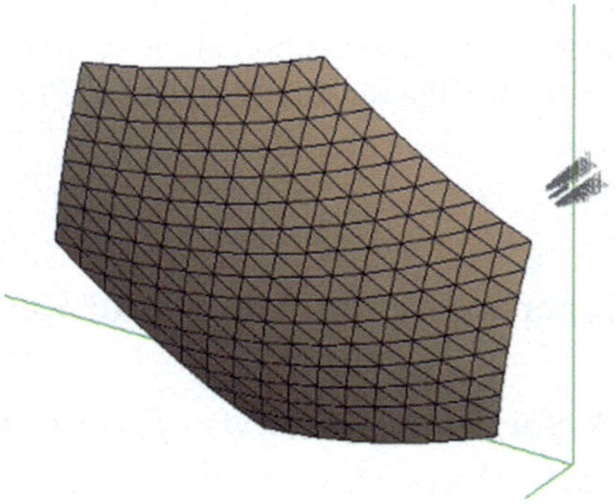

Fig. 5.10 Geometric diagram of mesh reflector

The equivalent model is used to analyze the mesh reflector, and the electrical performance analysis is performed. The far field patterns at the 0 and 90° of the mesh reflector antenna with an array feed are shown in Figs. 5.12 and 5.13. The antenna gain, side-lobe level and half power lobe width are shown in Table 5.1.

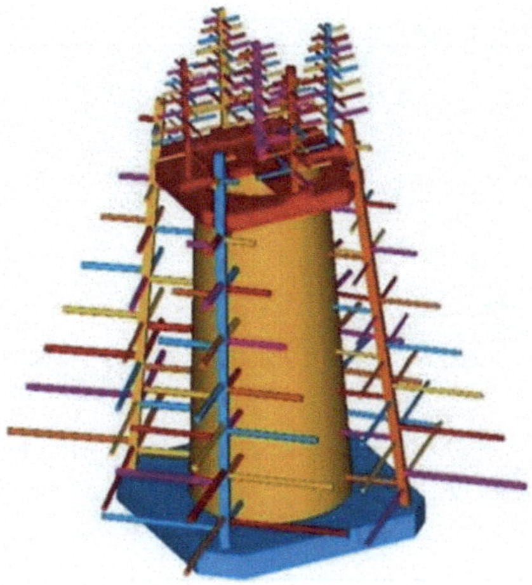

Fig. 5.11 Schematic diagram of array feed CAD model

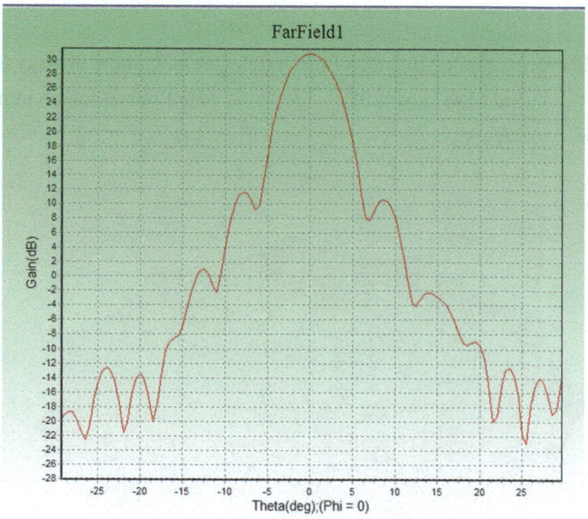

Fig. 5.12 Far field pattern of mesh reflector antenna with array feed at 0°

5.5.2 Electrical Performance Analysis of Radial Rib Deployable Antenna

The structural model of the radial rib mesh reflector antenna is also established by the SDADS and the electric dipole is employed as the initial excitation of the antenna. The electrical performance analysis of the radial rib mesh reflector antenna excited by a single feed is performed. The geometric model of the radial rib mesh reflector

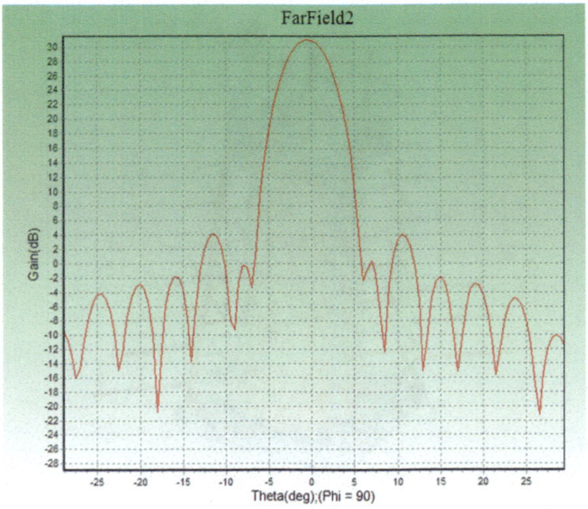

Fig. 5.13 Far field pattern of mesh reflector antenna with array feed at 90°

Table 5.1 Far field electrical performance of mesh reflector antenna with array feed

Gain/dB	Side-lobe level/dB	Half power lobe width/°
31.12	−19.38	5.0

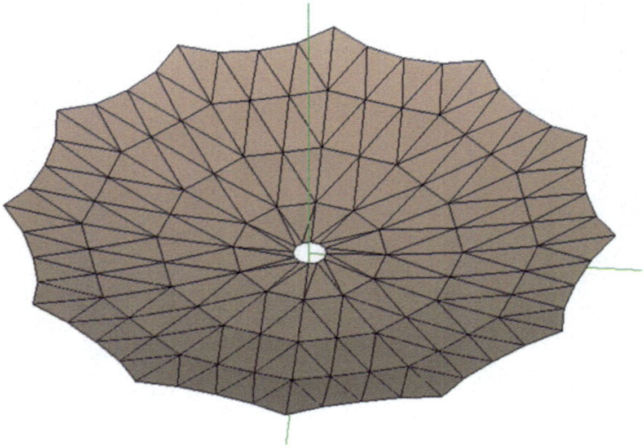

Fig. 5.14 Geometry model of radial rib mesh reflector antenna

antenna is shown in Fig. 5.14. The antenna aperture is 5 m, the operating frequency is 6 GHz, the focal length is 4 m, and a gold-plated molybdenum mesh is employed as the reflector. The antenna consists of 12 main ribs, each of which is divided into 4 segments.

The mesh reflector is treated with the equivalent model, and electrical performance analysis is performed. The far field patterns of the two planes (0 and 90°) of the radial rib mesh reflector antenna are shown in Figs. 5.15 and 5.16. The antenna gain, side-lobe level and half power lobe width are listed in Table 5.2.

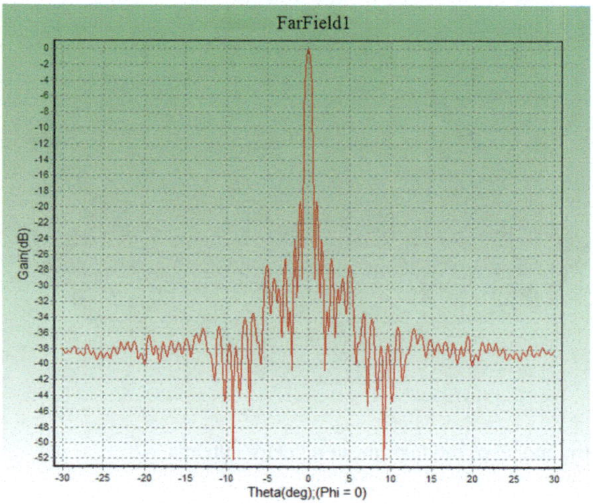

Fig. 5.15 Far field pattern of radial rib mesh reflector antenna at 0°

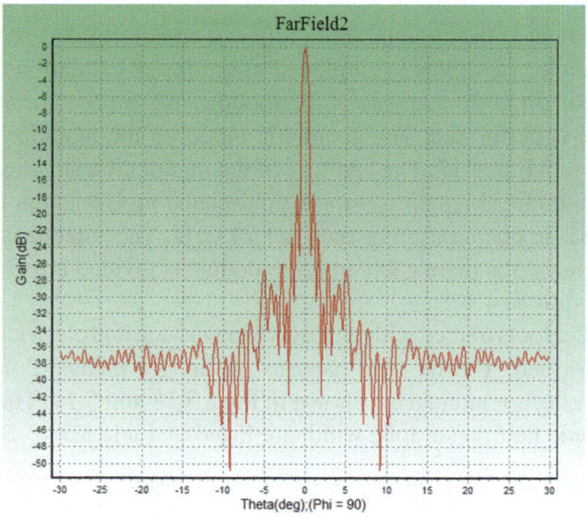

Fig. 5.16 Far field pattern of radial rib mesh reflector antenna at 90°

Table 5.2 Far field electrical performance of radial rib mesh reflector antenna

Gain/dB	Side-lobe level/dB	Half power lobe width/°
39.24	−19.24	0.6

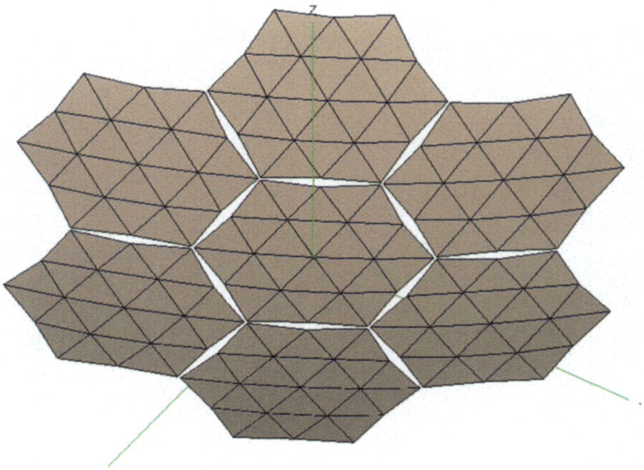

Fig. 5.17 Geometric model of modular mesh reflector antenna

5.5.3 Electrical Performance Analysis of Modular Reflector Antenna

The structural model of the modular mesh reflector antenna is also established by the SDADS and the electric dipole is employed as the initial excitation of the antenna. The electrical performance analysis of the modular mesh reflector antenna excited by a single feed is performed. The geometric model of the modular mesh reflector antenna is shown in Fig. 5.17. The aperture of the modular mesh reflector antenna is 4.81×5 m and the focal length is 4 m. A gold-plated molybdenum mesh serves as the reflector, and the operating frequency is 1.71 GHz. The antenna is divided into 2 rings and 7 modules, the main cable in each module is divided into 2 segments, and the sag-to-span ratio is 4%.

The mesh reflector is treated with the equivalent model, and electrical performance analysis is performed. The far field patterns of the two planes (0 and 90°) of the modular mesh reflector antenna are shown in Figs. 5.18 and 5.19. The antenna gain, side-lobe level and half power lobe width are listed in Table 5.3.

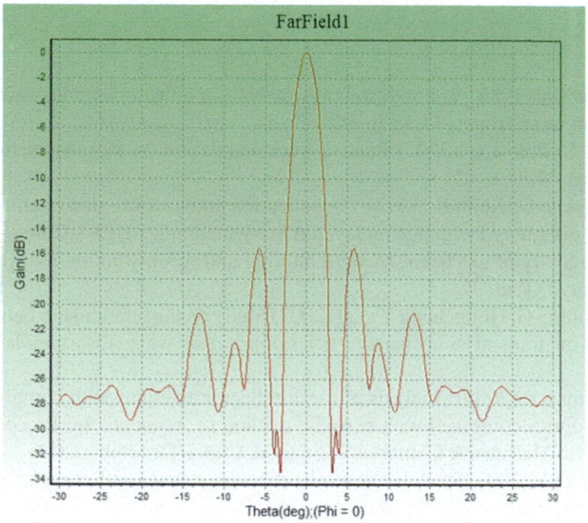

Fig. 5.18 Far field pattern of modular mesh reflector antenna at 0°

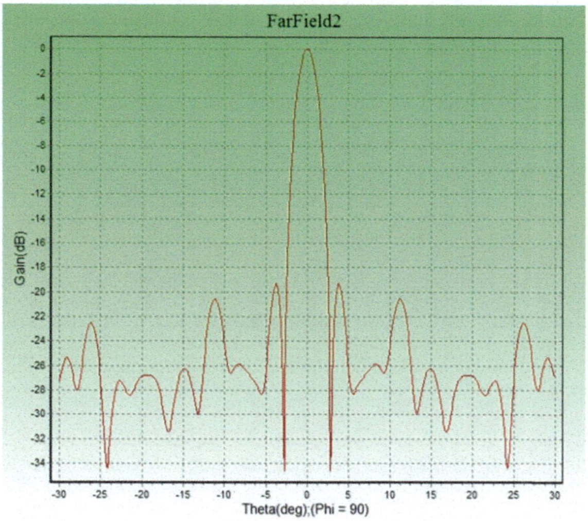

Fig. 5.19 Far field pattern of modular mesh reflector antenna at 90°

Table 5.3 Far field electrical performance of modular mesh reflector antenna

Gain/dB	Side-lobe level/dB	Half power lobe width/°
29.98	−15.52	2.4

References

1. Rahamt SY, Lee SW (1985) Vector diffraction analysis of reflector antennas with mesh surfaces. IEEE Trans Antennas Propag 33(1):76–90
2. Bolli P, Selleri S, Pelosi G (2002) Passive intermodulation on large reflector antenna. IEEE Antennas Propag 44(5):13–20
3. Bolin Z, Zhengping G, Kejin R (1993) The reflection and transmission coefficients for electromagnetic wave obliquely incidence on layered lossy medium. J UEST China 22(5):483–489
4. Bing W (2003) Study of electromagnetic scattering and reconstruction for Anisotropic media. Xidian University, Xi'an
5. Xueguan L, Huiping G, Hongcheng Y et al (2003) Study on the relationship between the medium parameters and the transmission characteristics of arbitrary wave-absorbing layer. Int J Microw 19(4):5–14
6. Miura A, Rahamt SY (2005) Analysis of mesh reflector antennas with complex mesh surfaces using physical optics combined with periodic method of moments. In: Proceedings of APCM 2005 Asia-Pacific Microwave Conference, Suzhou, China, December 2005
7. Miura A, Rahamt SY (2007) Spaceborne mesh reflector antennas with complex weaves: extended PO/periodic MoM analysis. IEEE Trans Antennas Propag 55(4):1022–1029

Chapter 6
Reflector Surface Precision Measurement and Adjustment of Cable-Truss Combined Structures

6.1 Introduction

To observe the universe more clearly, it is a good strategy to put an antenna in space. The Hubble (Fig. 6.1) space telescope in the USA is an example. In the harsh environment of space, it is difficult to maintain the surface precision of the antenna, which is one of the most crucial indicators to evaluate the performance of the antenna. Although the requirements of surface precision and structure stiffness have been considered in the design phase, deployable mesh reflector antennas inevitably introduce random errors in the phases of processing, manufacturing and assembly phases, which lead to the degradation of surface precision. To reduce the influence of manufacturing and assembling errors and improve the surface precision, it is necessary to measure and adjust the cable net surface of the antenna [1, 2]. The methods of measuring the surface of an object can be roughly classified into two types, i.e., contact and non-contact types. Due to the flexibility of cable net structure, any contact-type measurement would change, to some degree, the surface precision, so only the latter can be applied.

In the process of adjusting the surface precision of the cable truss combined structure, since there are some differences between the finite element analysis model established in the initial design phase and the physical model of the antenna. To reflect the actual situation of the physical antenna by using the finite element analysis model, the initial finite element model should be modified based on the measurement results for the physical antenna.

6.2 Photogrammetry of Surface Precision

The photogrammetry process of the surface of the cable truss combined structure largely consists of the arrangement of the photogrammetric environment (prepared

B. Duan et al., *Large Deployable Satellite Antennas*,
Springer Tracts in Mechanical Engineering,
https://doi.org/10.1007/978-981-15-6033-0_6

Fig. 6.1 Edwin Powell Hubble, 1889–1953, American astronomers, the founder of galaxies astronomy and the pioneer of observational cosmology, confirmed the existence of other galaxies outside the Milky Way, and discovered Hubble's law

in advance), the photograph of the measurement field and photo processing. The correct photogrammetric process is the prerequisite for ensuring the accuracy of measurement. Based on the Tianyuan 3D photogrammetry system and a 2m-diameter physical antenna, the photogrammetric process will be discussed in detail.

1. Arrangement of photogrammetric environment

There are two key points should be noted during the arrangement of photogrammetric environment. One is to properly arrange the target points, the coding points and the standard ruler, and the other is to correctly select the size of the target points and coding points. The quantitative relationship between the size and distance of the coding points is shown in Fig. 6.2.

When arranging, the photographic reflective targets should be pasted on the cable net nodes to be tested firstly, and then the standard ruler should be placed around the measured antenna. Finally, the coding points have to be placed, even and flat, around

Fig. 6.2 Quantitative relationship between the size of the coding point and distance

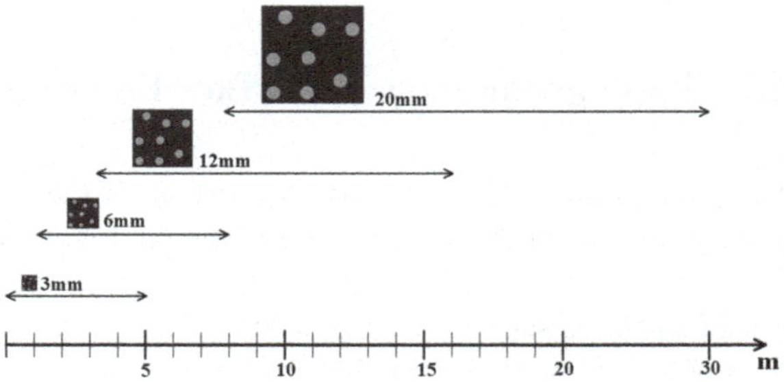

Fig. 6.3 Layout of the measurement environment

the standard ruler and the entire antenna. It should be noted that as many coding points as possible should be placed in the corners of the antenna, with a distance of 40–60 cm between the coding points. In summary, the layout of the measurement environment is shown in Fig. 6.3.

2. Photograph of measured object

During the photography process, the camera shoots around the measured object, which should be completely covered by the camera's photograph network. The photograph network schematic diagram for the antenna model is described in Fig. 6.4. In addition, the measurement distance is generally between 1.0 and 3.0 m, and the number of photos taken is related to the photograph position; the corners of the measured object should be shot sufficiently.

3. Photo processing

After obtaining the photos of the antenna model, the Digimetric, a three dimensional photogrammetry system software, will be used to calculate the three-dimensional spatial coordinates of targets [3]. The operation interface of the Digimetric software is given in Fig. 6.5. The obtained target distributions (including the top and left views) are shown in Fig. 6.6.

The measuring precision of the photogrammetric system is an evaluation criterion for measurement results, which is critical to the precision of the measurement results. The measuring precision of photogrammetric systems is usually related to system hardware, measurement distance and environmental factors. Taking the Tianyuan Photogrammetry System as an example, the measurement precision can be expressed as

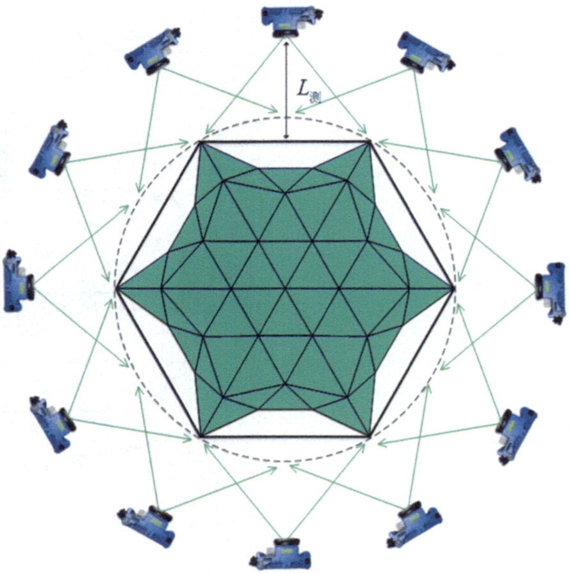

Fig. 6.4 Shooting network schematic diagram

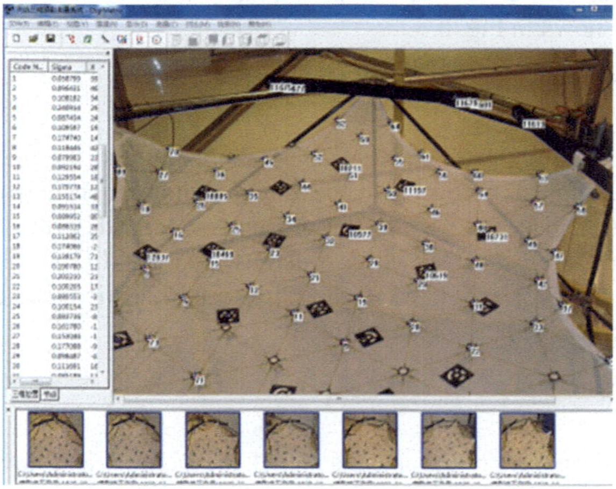

Fig. 6.5 Operation interface of the Digimetric software

$$\delta_{\text{precision}} = \zeta_0 + \xi \cdot L_{\text{measure}} + \Delta\zeta \tag{6.1}$$

where, ζ_0 is the basic error of the photogrammetric system; for the Tianyuan Photogrammetry System, $\zeta_0 = 20$ μm. $\xi = 10$ μm is the measurement distance factor, and L_{measure} is the measurement distance, i.e. the distance from the camera

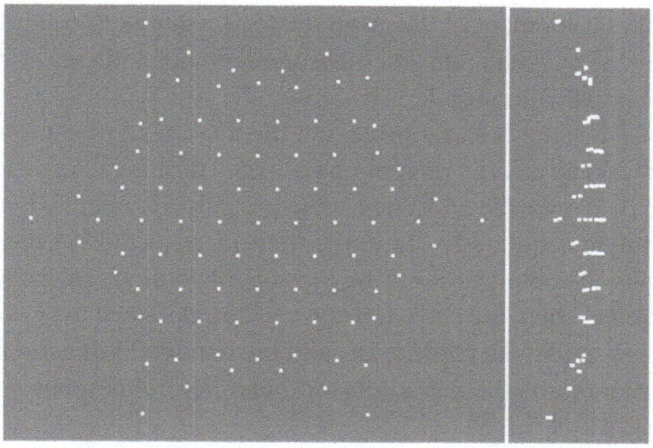

Fig. 6.6 Top view and left view of the target points

lens to the measured object. $\Delta\zeta$ is the value of the random error affected by environmental factors, the range of which is $[0,\ 20\ \mu\text{m}]$. In general, the ideal measuring environment is windless, with an indoor temperature of 20 °C, humidity 65% RH and indoor brightness 300 lx.

If measuring the 2 m-diameter model, the measuring distance L_{measure} is approximately 1.5 m. According to Eq. (6.1), the measuring precision $\delta_{\text{precision}}$ is taken as $[35,\ 55]\ \mu\text{m}$, which meets the requirement for the measurement of the antenna model whose surface precision is <0.8 mm. That is, the measured data are credible.

After obtaining the three-dimensional coordinates of the targets, the reflector precision is calculated as follows: Assuming that the paraboloid fitted equation is $ax^2 + by^2 + cxy + dx + ey + f = z$, the corresponding z coordinates could be calculated by substituting the x and y coordinates of the targets into the paraboloid fitted equation. Since the ideal z' is already known, the value of $\Delta z = z' - z$ gives the reflector precision as

$$\delta_{rms} = \sqrt{\frac{\sum_{i=1}^{n} \Delta z_i^2}{n}} \tag{6.2}$$

where n is the total number of targets.

6.3 Cable Tension Measurement

The cable tension of the support cable net is the crucial parameter for adjusting the reflector precision and checking the stiffness of the cable net structure. Accurate measurement of the cable tension of the cable net structure plays an important role in the successful development of the mesh reflector antenna. It should be mentioned that there are few non-contact tension measurement technologies at present; hence, a hand-held dynamometer is used.

According to the force-balance principle, the hand-held dynamometer can indirectly measure the cable tension by applying a load in the lateral direction of the cable, as shown in Fig. 6.7. Once the tensioned cable is clamped onto the dynamometer, the lateral load F can be acquired by the force transducer located in the roller, and the distance h can also be obtained according to the inherent geometrical relationships of the dynamometer. Then, the cable tension can be calculated by $T = F\big/(2\cos\alpha)$. The measuring process of the cable tension of the hand-held dynamometer is shown in Fig. 6.8. The range of the cable tension hand-held dynamometer used here is 0–55 N with an error of ± 1 N.

However, there are three problems in measuring cable tension by means of a hand-held dynamometer. Firstly, the initial position of the end point of the cable changes due to clamping; and the initial equilibrium state is broken, which will increase the deviation between the measured result and the actual tension of cables. Secondly, the tension loss caused by friction is neglected when calculating the cable tension. Thirdly, a cable net structure with a large diameter is assembled using many cables, which will take plenty of time to measure with hand-held dynamometers. Therefore, the hand-held dynamometer has certain limitations in the tension measurement of the cable net structure.

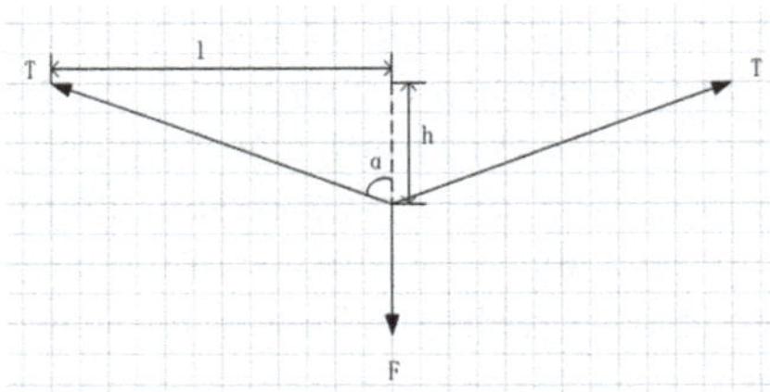

Fig. 6.7 Schematic diagram of hand-held dynamometer

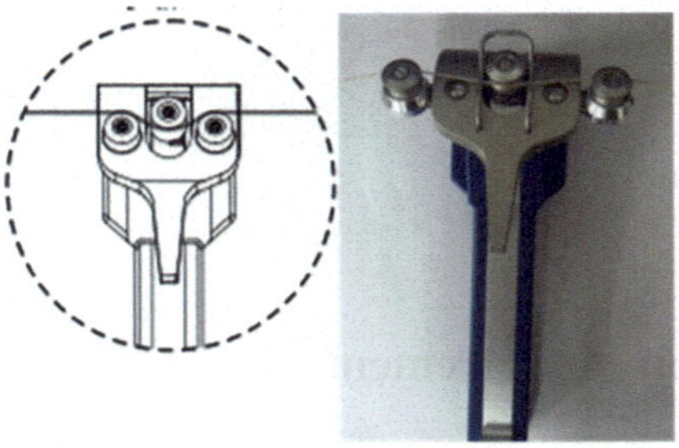

Fig. 6.8 Measuring process of the cable tension of the hand-held dynamometer

6.4 Surface Precision Adjustment Test

Before establishing the finite element model of the cable truss combined structure, the form-finding design of the cable net should be carried out to determine a reasonable initial length of the cables. The relevant design method is given in Sect. 3.3. According to the design parameters of the truss and initial lengths of the cables, a finite element model of the cable truss combined structure can be established to guide the surface precision adjustment.

First, a numerical simulation of the surface adjustment of the 2 m-diameter mesh reflector antenna shown in Fig. 6.9 is performed, and the finite element model is established according to the design parameters. To simulate the influence of manufacturing error, a random disturbance is applied to the initial length of the truss rods and cables of $l_{0p} = l_0 + 2.0\ \text{mm} \times \alpha$ where l_0 represents the initial length of the truss rods and cables and α is a random variable evenly distributed in the interval [−1, 1]. After the error disturbance is applied, the support cable-net will be in a new equilibrium position, and the surface precision will degrade. This enables us to adjust the surface precision.

To investigate the influence of the finite element model on the effectiveness of the surface precision adjustment, traditional adjustment and the model-based adjustment were simulated and compared, and each adjustment was repeated 5 times. In the simulation, the perturbation is applied to the finite element model first, and the resulting model is used to represent the physical model. The corresponding surface precision changes during the adjustment process are shown in Fig. 6.10. It can be seen that the convergence speed of the traditional adjustment is slower, and the model-based adjustment can speed up the convergence and improve the adjustment efficiency.

To further investigate the differences between the two methods, the variation of the length of cables during the adjustment process is analyzed and compared.

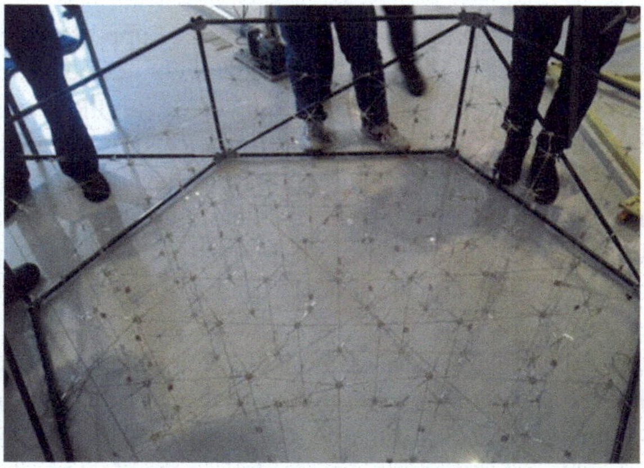

Fig. 6.9 Support cable net and targets of the 2 m-diameter deployable reflector antenna prototype

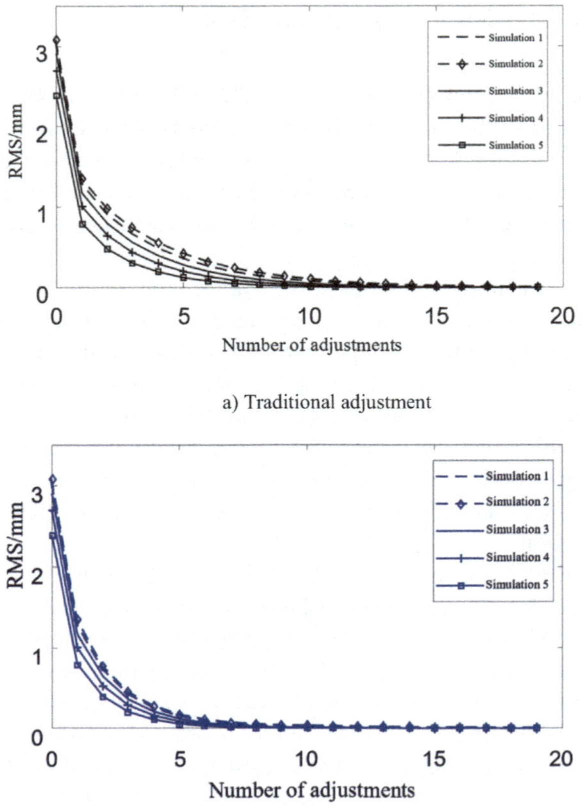

a) Traditional adjustment

b) Model-based adjustment

Fig. 6.10 Shape surface accuracy change in the adjustment process

Considering the geometric symmetry, one-sixth of the support cable net shown in Fig. 6.11 is taken for analysis. Each support node is connected with an adjustable cable, and the variation of adjustable cables connected with the nodes marked with an asterisk is given in Fig. 6.12. For the sake of clarity, only the first 10 variations in cable length are given here. It can be seen from the figure that both methods have some degree of over adjustment, that is, the adjustable cable is elongated at a certain adjustment step, and shortened in the next adjustment step. The degree of over adjustment is smaller for the model-based adjustment and the adjustment process converges more quickly.

To further verify the effectiveness of the method, a surface adjustment test was performed on the 2 m-diameter antenna model shown in Fig. 6.9. By adopting the above adjustment method and adjusting the length of the vertical cable, the nodes of the upper support cable net (reflective surface) fall on the ideal paraboloid.

Fig. 6.11 Numbered nodes of the one-sixth of the support cable net

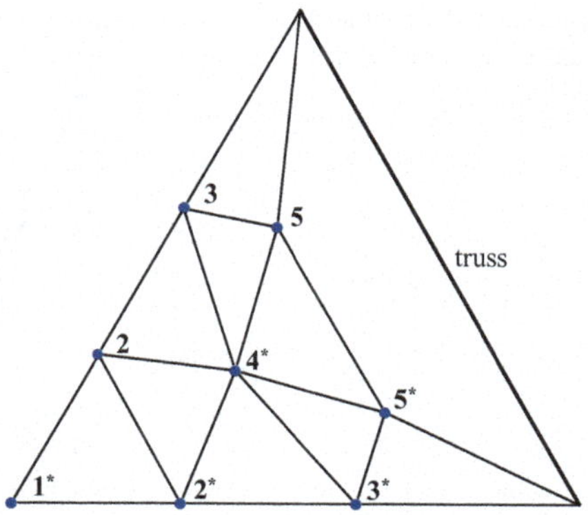

Fig. 6.12 Variation of cable length during the adjustment process

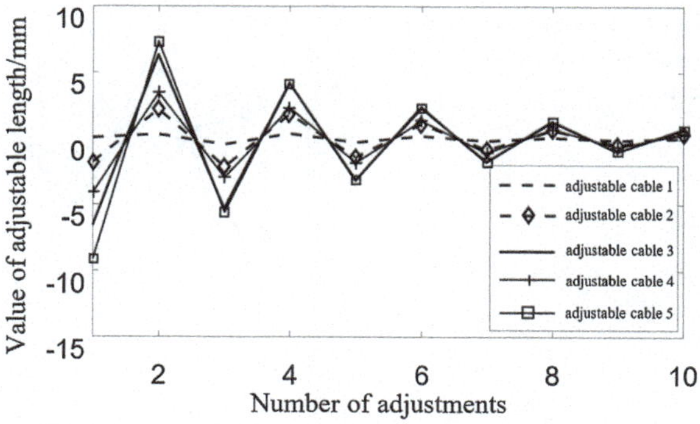

a) Traditional adjustment

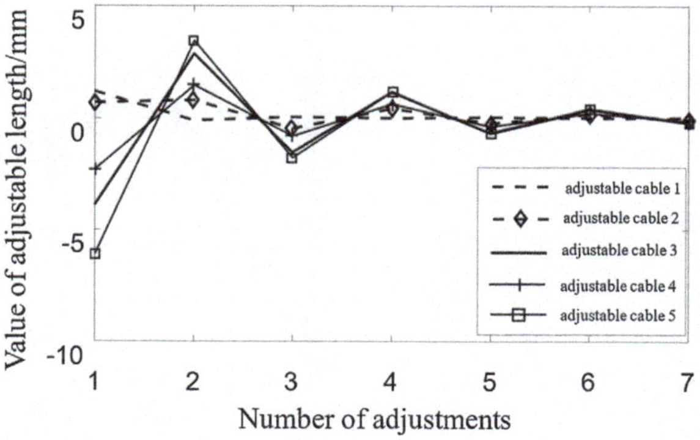

b) Model-based adjustment

Fig. 6.13 Variation of surface accuracy during adjustment test

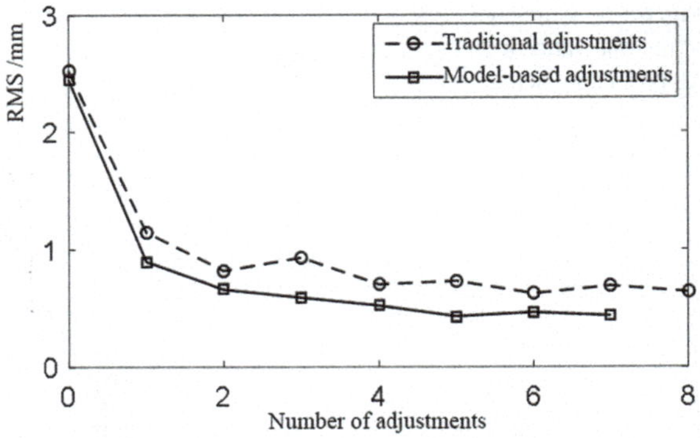

Fig. 6.14 Variation of cable length adjustment during two tests

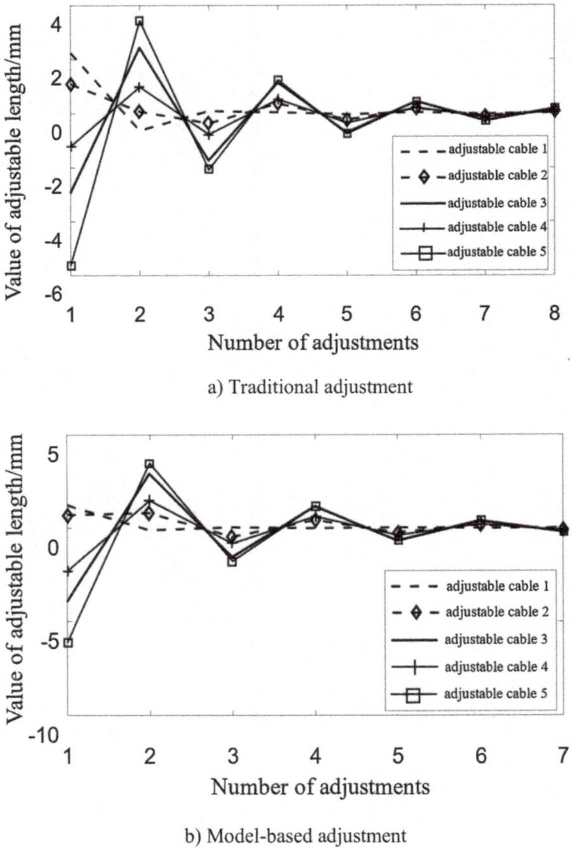

In the adjustment process, the positions of the nodes on the upper support cable net are obtained by photogrammetry. The cable length is adjusted by a screw device that is in series with the adjustable cable.

In the 2 m-diameter antenna model, the surface precision of the initial support cable net mounted on the rim trusses is 2.52 mm and it reaches to 0.60 mm after 8 iterations by the traditional adjustment method. Then a model-based adjustment is been performed after applying a random length disturbance of [−3, 3] mm to all adjustable cables. At that time, the surface precision of the initial mesh antenna is approximately 2.54 mm and converges to 0.42 mm after 7 model-based adjustments. The variation in the surface precision during the two different adjustment tests is shown in Fig. 6.13. It can be seen that not only is the adjustment efficiency improved based on the model-based adjustment, but also the surface precision of the mesh antenna improves. The changing value of the cable length adjustment is shown in Fig. 6.14. The trend of the cable length change during the two adjustment processes is basically the same, and the cable length overshoot is slightly smaller for the model-based adjustment.

References

1. Jin M, Yasaka T, Miura K (1990) Shape control of the tension truss antenna. AIAA Journal 28(2):316–322
2. Belvin WK, Edighoffer HH, Herstorm CL (1989) Quasistatic shape adjustment of a 15-meter-diameter space antenna. J Spacecr 26(3):129–136
3. Tabata M, Yamamoto K, Inoue T et al (1992) Shape adjustment of a flexible space antenna reflector. J Intell Mater Syst Struct 3(3):646–658

Chapter 7
Deployment Reliability Analysis of Cable Truss Combined Deployable Antenna

7.1 Introduction

Long March No. 4 sent the first black hole exploration satellite named HuiYan to space on June 15, 2017. Its goal is to explore a black hole and demonstrate the evaporation theory proposed by the great scientist Stephen William Hawking (Fig. 7.1). It is hoped that γ-ray storm and gravitation wave will be explored by means of X-rays. The information obtained by the satellite could not be sent back to Earth without a satellite antenna. Furthermore, the normal operation of the satellite antenna depends upon the successful deployment of the antenna.

However, confirming that an antenna has successfully deployed in space is not easy; the antenna may experience varied and complicated failure cases during deployment for many reasons, such as failure of the cable management mechanism, an insufficient dynamic driving moment, a broken cable, a sliding telescopic rod sliding, a rotating joint jam, and sticking synchronous gears. Any failure will lead to unsuccessful deployment of the antenna, which would cause the satellite to be out of operation [1]. It is necessary to carry out reliability analysis and prediction for spaceborne mesh antennas during deployment. To be as consistent as possible across the whole book, this chapter considers the AstroMesh deployable antenna (Fig. 3.11) as an example. However, the model, theory and method are valid for other cable truss combined deployable antennas.

Research on spaceborne mesh deployable antennas is currently focused on mesh adjustment, cable net form-finding, dynamics control and analysis during deployment, spatial on-orbit thermal analysis, and electronic performance analysis, and there are already many related research results. Masawa et al. [2] estimated and discussed the reliability of the spaceborne antenna deployment mechanism. Gerhard and Wolffgang [3] predicted the reliability of spatial deployable mechanisms by the Monte Carlo method. Hu et al. [4] conducted a reliability analysis on the motion function of the rotating joint in an umbrella-shaped deployable antenna mechanism. Based on the unascertained theory, Zhu and Chen [5] established a reliability analysis

B. Duan et al., *Large Deployable Satellite Antennas*,
Springer Tracts in Mechanical Engineering,
https://doi.org/10.1007/978-981-15-6033-0_7

Fig. 7.1 Stephen William Hawking, 1942–2018, a famous British theoretical physicist and cosmologist, one of the greatest contemporary physicists of the twentieth century. Put forward the singular theorem and black hole area theory of general relativity; The first person to propose the joint interpretation of cosmology theory by general relativity and quantum mechanics

model for the rotating joints of spaceborne deployable antennas, and predicted the reliability of the antenna mechanism during deployment. Chen et al. [6, 7] studied the deployment possibilities and gear reliability of AstroMesh antennas and calculated the reliability of antenna deployment by the probabilistic reliability method. Combined with DS theory and interval probability theory, Lin et al. [8] established a reliability analysis model with the interval as the variable for the AstroMesh deployable antenna prototype. Zhang and Chen [9] established a non-probabilistic reliability analysis model for the motion of the deployment mechanism for an AstroMesh deployable antenna, and analyzed and predicted the non-probabilistic reliability during the deployment process.

In fact, the probabilistic reliability analysis method requires knowing the probability density distribution function of the random parameter, which is based on a large amount of sample data. Moreover, the probability reliability model strongly depends on the original data, and a small deviation of the model may cause large deviation and even errors in the results [10–12]. Due to the limited information, it is difficult to obtain accurate probability distributions for some uncertain variables, and only the intervals of their changes can be given. These uncertain variables are called unknown but bounded variables (interval variables). They play a decisive role in the accuracy of reliability analysis, which scientifically describes the uncertain information based on the actual point of view. Large spaceborne mesh antennas are aerospace products with complex structures and a high cost and are often produced in a single piece lacking reliability-related statistical data. For many variables, there is little information and few samples. It is no longer a reasonable choice to use traditional probabilistic reliability analysis methods to predict its reliability. Therefore, we consider a reliability analysis method based on both interval parameters and probability parameters.

At present, scholars have conducted some related discussions and research on the reliability problem of combining interval and probabilities. For example, E1ishakeoff and Colomhi [13] first explored the hybrid problem of the uncertainty probability model and convex model and played a fundamental role in the study of probabilistic and convex set based hybrid reliability methods. Guo and Lv [14] proposed a probabilistic non-probability hybrid model but is only suitable for the reliability analysis of simple problems. Luo et al. [15] transformed the solution of the lower limit of reliability into a single-layer optimization problem by a mathematical programming method, but it is not suitable for finding the upper limit of reliability. Jiang et al. [16] considered the interval variable as a probability variable that obeys a uniform distribution and transformed the interval and probability hybrid reliability problem into a probability reliability problem, but the method is not able to deal with the actual variation range of reliability. References [17–19] transformed the problem of the coexistence of probability and interval variables into two-level nested optimization problems and successfully applied them to the reliability analysis of automobile bodies.

This chapter, based on [20], introduces a reliability analysis method based on interval and probabilities for spaceborne mesh antenna deployment.

7.2 Spaceborne Mesh Antenna Deployment Failure Tree and Bottom Event Probability

FTA (failure tree analysis) is a systematic method that can effectively determine the basic shortcomings of structures or products, the reason for defects, and the probability of occurrence and thus serve as the basis for reliability analysis. This chapter analyzes the deployment failure of the spaceborne mesh deployable antenna by this method.

7.2.1 Establishment of Failure Tree Model

The spaceborne antenna experiences three main stages of unlocking, deployment and positioning from the process of the stowed state to the deployed state. In the unlocking phase, the antenna is released from the self-locking position by torsion springs; during deployment, the antenna is gradually deployed under the motor driving force. In the positioning phase, the antenna is locked in the working state under the action of the limit switch. Figure 7.2 shows both the stowed and deployed states of the antenna.

The top, intermediate and bottom events of antenna deployment failure are determined according to the working environment of the antenna in space and the principle of mechanism deployment. The top event is the deployment failure of the spaceborne mesh antenna, and the events that directly and indirectly cause the failure of the spaceborne mesh antenna are intermediate events and bottom events. If we can establish a corresponding logical relationship between each bottom event and an intermediate

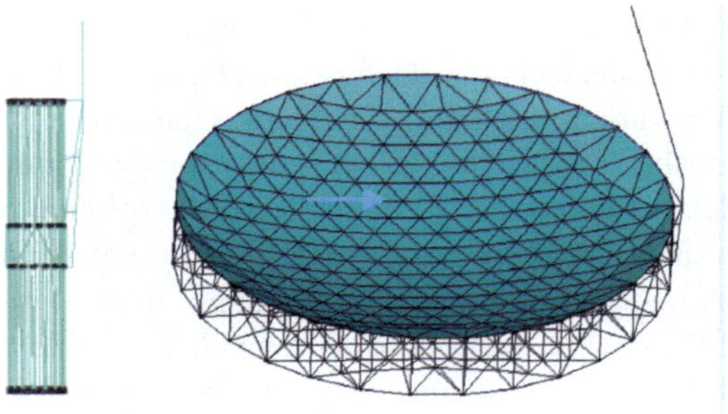

(a) stowed state (b) deployed state

Fig. 7.2 Stowed and deployed states of AstroMesh antenna. **a** Stowed state. **b** Deployed state

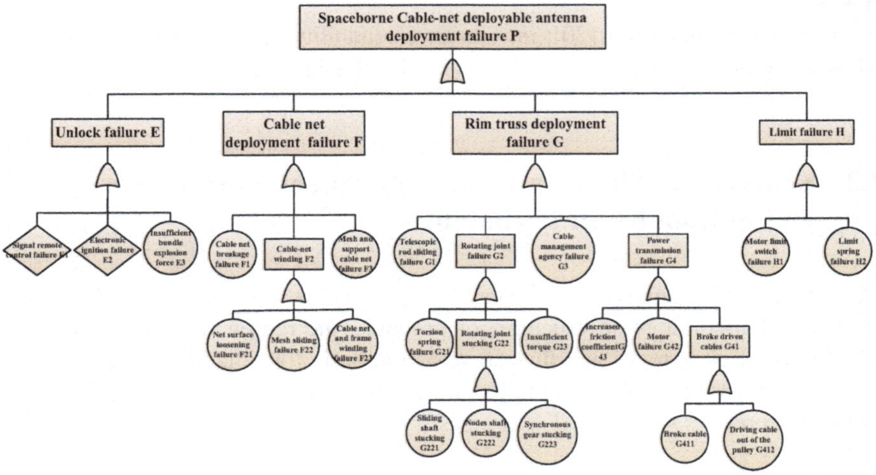

Fig. 7.3 Failure tree model of spaceborne deployable mesh antenna

event, the failure mode of the top event failure can be determined. The failure tree model is shown in Fig. 7.3.

7.2.2 *Classification of Bottom Events*

According to the methods of determining reliability, the bottom events of the failure tree model of spaceborne deployable AstroMesh antenna can be divided into three categories:

(1) Reliability is obtained by calculation for the following events: torsion spring failure G21, sliding shaft sticking G221, nodes shaft sticking G222, synchronous gear sticking G223, insufficient torque G23, and broken cable G411.

(2) Reliability is obtained by experiment for the telescopic rod sliding failure G1.
(3) Reliability is obtained by surveys, expert opinions and references for the following events: signal remote control failure E1, electronic ignition failure E2, insufficient bundle explosion force E3, cable net breakage failure F1, cable net surface loosening failure F21, limit spring failure H2, cable net and frame winding failure F23, mesh and support cable net failure F3, motor failure G42, motor limit switch failure H1, mesh sliding failure F22, cable management agency failure G3, and driving cable out of the pulley G412.

7.2.3 Determining Bottom Events for Deployment Reliability Analysis

Table 7.1 shows the bottom events and the corresponding failure probability of the AstroMesh deployable antenna.

Table 7.1 Bottom events and failure probability of perimeter truss of AstroMesh antenna

Bottom events	Failure probability	Remarks
E1 (Signal remote control failure)	10^{-6}–10^{-5}	Satellite Measurement and Control Center in Xi'an
E2 (Electronic ignition failure)	10^{-5}–10^{-4}	China Academy of Space Technology, headquarters
E3 (insufficient bundle explosion force)	10^{-5}–10^{-4}	China Academy of Space Technology, headquarters
F1 (Cable net breakage failure)	10^{-6}–10^{-5}	China Academy of Space Technology, Xi'an Branch
F21 (Net surface loosening failure)	10^{-9}–10^{-7}	China Academy of Space Technology, Xi'an Branch
F22 (Mesh sliding failure)	10^{-6}–10^{-5}	China Academy of Space Technology, Xi'an Branch
F23 (Cable net and frame winding failure)	10^{-6}–10^{-5}	China Academy of Space Technology, headquarters
F3 (Mesh and support cable net failure)	10^{-6}–10^{-5}	China Academy of Space Technology, Xi'an Branch
G3 (Cable management agency failure)	10^{-6}–10^{-5}	China Academy of Space Technology, Xi'an Branch
G412 (Driving cable out of the pulley)	10^{-8}–10^{-7}	China Academy of Space Technology, Xi'an Branch
G42 (Motor failure)	10^{-7}–10^{-6}	Xi'an Micro-Motor Institute
H1 (Motor limit switch failure)	10^{-6}–10^{-5}	China Academy of Space Technology, headquarters
H2 (Limit spring failure)	10^{-6}–10^{-5}	China Academy of Space Technology, headquarters

7.3 Interval and Probability Hybrid Reliability Model

7.3.1 Probabilistic Reliability Model

When performing structural reliability analysis, the most common method is to establish a probability model for uncertainties and use probabilistic reliability methods for analysis. Assume that $g(X_1, X_2, \ldots X_n)$ is a limit state function of the structure and $\mathbf{X} = \{X_1, X_2, \ldots X_n\}$ is an n-dimensional random vector of the structure, which are independent of each other. After converting the basic random vector $\mathbf{X}$ into a separate standard normal space U, the n-dimensional space U could be divided into two parts by surface $G(\mathbf{u}) = 0$, the security domain and the failure domain. It can be described as shown in Fig. 7.4.

According to the theory of probability reliability, the reliability of an uncertain structure can be determined by the following integral equation:

$$R = \Pr\{g(\mathbf{X}) \geq 0\} = \int\limits_{G(\mathbf{u}) \geq 0} f_{\mathbf{u}}(\mathbf{u}) \mathrm{d}\mathbf{u} \tag{7.1}$$

where $f_{\mathbf{u}}(\mathbf{u})$ is the joint probability density function of $\mathbf{u}$.

According to the reliability indicator method, Eq. (7.1) can be transformed into an optimization problem:

$$\begin{cases} \beta = \min \|\mathbf{u}\| = \min \sqrt{\sum\limits_{i=1}^{n} u_i^2} \\ \text{s.t.} \;\; G(\mathbf{u}) = 0 \end{cases} \tag{7.2}$$

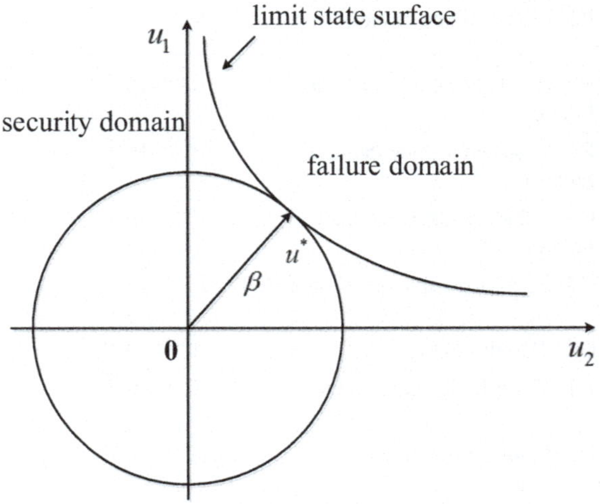

Fig. 7.4 Geometric interpretation of reliability indicators

The optimal solution $\mathbf{u}^*$ is the most likely failure point and $\beta = \|\mathbf{u}^*\|$ is the reliability indicator whose geometric meaning is to the shortest distance from the origin of the coordinate system to the failed surface (limit state surface) in normalized normal space U. Through the reliability indicator, the reliability of an uncertain structure can be calculated as $R = \Phi(\beta)$.

7.3.2 Interval and Probability Hybrid Reliability Model

In traditional reliability analysis, the uncertain parameters are random variables. To determine the probability density distribution function of the random variables, a large amount of experimental or statistical data is necessary. This is nevertheless difficult to obtain in practice. For this reason, hypothetical probability distribution is usually assumed in the case of insufficient data to describe the distribution of uncertainties. Unfortunately this method often leads to unreliable calculations.

In most cases, uncertainties in the structural system have both random variables and interval variables, which cannot be analyzed in a traditional way. To make full use of the known information and avoid the traditional unnecessary assumptions, random variables are described by both probability and interval variables. In the case of the probabilistic and interval hybrid models, a new reliability analysis method should be sought. Denote $\boldsymbol{X}$ and $\boldsymbol{Y}$ as the random variable and interval variables, respectively, where $X_i \sim \left(\mu_{X_i}, \sigma_{X_i}\right)$, $i = 1,2,\ldots, n$, $Y_i \in \left[Y_i^L, Y_i^R\right]$, $i = 1,2,\ldots, m$. The limit state function created by the system failure criteria is $g(\mathbf{X}, \mathbf{Y})$. Normalize $\boldsymbol{X}$ and $\boldsymbol{Y}$ to $\boldsymbol{u}$ and $\boldsymbol{v}$ and then convert the limit state function into $G(\mathbf{u}, \mathbf{v})$. The normalization process of the random variable $\boldsymbol{X}$ and the interval variable $\boldsymbol{Y}$ is as follows

$$u_i = \frac{X_i - \mu_{X_i}}{\sigma_{X_i}}, v_i = \frac{Y_i - Y_i^c}{Y_i^R}, \quad i = 1, 2, \ldots, n \tag{7.3}$$

where

$$Y_i^c = \frac{Y_i^L + Y_i^R}{2}, \quad Y_i^r = \frac{Y_i^R - Y_i^L}{2}.$$

For hybrid reliability models with both random and interval variables, the limit state equation $G(\mathbf{u}, \mathbf{v}) = 0$ represents a cluster of failure surfaces in the normalized U space. The entire U space is divided into three parts: the security, critical and failure domains. Figure 7.5 shows a case with two random variables u_1,u_2 and one interval variable v, where β_m is the reliability indicator of the hybrid model and $\bar{\beta}_m$ and $\underline{\beta}_m$ are the upper and lower limits of the reliability indicator, respectively [20].

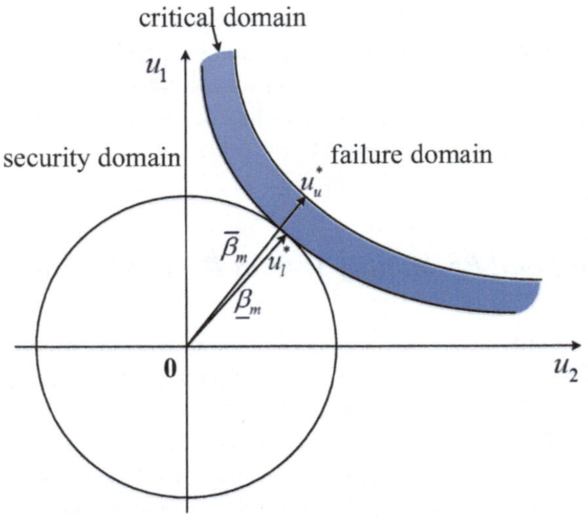

Fig. 7.5 Description of reliability indicators under hybrid model

In the random variable space, the reliability of the hybrid reliability model can be defined as $P_m = \Pr\{G(\mathbf{u}) > 0\}$, and the structural reliability should also be an interval due to the existence of interval variables; i.e. $P_m \in \left[\underline{P}_m, \bar{P}_m\right]$.

If the interval variable $\mathbf{v}^*$ exists, the expression of the reliability indicator, according to the geometric meaning of the reliability indicator, is [15]

$$\beta_m = \operatorname{sgn}\left(G\left(0, \mathbf{v}^*\right)\right) \cdot \sqrt{\mathbf{u}^{\mathrm{T}}\mathbf{u}} \tag{7.4}$$

where $\operatorname{sgn}(G(0, \mathbf{v}^*))$ is a symbolic function of the random variable of the structure when it takes a nominal value and is defined as

$$\operatorname{sgn}\left[G\left(0, \mathbf{v}^*\right)\right] = \begin{cases} 1\,, & G\left(0, \mathbf{v}^*\right) > 0 \\ 0\,, & G\left(0, \mathbf{v}^*\right) = 0 \\ -1\,, & G\left(0, \mathbf{v}^*\right) < 0 \end{cases} \tag{7.5}$$

If $G(0, \mathbf{v}^*) > 0$, then $\beta_m > 0$, and the structure will be in a safe state; the larger β_m is, the more stable the structural safety state is and the more reliable the structure will become. If $G(0, \mathbf{v}^*) < 0$, then $\beta_m < 0$, and the structure will be in a failed state; and similarly, the smaller β_m is, the more stable the structural failure state and the poorer the structural reliability. If $G(0, \mathbf{v}^*) = 0$, then $\beta_m = 0$, and the structure will be in a critical state; that is, the structure may be reliable or may fail. When the interval variable takes a certain value, a slight change may cause the state of the structure to change. In this situation, the failure and safety states of the structure are not robust.

Figure 7.5 shows that the lower limit of the reliability indicator is

$$\underline{\beta}_m = \text{sgn}\left(G\left[0, \mathbf{v}_l^*\right]\right) \cdot \sqrt{\mathbf{u}_l^{*\text{T}} \mathbf{u}_l^*} \tag{7.6}$$

where the parameter $\underline{\beta}_m$ can be obtained by the two nested optimization models below. The outer layer optimization aims to find the MPP (most probable failure point) $\mathbf{u}_l^*$ when the interval variable takes the value $\mathbf{v}_l^*$; its optimization model is

$$\begin{aligned} &\min_u \ \left\{\mathbf{u}^\text{T}\mathbf{u}\right\} \\ &\text{s.t.} \ \ G\left(\mathbf{u}, \mathbf{v}_l^*\right) = 0 \end{aligned} \tag{7.7}$$

The inner layer optimization aims to find the worst case point $\mathbf{v}_l^*$, and its optimization model is

$$\begin{aligned} &\min_v \quad G(\mathbf{u}, \mathbf{v}) \\ &\text{s.t..} \quad v_i^L \le v_i \le v_i^R \quad (i = 1, 2, \ldots, N) \end{aligned} \tag{7.8}$$

The lower limit of the structural reliability can be obtained by the reliability indicator $\underline{\beta}_m$ as

$$\underline{P}_m = \phi\left(\underline{\beta}_m\right) = \int_{-\infty}^{\underline{\beta}_m} \frac{1}{\sqrt{2\pi}} \exp\left(-\frac{1}{2}u^2\right) \text{d}u \tag{7.9}$$

Similarly, if the problem in Eq. (7.8) is transformed into a maximum value problem, the best case point $\mathbf{v}_u^*$, the upper boundary u_u^* of the MPP point, and the upper limits of the reliability indicator $\bar{\beta}_m$ and the reliability $\bar{P}_m$ can also be obtained.

The geometric meaning of the lower limit of the reliability indicator $\bar{\beta}_m$ is the shortest distance from the origin of the coordinate system to the failure surface, on which the worst case point is located in the standardized U space. The upper limit $\bar{\beta}_m$ represents the shortest distance from the origin of the coordinate system to the failure surface, on which the best case point is located. If all uncertain variables are random variables, the hybrid reliability model reduces to the traditional probability reliability model, and the reliability indicator reduces into the traditional probability reliability indicator.

7.4 Reliability Analysis of Critical Bottom Events

7.4.1 Uncertainty Classification Involved in Reliability Analysis of Critical Bottom Events

The uncertainties involved in the calculation of the bottom event reliability for each critical part of the spaceborne mesh antenna are shown in Table 7.2. Some uncertainties, such as the diameter of the hinge pin and the inner diameter of the bracket hole, are often given in the form of nominal values with upper and lower tolerances, and it is easy to determine the range of those uncertainties; therefore, they are often regarded as interval variables [21, 22]. The other uncertainties, such as the coefficient of friction, verticality, deflection angle, the tension of the driving cable and the strength of the driving cable are generally highly random; hence, these uncertainties are usually regarded as random variables.

The corresponding limit state functions are established for each critical bottom event failure condition, and reliability is evaluated by the interval probability hybrid reliability model analysis method described in Sect. 7.2. Take the bottom event "nodes shaft sticking" as an example to illustrate the analysis process.

Table 7.2 Classification of uncertainties in the bottom event

	Torsion spring failure G21	Pulley shaft locked G221	Node shaft locked G222	Synchronous gear rotation stuck G223	Driven cables broken G411	Insufficient torque G23
Random variables	Coefficient of friction f	Verticality Δ_{A1} and Δ_{B1}	Verticality Δ_{A2} and Δ_{B2}	Deflection angle of the gear axis θ_1,θ_1 and the center distance of the two gears a	The strength of driven cable r, the force of driven cable F	Friction coefficient between node shaft and bushing f_1, Friction coefficient between pulley shaft and bushing f_2, the force of driven cable T
Interval variables	The length of spring twist arm l, The radius of pin r	The diameter of pulley pin a_1, the diameter of bracket hole b_{A1}, the inner diameter of bushing b_{B1}	The diameter of nodal pin a_2, the diameter of bracket hole b_{A2}, the inner diameter of bushing b_{B2}	The diameter of addendum circle d_a和 The diameter of dedendum circle d_f	Diameter of driven cable d	The radius of pins R_1 and R_2

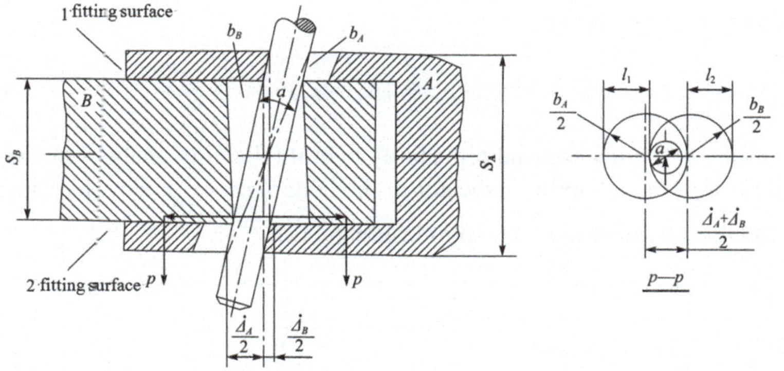

Fig. 7.6 The fit between the positioning pin and hole with double fitting surface

7.4.2 *Reliability Analysis of the Bottom Event "Nodes Shaft Sticking"*

The rotating joint mainly includes the node pin and pulley shafts, whose fitting principles are as shown in Fig. 7.6. For some important fits between components, such as the fits of the components of the rotating joints on rockets, satellites, and airplanes, a high fitting precision is often required. Hence, the problem of misalignment and fits should be solved first during the process of development. The main factors influencing fitting failure include verticality errors of the shaft and the hole in the axial direction, the coordination accuracy of the central axis of the shaft and hole, etc.

To ensure that the pin is smoothly inserted into the hole and can be freely rotated (Fig. 7.6), the following formula must be satisfied at fitting surfaces 1 and 2:

$$\bar{b}_A - \bar{a} - \left[\dot{a} + \frac{S_A(\dot{\Delta}_B - \dot{b}_B) + S_B(\dot{\Delta}_A - \dot{b}_A)}{S_A + S_B}\right] \geq 0 \tag{7.10}$$

and

$$\bar{b}_B - \bar{a} - \left[\dot{a} + \frac{S_A(\dot{\Delta}_B - \dot{b}_B) + S_B(\dot{\Delta}_A - \dot{b}_A)}{S_A + S_B}\right] = 0 \tag{7.11}$$

where $\bar{b}_A$ and $\bar{b}_B$ are the mean values of the apertures of components A and B, respectively, $\dot{b}_A$ and $\dot{b}_B$ are the random variables of the apertures of components A and B after centralization, respectively, and $\dot{\Delta}_A$ and $\dot{\Delta}_B$ are the verticality errors of the apertures of components A and B along the axial direction, respectively. Because $\bar{b}_A = \bar{b}_B$, the limit state function of the "nodes shaft sticking" rotation is

$$g = \bar{b}_A + \left(\frac{\dot{b}_A + \dot{b}_B}{2} - \frac{\dot{\Delta}_A + \dot{\Delta}_B}{2}\right) - a \tag{7.12}$$

It can be standardized as

$$G(\delta_{\dot{\Delta}_A}, \delta_{\dot{\Delta}_B}, \delta_a, \delta_{\dot{b}_A}, \delta_{\dot{b}_B}) = 0 \tag{7.13}$$

According to the interval and probability hybrid reliability analysis method, the lower limit $\underline{\beta}$ of the reliability indicator of the bottom event "nodes shaft sticking" can be obtained by the following optimization model:

Outer layer optimization

$$\begin{aligned} &\text{Find} && \delta_{\Delta_A}, \delta_{\Delta_B} \\ &\underset{(\delta_{\Delta_A}, \delta_{\Delta_B})}{\text{Min}} && \sqrt{\delta_{\Delta_A}{}^2 + \delta_{\Delta_B}{}^2} \\ &\text{S.T.} && G(\delta_{\Delta_A}, \delta_{\Delta_B}, \delta_a, \delta_{b_A}, \delta_{b_B}) = 0 \end{aligned}$$

$\delta_a, \delta_{b_A}, \delta_{b_B}$ (passed from inner layer to outer layer); $\delta_{\Delta_A}, \delta_{\Delta_B}$ (passed from outer layer to inner layer) (7.14)

Inner layer optimization

$$\begin{aligned} &\underset{(\delta_a, \delta_{b_A}, \delta_{b_B})}{\text{Min}} && G(\delta_{\Delta_A}, \delta_{\Delta_B}, \delta_a, \delta_{b_A}, \delta_{b_B}) \\ &\text{S.T.} && |\delta_a| \le 1 \\ & && |\delta_{b_A}| \le 1 \\ & && |\delta_{b_B}| \le 1 \end{aligned}$$

The upper limit of the probability of failure of a set of rotating joints is

$$\underline{P}_{fxz}\phi\left(-\underline{\beta}_{xz}\right) = 1 - \int_{-\infty}^{\underline{\beta}_{xz}} \frac{1}{\sqrt{2\pi}} \exp\left(-\frac{1}{2}u^2\right) \mathrm{d}u \tag{7.15}$$

By changing the objective function of the inner layer optimization in Eq. (7.14) to the max function, the lower limit of the failure probability of the bottom event "nodes shaft sticking" can be obtained.

There are N sets of rotating joints in the deployable antenna structure and the joints are independent of each other. The failure probability of the bottom event "nodes shaft sticking" can be obtained by

$$P_{fxz} = \left[1 - \left(1 - \underline{P}_{fxz}\right)^N, 1 - \left(1 - \bar{P}_{fxz}\right)^N\right] \tag{7.16}$$

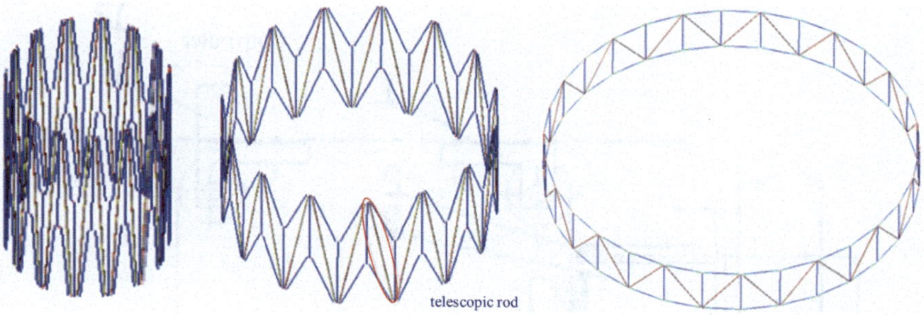

Fig. 7.7 Motion scheme of telescopic rod during deployment of the periphery truss

Similarly, the failure probability of other key bottom events can be determined by the interval and probability hybrid reliability analysis method.

7.5 Experimental Device and Test Method for Telescopic Rod Sliding

It is necessary that the telescopic rod slides smoothly without jamming for the smooth deployment of a deployable antenna with a rim truss. The motion of the telescopic rod during the deployment of the perimeter truss of the AstroMesh antenna is shown in Fig. 7.7.

To determine the reliability of the telescopic rod sliding without jamming, the motion test of the telescopic rod during the deployment process can be carried out by designing a corresponding experimental device considering the easy movement of the telescopic rod. The designed experimental device (Fig. 7.8) is simple in structure and reasonable in design and can be used to find the reliability of the telescopic rod of the spaceborne mesh deployable antenna [23].

The specific method and flow chart of performing the reliability test on the telescopic rod are given in Fig. 7.9.

After the experiment, the probability distribution and the distribution parameters of the random variables can be acquired by analysing the experimental data according to the moment estimation method, and then the failure probability of the telescopic rod sliding can be obtained from the probability distribution.

7.6 Failure Probability Calculation for AstroMesh Antenna

The failure probability interval of the key bottom events can be calculated by the method given in Sect. 7.3. For the bottom event of "sliding failure of the telescopic rod", the failure probability estimation interval can be obtained by the experimental

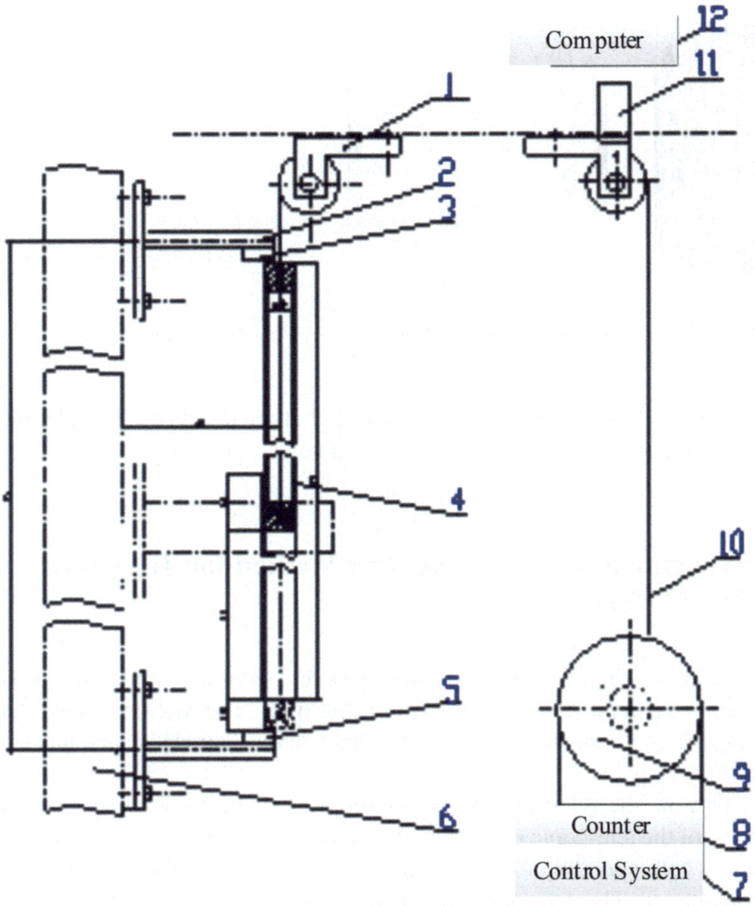

Fig. 7.8 Design ofthe telescopic rod sliding reliability experimental device. 1. Fixed pulley component, 2.Switch block, 3. Upper limit switch, 4. Telescopic rod, 5. Lower limit switch, 6. Fixtures, 7. Control system, 8. Counter, 9. DC stepper motor, 10. Wire rope, 11. Force gauge, 12. Computing unit

device and testing method in Sect. 7.4. For the third type of bottom event, the failure probability estimation interval of other bottom events is obtained by analyzing the corresponding data given by the relevant organization and by consulting the relevant experts. The specific data are listed in Table 7.3.

According to the failure tree analysis method, the failure probability of the top event "deployment failure of the spaceborne mesh antenna" can be calculated from the logical relationship of events at various levels. Considering the complex and unknown relationships between events, we assume that all events of each level in the spaceborne mesh antenna failure tree are independent of each other. The relationship between every event at each level in the failure tree is "or". The failure probability estimation interval of the bottom event is obtained through various methods, and

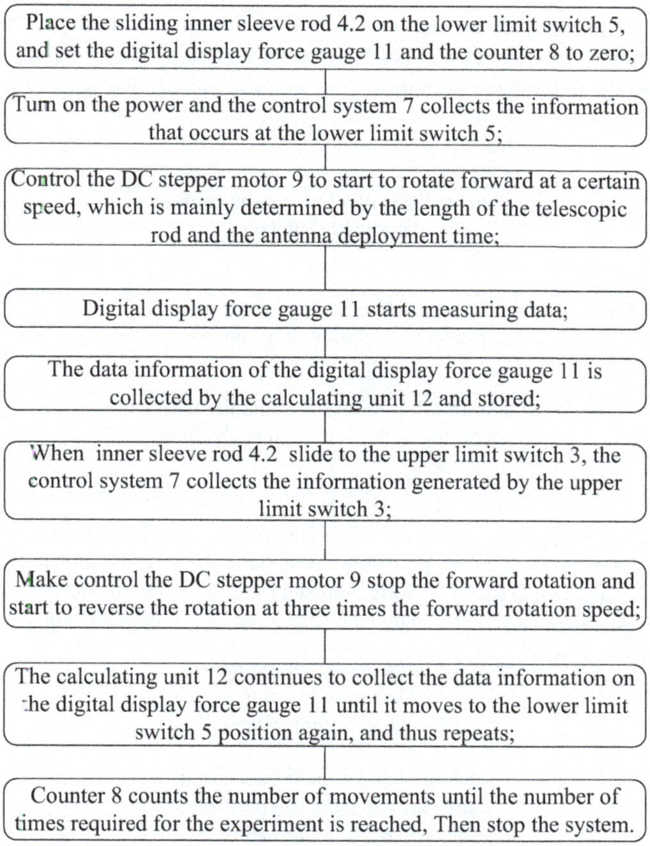

Fig. 7.9 Method and flow chart for telescopic rod sliding reliability test

Table 7.3 Failure probability in deployment failure tree for a spaceborne mesh antenna

Bottom event	Failure probability	Bottom event	Failure probability
E1	10^{-6}–10^{-5}	G221	2.50×10^{-24}–2.97×10^{-4}
E2	10^{-5}–10^{-4}	G222	9.00×10^{-26}–1.28×10^{-4}
E3	10^{-5}–10^{-4}	G223	2.66×10^{-11}–7.63×10^{-4}
F1	10^{-6}–10^{-5}	G23	3.93×10^{-5}–3.95×10^{-5}
F21	10^{-9}–10^{-7}	G3	10^{-6}–10^{-5}
F22	10^{-6}–10^{-5}	G411	1.30×10^{-21}–1.00×10^{-19}
F23	10^{-6}–10^{-5}	G412	10^{-8}–10^{-7}
F3	10^{-6}–10^{-5}	G42	10^{-7}–10^{-6}
G1	7.33×10^{-7}–3.06×10^{-6}	H1	10^{-6}–10^{-5}
G21	3.03×10^{-16}–3.53×10^{-13}	H2	10^{-6}–10^{-5}

then the failure probability estimation interval of the top event "deployment failure of the spaceborne mesh antenna" is obtained by calculation step by step according to the method of interval mathematics. The specific calculation flow chart is shown in Fig. 7.10.

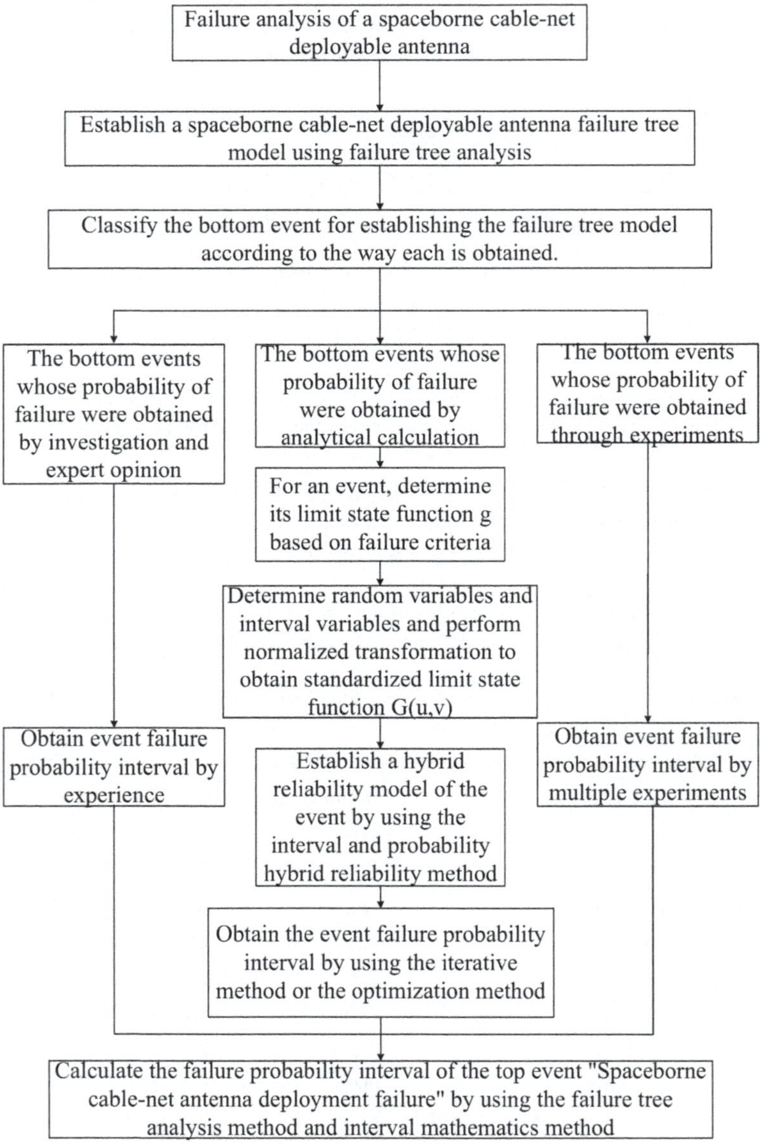

Fig. 7.10 Flow chart of deployment reliability calculation for spaceborne mesh antenna

The failure probability of the rotating joint sticking G22 is obtained by calculating the probabilities of the sliding shaft sticking G221, nodes shaft sticking G222 and synchronous gear sticking G223. The specific calculation formula is

$$P_f(G_{22}) = [1, 1] - \big([1, 1] - p_f(G_{221})\big)([1, 1] - p(G_{222}))([1, 1] - p(G_{223})) \tag{7.17}$$

The failure probability of rotating joint failure is obtained by calculating the torsion spring failure G21, the rotating joint sticking G22 and insufficient torque G23. The specific calculation formula is

$$P_f(G_2) = [1, 1] - ([1, 1] - p(G_{21}))([1, 1] - p(G_{22}))([1, 1] - p(G_{23})) \tag{7.18}$$

In the same way, $p(G_1)$, $p(G_3)$ and $p(G_4)$ can be acquired, and then we have

$$P_f(G) = [1, 1] - ([1, 1] - p(G_1))([1, 1] - p(G_2))([1, 1] - p(G_3))([1, 1] - p(G_4)) \tag{7.19}$$

Furthermore, $p(E)$, $p(F)$ and $p(H)$ can also be calculated and then the failure probability P of the top event "deployment failure of the paceborne mesh antenna" is

$$\begin{aligned} P_f(P) &= [1, 1] - ([1, 1] - p(E))([1, 1] - p(F))([1, 1] - p(G))([1, 1] - p(H)) \\ &= 1 - \prod_{i=1}^{n} ([1, 1] - p(i)) \end{aligned} \tag{7.20}$$

Substituting the data into Eq. (7.20) gives the failure probability estimation interval of the top event "deployment failure of the spaceborne mesh antenna" as $P_f(P) = \left[2.56 \times 10^{-4}, 1.663 \times 10^{-3}\right]$, and the reliability is $P_r(P) = [0.998337, 0.999744]$.

Compared with the traditional method whose reliability result is a single value, the method proposed in this chapter makes full use of the known information and considers the existence of the interval variable, which leads to the results of calculated reliability as an interval and has great meaning in engineering research.

7.7 Calculation of the Importance of AstroMesh Antenna Bottom Event

The importance is a key value to describe the contribution of the basic events in the failure tree of system failure. Importance analysis has different meanings and calculation methods. In engineering, the probability importance and key importance are commonly used. The probability importance is used to measure the degree of the effect of the basic event states on the probability of system failure. Its physical meaning is the probability that system failure occurs when only the *j*th basic event occurs and it can be expressed as [24]

$$I_j = \frac{\partial G(P)}{\partial P_j} \tag{7.21}$$

where I_j is the probability importance when the *j*th bottom event occurs, P_j is the probability of failure of the *j*th bottom event, and *G(p)* is the top event probability expression.

According to the spaceborne mesh antenna deployment failure tree, the probability importance can be defined as

$$I_j = \prod_{i=1, i \neq j}^{n} (1 - P_i) \tag{7.22}$$

Considering both the probability importance of the basic event and the effect that the probability of occurrence of the basic event itself on system failure, the key importance of the bottom event is defined as

$$K_j = \frac{P_j I_j}{G(P)} \tag{7.23}$$

where K_j is the key importance of the *j*th bottom event. The failure probability of the spaceborne deployable mesh antenna bottom event obtained here is an interval, and to ensure that the calculated result is robust and reliable, the maximum failure probability of the bottom event is used in the calculation of the probability importance and critical importance. The specific calculation results of probability importance and key importance are described in Table 7.4.

As seen from Table 7.4, a few basic bottom events that have the greatest impact on the deployment process of the spaceborne mesh deployable antenna are synchronous gear sticking (G223), sliding shaft sticking (G221), insufficient torque (G23), nodes shaft sticking (G223) and so on, which should be emphasized during the engineering design.

For spaceborne mesh deployable antennas, there are many uncertain variables involved in deployment reliability analysis due to the complexity of the structure, and some of the uncertainties are difficult to obtain accurate probability information in most cases. If these uncertainties are assumed to be random variables, unreliable calculation results could be obtained by traditional probabilistic methods. Therefore, this chapter focuses on the interval and probability hybrid reliability analysis method, by which the analysis and evaluation of the deployment reliability of spaceborne mesh antennas are carried out. Through theoretical analysis, numerical simulation and physical verification, the following results can be obtained.

Table 7.4 Probability importance and key importance of bottom events

Bottom event	Probability importance	Key importance
E1 (Signal remote control failure)	0.998354054	0.006028963
E2 (Electronic ignition failure)	0.998443914	0.060295056
E3 (Insufficient bundle explosion force)	0.998443914	0.060295056
F1 (Cable net breakage failure)	0.998354054	0.006028963
F21 (Net surface loosening failure)	0.998344170	6.0289E-05
H2 (Limit spring failure)	0.998354054	0.006028963
F23 (Cable net and frame winding failure)	0.998354054	0.006028963
F3 (Mesh and support cable net failure)	0.998354054	0.006028963
G42 (Motor failure)	0.998345068	0.000602891
H1 (Motor limit switch failure)	0.998354054	0.006028963
F22 (Mesh sliding failure)	0.998354054	0.006028963
G3 (Cable management agency failure)	0.998354054	0.006028963
G412 (Driving cable out of the pulley)	0.998344170	6.0289E-05
G1 (Telescopic rod sliding failure)	0.998347125	0.00184485
G21 (Torsion spring failure)	0.998344080	6.0289E-06
G221 (Sliding shaft stucking)	0.998640525	0.179026365
G222 (Nodes shaft stucking)	0.998472304	0.077439443
G223 (Synchronous gear stucking)	0.999106559	0.460459773
G23 (Insufficient torque)	0.998527999	0.111072846
G411 (Broken cables)	0.998344071	6.0289E-07

First, the hybrid reliability analysis method is used to obtain the estimation interval of antenna deployment reliability, which can make full use of the known information and avoid some artificial assumptions. The problem of the coexistence of both random variables and interval variables of the spaceborne mesh antenna in reliability analysis can be solved by this method and the calculated results are more valuable.

Second, for the deployment failure tree of the spaceborne mesh antenna, the bottom events can be divided into three categories according to the method of obtaining the reliability, and the failure probability intervals can be obtained by different methods for different kinds of bottom events.

Third, the failure probability interval of telescopic rod sliding failure is determined by designing a telescopic rod sliding experimental device and a test method; the result is $[7.33 \times 10^{-7}, 3.06 \times 10^{-6}]$.

Finally, by analyzing the of importance of the basic bottom event of the spaceborne mesh antenna failure tree, it can be seen that the bottom events such as synchronous gear sticking, sliding shaft sticking, insufficient torque, and node shaft sticking, have the greatest impact on the reliability of the deployment of the spaceborne mesh antenna and are the weak links in the antenna deployment process.

In this chapter, the reliability of the spaceborne mesh antenna is analyzed and studied in detail. The reliability of the antenna is evaluated and predicted with the hybrid reliability analysis method, which provides a reference for the development and engineering application of antennas.

References

1. Chen JJ, Cao HJ, Zhang JG (2014) An opinion about reliability of large spaceborne deployable antenna structure system. J Eng Mech 31(2):25–33 (In Chinese)
2. Misawa M, Yasaka T, Miyake S (1989) Analytical and experimental investigations for satellite antenna deployment mechanisms. Spacecraft 26(3):181–187
3. Gerhard SS, Wolffgang HT (1998) Practical procedures for reliability estimation of spacecraft structures and their components. AIAA 36(8):1509–1515
4. Hu TB, Chen JJ, Zhang JG (2005) Reliability analysis of rotating joint motion function of umbrella antenna. J Space Sci 25(6):552–557 (In Chinese)
5. Zhu ZQ, Chen JJ (2010) Motion reliability analysis of rotation joint in umbrella antenna. Int J Perform Eng 6(3):233–242
6. Chen JJ, Zhang JG, Duan BY (2005) Analysis of the deployment possibilities of large-scale spaceborne antennas. J Astronaut 26(SI):130–134 (In Chinese)
7. Chen JJ, Zhang JG, Duan BY (2005) Reliability analysis of synchronization of synchronous gear system in large spaceborne antenna deployment mechanism. J Xidian Univ 32(3):1–5 (In Chinese)
8. Lin LG, Chen JJ, Ma J (2010) Interval analysis method for large-scale spaceborne antenna deployment system failure tree. Mech Strength 32(1):68–73 (In Chinese)
9. Zhang JG, Chen JJ (2006) Reliability analysis of the deployment mechanism of a large satellite antenna based on the non-probabilistic model. In: The 1st international symposium on systems and control in aerospace and astronautics, Harbin, China, pp 1109–1114
10. Ben HY (1994) A non-probabilistic concept of reliability. Struct Saf 14(4):227–245
11. Elishakoff I (1995) Essay on uncertainties in elastic and viscoelastic structures: from A M Freudenthal's criticisms to modem convex modeling. Comput Struct 56(6):871–895
12. Schettrkogel A, Elishakoff I (1985) On the ergodicity assumption in an applied mechanics problem. J Appl Mech 52(1):133–136
13. Elishakoff I, Colomhi P (1993) Combination of probabilistic and convex models of uncertainty when scarce knowledge is present on acoustic excitation parameters. Comput Methods Appl Mech Eng 104(2):187–209
14. Guo SX, Lv ZZ (2002) Probabilistic and non-probabilistic hybrid models for structural reliability analysis. Mech Strength 24(4):524–526 (In Chinese)
15. Luo YJ, Kang Z, Alex L (2009) Structural reliability assessment based on probability and convex set mixed model. Comput Struct 87(21):1409–1410
16. Jiang C, Lu GY, Han X et al (2012) A new reliability analysis method for uncertain structures with random and interval variables. Int J Mech Mater Des 8(2):169–182
17. Du XP (2007) Interval reliability analysis. In: Proceedings of the ASME 2007 international design engineering technical conferences & computers and information in engineering conference, Las Vegas
18. Qiu ZP, Wang J (2010) The interval estimation of reliability for probabilistic and non-probabilistic hybrid structural system. Eng Fail Anal 17(5):1142–1154
19. Jiang C, Li WX, Han X et al (2011) Structural reliability analysis based on random distributions with interval parameters. Comput Struct 89(23):2292–2302
20. Liu SJ, Duan BY, Yang DW (2016) Gear's anti-sticking analysis of spaceborne deployable antenna based on interval and probability. J Xidian Univ 43(3):57–72 (In Chinese)

21. Ferson S, Joslyn CA, Helton JC et al (2004) Summary from the epistemic uncertainty workshop: consensus amid diversity. Reliab Eng Syst Saf 85(1):355–369
22. Du X, Sudjianto A, Huang B (2005) Reliability-based design under the mixture of random and interval variables. J Mech Des 127(6):1068–1076
23. Xidian University (2014) Reliability test device and method for peripheral truss type deployable antenna telescopic rod. CN104101493A. Accessed 15 Oct 2014 (In Chinese)
24. Chen JJ, Zhang JG, Duan BY (2005) Failure tree analysis of deployment system of large spaceborne antenna. Mech Des Res 21(3):7–8 (In Chinese)

Chapter 8
Prototype Development and Experiment

8.1 Introduction

Karl Guthe Jansky (Fig. 8.1) was the inventor of the world's first radio telescope, which he used to discover radio waves from the center of the Milky Way galaxy in 1932, marking the birth of radio astronomy. The importance of test equipment in scientific research is self-evident. To demonstrate the proposed mathematical model and method, a 2 m-diameter deployable mesh antenna prototype was produced.

8.2 Experimental Prototype Development

The manufacturing of the prototype includes three main tasks: index and error assignment, design and manufacture, and special equipment development.

8.2.1 Index and Error Assignment

In practical applications, the ratio of focus to diameter of a paraboloid is generally 0.3–0.8. Considering the demand of other types of tests (such as electrostatic forming membrane antennas), the ratio of focus to diameter of the upper and lower cable net are taken as 0.8 and 1.0 respectively. The overall height of the antenna can be calculated by Eqs. (3.26) and (3.102). The basic design indicators of the prototype can be obtained as required (Table 8.1).

Support cable net surface errors mainly include design and manufacturing errors. The design error is composed of two parts: one is the geometrical error introduced when the reflector surface formed by the cable net approximates the ideal parabolic reflector surface, which depends on the layout of the cable and the size of the grid.

B. Duan et al., *Large Deployable Satellite Antennas*,
Springer Tracts in Mechanical Engineering,
https://doi.org/10.1007/978-981-15-6033-0_8

Fig. 8.1 Karl Guthe Jansky, 1905–1950, radio engineer, radio astronomer, and inventor of the first radio telescope. His discovery of radio waves from the Milky Way galaxy in 1932 marked the birth of radio astronomy

Table 8.1 Basic design indicators of the prototype

Name	Value
Antenna diameter	2 m
Truss number	6 sides
Focal length	Upper mesh surface is 1.6 m, lower mesh surface is 2 m
Antenna height	0.6 m
Antenna mass	Less than 8 kg
Base frequency	Greater than 1 Hz
Surface precision	Better than 1 mm

The second is the error between the position of each node of the front cable net obtained in the form-finding design of the cable-truss combined structure (taking into account the deformation of the truss) and the ideal position, which is known as the mechanical design error. Taking various factors into consideration, the error assignment of the support structure of the 2 m-diameter prototype is presented in Table 8.2.

In the overall error assignment, the surface error of the 2 m-diameter prototype is required to be no more than 1 mm. With reference to the actual error situation of similar cable net antennas, the error assignment of the prototype is as follows: the design error is 0.43 mm, the manufacturing error is 0.87 mm, and the remaining error is 0.25 mm. The design error is divided into the geometric design and mechanical design errors, which are 0.40 and 0.15 mm, respectively. The manufacturing error is divided into cable cutting, mesh cutting and the assembly error of the cable net, mesh and truss.

Table 8.2 Error assign scheme for 2 m prototype support structure

Error source		Error value/mm	Ratio (%)
Design error	Geometric design	0.40	16
	Mechanical design	0.15	2
Manufacturing error	Cable cutting	0.10	1
	Mesh cutting	0.10	1
	Cable-net assembly	0.30	9
	Mesh assembly	0.40	16
	Truss assembly	0.70	49
Other error		0.25	6
Total error		1	100.0

8.2.2 *Design and Manufacture*

Regarding the design and manufacture of the prototype, the material selection and the manufacturing process of the devices and parts have to be addressed [1].

1. Material selection

According to the diameter, quality and stiffness requirement of the prototype, based on the mechanical performance analysis and taking into account the cost and processing capacity, the material and geometric parameters of the selected components are shown in Table 8.3.

2. Design and manufacture of truss structures

According to the structural features and functional requirements of the prototype, the truss structure includes the components such as the rod, joint, support cable net connection node, synchronous structure and driving structure.

(1) Rod

The rods mainly include cross, vertical and diagonal rods. The rod is an important part of the unit and truss. The diagonal rod is an extendable link that is composed of thick and thin rods connected with a sleeve. To meet the extendable limit constraint of the thick and thin diagonal rods, an aluminum locating pin is mounted between the thick and thin rods.

(2) Joint

Joint design is of great importance compared with the overall structure of the space truss. Joints are the key determinant of whether the structure can be deployed smoothly and locked in space. The joints of the hoop truss structure must be able

Table 8.3 Component parameters

Component	Material	Density/(g/cm^3)	Modulus/(GPa)	Passion ratio	Parameters/mm
Cross rod	Carbon fiber	1.8	118	0.3	Length: 895 outer diameter: 18 wall thickness 1.2
Vertical rod	Carbon fiber	1.8	118	0.3	Length: 535 outer diameter: 18 wall thickness 1.2
Thick obique rod	Carbon fiber	1.8	118	0.3	Length: 915 outer diameter: 16 wall thickness 1.2
Thin diagonal rod	Carbon fiber	1.8	118	0.3	Length: 915 outer diameter: 12 wall thickness 1.2
Support cable net	Kevlar	1.44	105.8	0.3	Cable cross diameter is 1
Three-dimension joint	Aluminum	2.8	71	0.36	Envelope size 72.6 × 41.7 × 67.0
Five- dimension joint	Aluminum	2.8	71	0.36	Envelope size 121.2 × 54.6 × 77.0
Synchronous gear	45# steel	7.85	210	0.31	Envelope size 38.0 × 38.0 × 16.0
Torsion spring	Carbon spring wire	7.81	202	0.3	Envelope size 23.0 × 7.0 × 10.4

to ensure that the rod moves freely during deployment. The friction between the rod and joint is small and there is no easy bending deformation. When the structure collapses, the rods are closely compressed for storage. In addition, the joints need to have enough strength to overcome the tension caused by the rod and local bending, shearing, friction and so on.

The joint design should start from the function of the joint to choose the overall design, and then the joint should be combined with the overall structure for joint consideration. Two types of connectors are required in the truss structure.

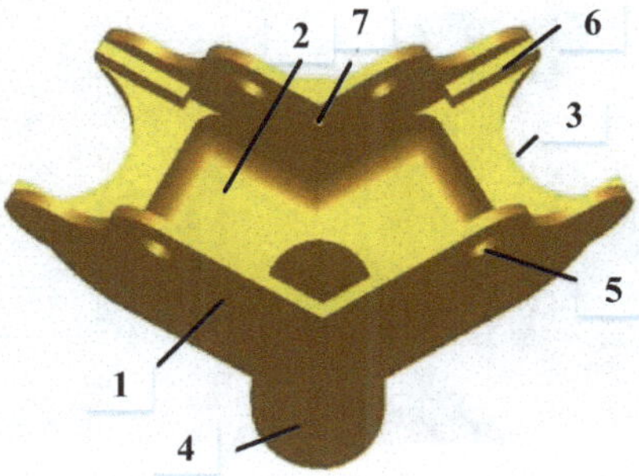

Fig. 8.2 Three-dimensional connector housing

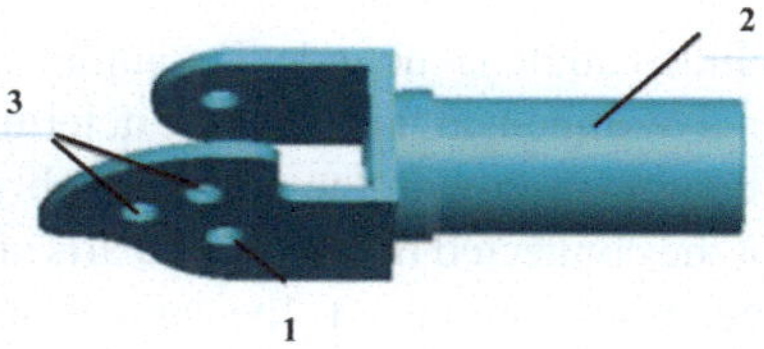

Fig. 8.3 Cross rod connector

(1) Three-dimensional joint

This consists of two connectors for the outer casing and the connecting cross rod. The specific structure is shown in Figs. 8.2 and 8.3. The outer casing is composed of two symmetrical parts, and the angle is the indexing angle of two adjacent units. Since the 2 m-diameter prototype is composed of 6-unit parallelograms, the indexing angle here is 120°. The joint and cross rod are assembled with the cross rod connecting piece. The cross rod connecting piece (Fig. 8.3) has the left end-1 and the outer casing connected by the pin shaft, and the right end-2 is fixed with the carbon fiber cross rod. Considering that the carbon fiber punching is easy to crack, it is fixed by a glue connection, and part-3 is connected with a bevel gear through the pin.

(2) Five-dimensional joint

This includes the outer casing and the connecting piece connecting the cross rod and the diagonal rod. The specific structure is shown in Fig. 8.4. Similar to the three-dimensional joint, the outer casing is composed of two symmetrical ends, and the

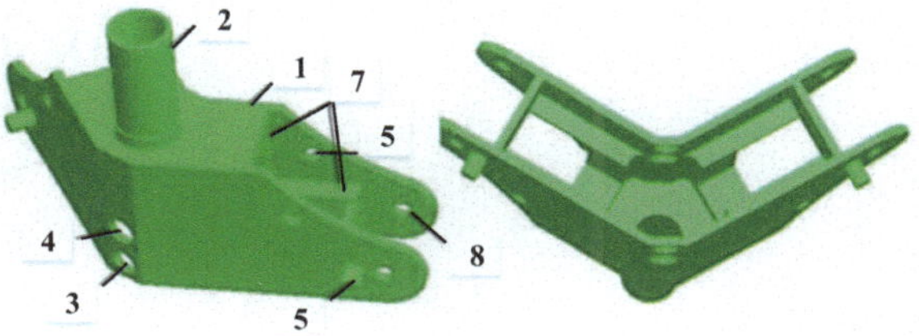

Fig. 8.4 Five-dimensional connector housing

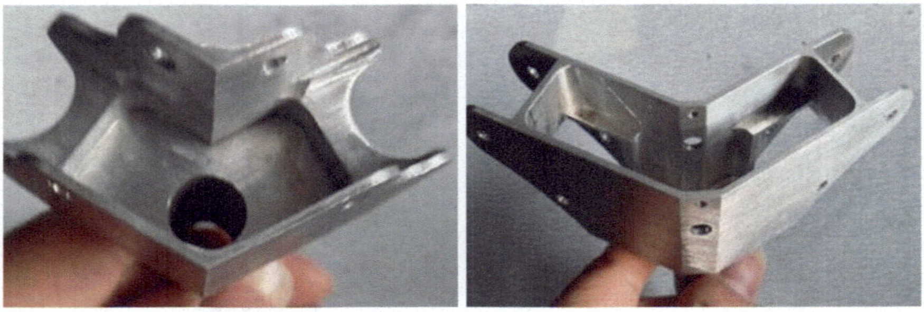

Fig. 8.5 Physical drawings of three-dimensional and five-dimensional joints

included angle is the indexing angle of two units. Figure 8.5 is the physical diagram of a five-dimensional joint. The determination of the joint structure size is carried out in combination with the post-assembly motion simulation, the purpose is to ensure that the connected or contained parts are mountable and so that interference does not occur when moving. At the same time, the rods are near to each other and close in place with interference.

(3) Connection node of support cable net

There are a group of devices (Fig. 8.6) in the support cable net. For each of these device, several cables are connected and fixed to it. A gap is ensured between the upper and lower support plates. The upper and lower support plates are evenly provided with corresponding positioning holes. Each pair of corresponding upper and lower positioning holes are connected with a cylindrical pin for winding the cable, and the upper and lower end sleeves of the cylindrical pin are connected in the upper and lower positioning holes, respectively.

Fig. 8.6 Connecting node

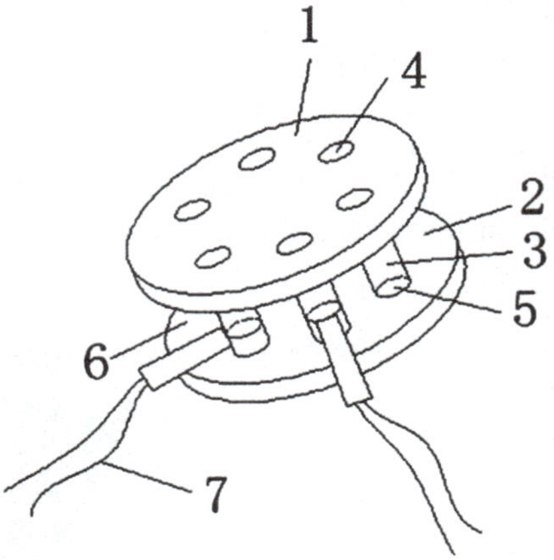

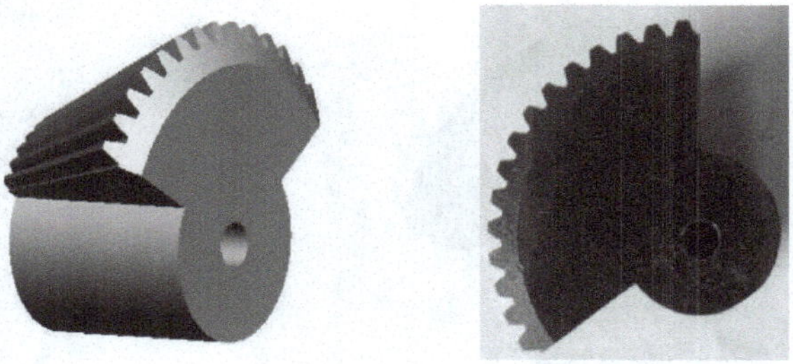

a)design model b) product picture

Fig. 8.7 Synchronous bevel gear. **a** Design model. **b** Product picture

(4) Synchronous structure

The gear drive is used as a synchronous drive connection for the joint of the deployable antenna structure. It is made up of a pair of bevel gears with the same design parameters and is assembled in the three-dimensional joint to be fastened with the portion connecting the cross rod to achieve the synchronous deployment of two adjacent quadrilateral units. Figure 8.7 shows the design and physical diagram of the bevel gear involved. The actual meshing portion is only 1/4 of the circumference, so the design cuts approximately 2/3 on a complete circumferential gears. Table 8.4 shows the bevel gear design parameters.

(5) Driving structure

The driving structure includes a driving torsion spring, a driving cable, a pulley driven by the driving cable, and a pulley baffle that prevents the drive cable from falling off the pulley (Fig. 8.8).

3. Manufacture of the support cable net and mesh

There are two main parts: the support cable net and the reflective mesh. The first part is considered first.

Table 8.4 Bevel gear design parameters

Name	Value
Modulus	$m = 1$
Number of teeth	$z = 43$
Pressure angle	$\alpha = 20°$
Base diameter/mm	$D_b = D\cos\alpha = 40.41$
Tip diameter/mm	$D_a = mz + 2\text{m}\cos\delta = 44.73$
Root diameter/mm	$D_f = mz - 2.4\ \text{m}\cos\delta = 40.92$
Reference diameter/mm	$D = mz = 43$
Pitch cone angle/°	$\delta = 30°$

Fig. 8.8 Pulley, pulley baffle and assembly drawing

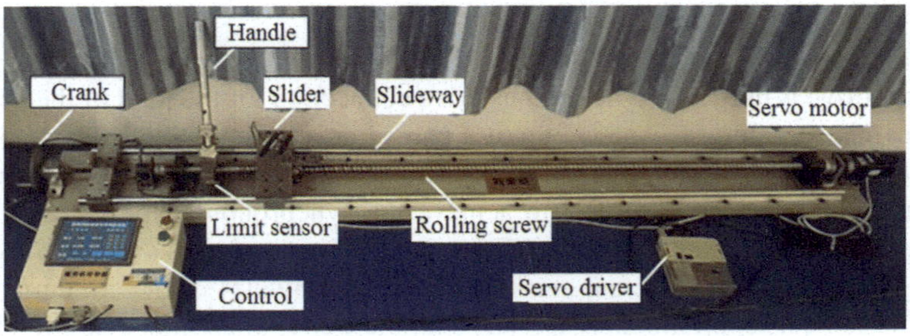

Fig. 8.9 Cable cutting device

(1) Cutting the pre-tensioned cable

To reduce the machining error of the support cable section during production of the support cable net, a special cutting machine for the cable segment was developed [2]. The device cuts the required pre-tensioned length of the cable to ensure that the cable segment reaches the required length when it is subjected to its design pre-tension. The main components of the cable cutting machine are shown in Fig. 8.9. The maximum length of the cable can be cut up to 2 m, and the tension and length are 0.01 N and 0.01 mm, respectively.

(2) Assembly cable net

The assembly of each cable segment of the support cable net structure is performed with a special button as shown in Fig. 8.10a, and a button on the cable net surface is assembled as shown in Fig. 8.10b. Finally, the cable segment can be connected according to topological relationship of the upper and lower support cable nets to assemble of the support cable net structures, as shown in Fig. 8.10c.

(3) Mesh cutting and sewing

For the mesh installed by cutting and sewing, the basic process is as follows:

(a) Rivets, buttons and cable (b) Buttons after assembly (c) Support cable net after assembly

Fig. 8.10 Assembly process of the cable net. **a** Rivets, buttons and cable. **b** Buttons after assembly. **c** Support cable net after assembly

(1) Mesh shape calculation

After the form-finding analysis of the structure composed of the support cable net and mesh (where the mesh is treated as a membrane), each triangular facet in the final configuration is in the prestress shape. The actual stacking size can be obtained after the prestress on the mesh is released. Here, the calculation is carried out with triangular membrane elements. The classification of the triangular size and shape in the final configuration of the support cable net and mesh structure is shown in Fig. 8.11, and the corresponding cutting dimensions are shown in Table 8.5.

(2) Cutting of the mesh

The size of the triangular mesh obtained in step (1) is a net size, and a certain sewing allowance is reserved when the wire is sewn to facilitate sewing the mesh boundary. The reserved sewing allowance is shown with the dotted line in Figs. 8.12a, and b

Fig. 8.11 Classification of the shape of a flat triangular mesh

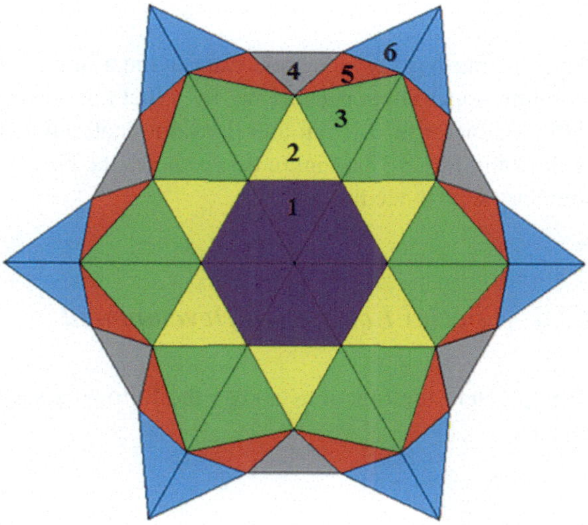

Table 8.5 Cutting dimensions of triangular mesh

Number	Deployed area/mm^2	Area without stress/mm^2	Retraction/mm^2
1	42,799	38,742	4057
2	44,173	39,985	4188
3	57,203	51,780	5423
4	23,528	21,297	2231
5	20,482	18,541	1941
6	27,961	25,310	2651

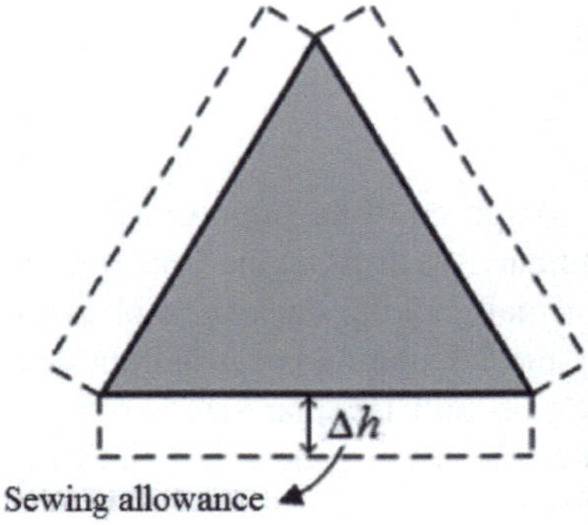

(a) Cutting size of the mesh

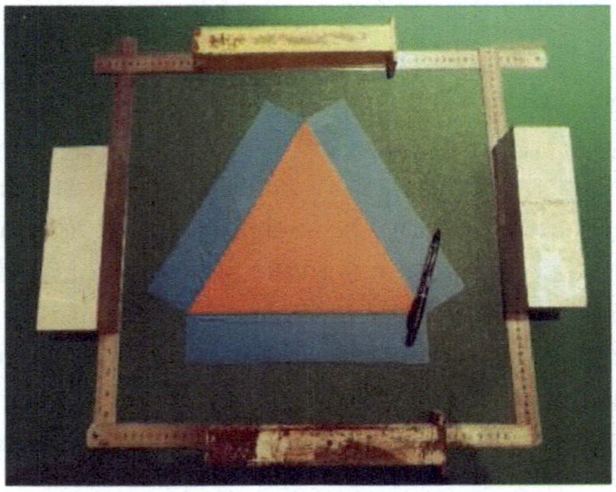

(b) Mark of the mesh cutting size

Fig. 8.12 Mesh cutting. **a** Cutting size of the mesh. **b** Mark of the mesh cutting size

shows the process of marking the mesh cutting size.

(3) Sewing of mesh

When the mesh is sewn, the boundary of the triangular flat mesh is sewn to the corresponding upper support cable net by a locking needle. A soft metal monofilament with a diameter of 0.1 mm and the same material as the mesh is used for sewing, as shown in Fig. 8.13a. Figure 8.13b shows the process of sewing a triangular mesh boundary with a needle.

8.2.3 Special Equipment Development

To implement the functions above, the following special devices must be made in advance.

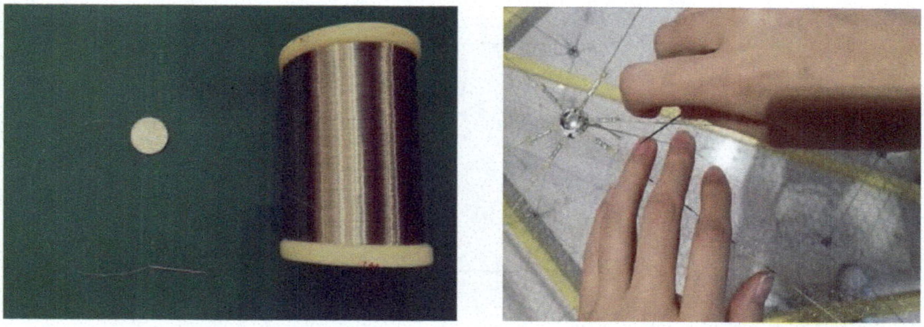

(a) Soft metal monofilament (b) Sewing of triangular mesh border

Fig. 8.13 Sewing process of the mesh

1. Gravity-compensating device design

Considering that the space deployable antenna operates in a microgravity or zero-gravity environment and is assembled and tested in a gravity environment, it is necessary to design a gravity-compensating device to eliminate the influence of gravity on the deployment of the antenna so that the ground test can accurately reflect the mechanical properties of the antenna in the space environment. For antennas with smaller diameters, a mechanical suspension is generally used to create an approximate zero-gravity environment; in other words, a force equivalent to the magnitude of gravity and in the opposite direction is applied to some components or positions of the structure. These forces are mainly obtained by suspending some of the antenna nodes from cables or struts. Gravity-compensating device design needs to fully consider the requirements of the mechanical precision, stiffness and size, installability and operability of the antenna. At the same time, compensation needs to be realized according to the deployment mode and trajectory of the antenna during the whole deployment process.

For the structural characteristics of the 2 m-diameter antenna prototype, the cable suspension method is selected. Gravity compensation for the antenna is performed with a pulley and an appropriate weight. At the same time, based on the kinematics analysis of the space deployable antenna, the deployment trajectory of the 2 m-diameter prototype is analyzed (see Fig. 8.14). The vertices of each truss unit move along the linear trajectory shown by the dashed line. Guide rails are erected along the trajectory so that the pulleys can slide along the guide rails while the antenna is deployed. It should be noted that to reduce the influence of the friction between the pulley and the guide rail during deployment, relatively smooth linear bearings are necessary.

2. Hardware of deployment control system

The deployment control system consists of a cable winding mechanism and controller (Fig. 8.15).

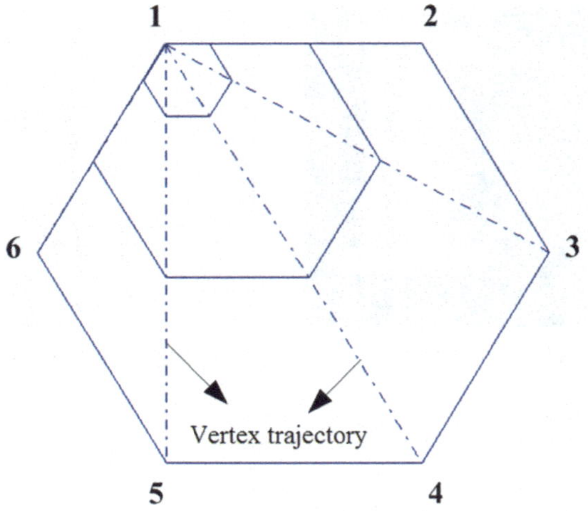

Fig. 8.14 Top view of deployment

(a) Cable reel mechanism

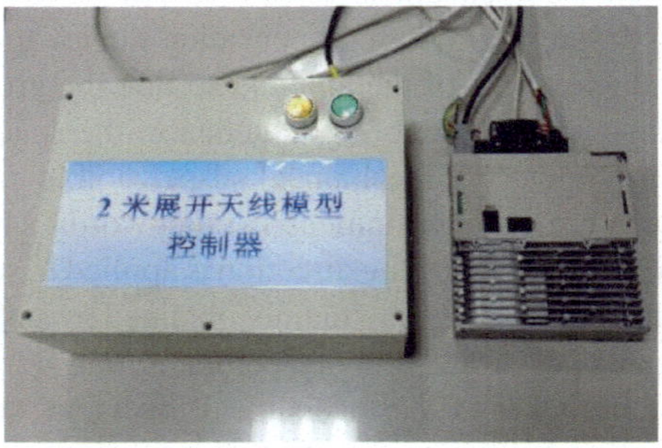

(b) The controller

Fig. 8.15 Control system of deployment

Based on the driving cable tension range obtained by the deployment dynamics analysis, the servo motor parameters selected are shown in Table 8.6. Additionally, according to the size of the motor body, a reel with a diameter of 60 mm is designed to accommodate the driving cable. Truss units may have an unsynchronized deployment problem, so it is necessary to detect whether each truss unit is deployed into the correct position by installing a limit sensor. The sensor signals are sent to the control system. Once all the units are in the required position, the control system stops the motor from running on the cable.

The controller consists of a control board (Fig. 8.16) and driver (YASKAWA's SGDV-2R8A01A series). The external power supply (220 V) supplies the power for each control board power supply (12 V) through a power switch, and the control board can be interrupted at any time by suspending the switch. The chip of the control board is composed of an ARM (LPC2214) and FPGA (EP2C5T114C8N).

Table 8.6 Servo motor parameter

Name	Value
Type	YAMAHA-SGMJV-04A
Size	128.5 mm × 60 mm × 60 mm
Weight	1.9 kg
Rated output power	400 W
Rated current	1.6 A
Rated speed	3000 r/min
Maximum speed	6000 r/min
Rated torque	1.27 N·m
Rated radial load	245 N

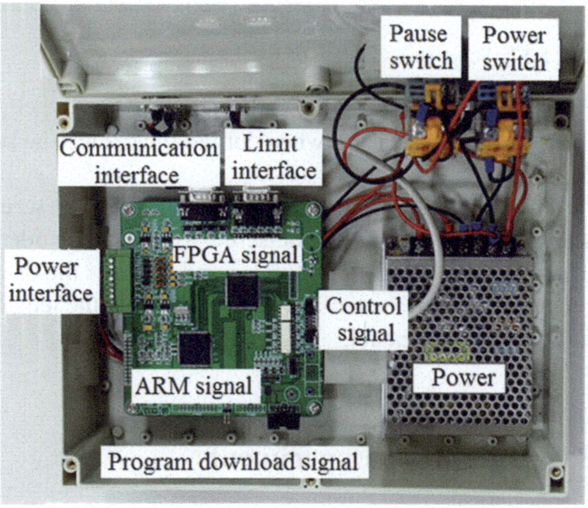

Fig. 8.16 Control board layout

The computer communicates with the control board by a communication interface (RS-422). After the ARM parses the command, it generates a control signal through the FPGA and sends it to the driver. The driver can drive the motor to perform the corresponding actions. When the mechanical interface reaches a limit, the generated electrical signal will be transmitted to the control board through the limit interface, thereby generating a stop section.

3. Cable cutting device

Machining error is an unavoidable source of profile accuracy errors in mesh antennas, especially cutting and assembly error in cable segments. The traditional method generally cuts the cable based on the original length data. Considering the nonlinearity of the initial stretched section of a Kevlar cable, this method tends to produce large deviations. Therefore, a special cutting device for cable cutting (Fig. 8.9) is developed. The device can cut the cable under a constant length and constant force

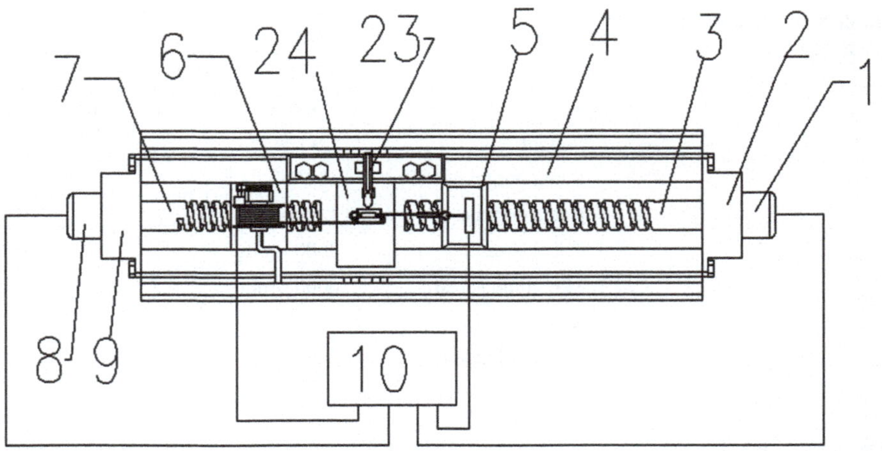

Fig. 8.17 Schematic diagram of special cutting device

to ensure high precision when cutting the cable. The schematic diagram is shown in Fig. 8.17.

The special cutting device includes a base, a screw, a motor, a controller, and sliders. The motor drives the screw to move the slider in the guide rail of the base. The controller controls the motor and receives the data fed from the force sensor to precisely control the tension of the cable and the length and then displays the length and force on the panel.

8.3 Prototype Experiments

The purpose of prototype development is to confirm the proposed design methods, and it may be able to serve large deployable antennas in orbit in the future. Therefore, once the prototype is designed and completely developed, a large number of experiments are needed. Via comparing the simulation and experiment results, one can find problems with the mathematical models and design methods add address them. The experiments consist of statics, dynamics, deployment mechanisms, and reliability tests, and so on.

8.3.1 Statics Experiment

The abovementioned 2 m-diameter prototype is tested for reflector surface precision, and the test method is in accordance with the photogrammetry method described in Sect. 6.2.

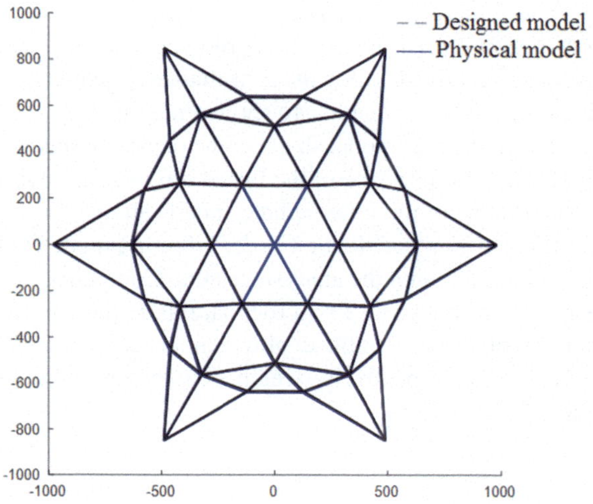

Fig. 8.18 Model comparison chart

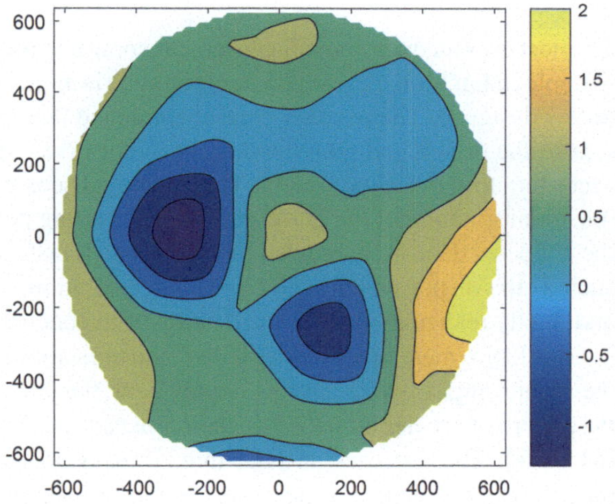

Fig. 8.19 Z-coordinate error of physical model

The comparison between the prototype design and the physical models is shown in Fig. 8.18. Figure 8.19 is a contour plot of the Z-coordinate error of the front cable net nodes of the physical model. The surface precision of the physical model reached 1 mm as required by the index.

8.3.2 Natural Frequency Experiment

The model test uses the DEWETRON modal tester, and an acceleration sensor is attached to the antenna joint to measure the modality of the antenna prototype by hammering.

The finite element models of the deployed state, the furled state and the intermediate states (in which the deployment angles are 30° and 70°) of the deployment are established for the 2 m-diameter antenna, and natural-frequency analysis is performed. The results are shown in Table 8.7.

The first-order modal shape of the deployed state swings around the fixed rod, the first-order modal shape of the furled state swings at the upper end, and the first-order modal shape of the intermediate state is almost consistent with the deployed state.

It is obvious that the first-order vibration modes of the deployed and furled states are consistent with the analysis results. Comparing the experimental results with the simulation results, the first-order natural frequency errors of the deployed, furled, and intermediate states of the deployment process are found to be 4.74%, 1.99%, 4.99%, and 6.18%, respectively, which demonstrates the accuracy of the finite element model.

8.3.3 Deployment Experiment

This test uses an open-loop displacement control strategy. First, the dynamics model of deployment for the 2 m-diameter prototype is modeled, and the deployment trajectory is designed. The position control command that the controller can recognize, is sent to the driver, and then the driving motor drives the reel to deploy the antenna according to the design motion. The positional sensor measures the dynamic coordinate information of the corresponding measuring point during deployment. Then, the collected data are processed by coordinate transformation to obtain information, such as the displacement, velocity and acceleration of each node of the prototype, and finally realize dynamic visualization of the deployment.

The deployment time is set to be 46 s, and the sampling frequency is set to 100 Hz in the monitoring software. During deployment, the motor speed is driven according to the pattern of acceleration (0–5 s), then constant speed (5–41 s), and then deceleration (41–46 s). The test uses an NDI three-dimensional motion measurement system, including 3D investigator measuring points and Optotrak monitoring software. The measuring points are selected at the rod and the center of the rod of the first truss unit (Fig. 8.20). The displacement curve with the time of measurement point-1 (test measurement and simulation analysis data) is shown in Fig. 8.21a–c, and the two curves are basically consistent.

It can be seen from the error curve that the measured points have relatively large errors at the beginning of the movement in the X and Y directions. The reason is that the antenna is driven by the torsion spring to deploy to a certain angle at the beginning. When the prototype is made, the torsion spring force is slightly larger than the design value, thus causing a large initial velocity and a deviation. During the steady-motion phase, errors in all directions are controlled within ±10 mm.

The velocity curve of point-1 in the test is shown in Fig. 8.21d, and the overall trend of the measured velocity curve is consistent with the simulation result. However,

Table 8.7 Modal analysis result

State of antenna	First-order natural frequency/Hz			Second-order natural frequency/Hz			Third-order natural frequency/Hz		
	Analysis	Test	Error (%)	Analysis	Test	Error (%)	Analysis	Test	Error (%)
Furled state	3.826	**3.750**	1.99	5.792	**5.375**	7.2	8.265	**7.750**	5.08
Intermediate state (30°)	1.763	**1.675**	4.99	3.987	**4.025**	0.9%	10.589	**10.125**	4.38
Intermediate state (70°)	1.812	**1.700**	6.18	3.812	**3.850**	1.00	10.834	**10.375**	4.24
Deployed state	1.837	**1.750**	4.74	3.707	**3.750**	1.16	11.087	**10.625**	4.17

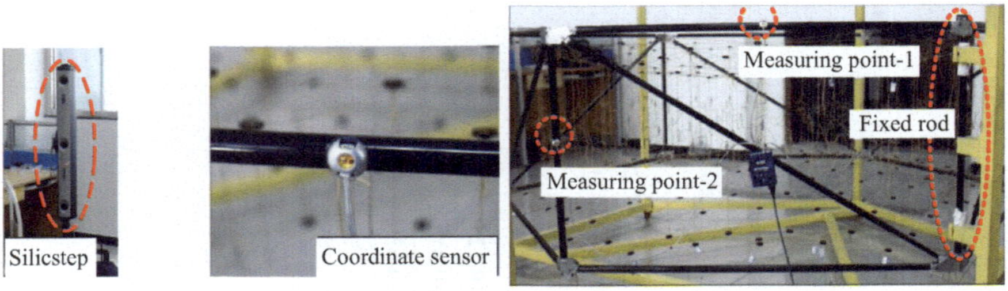

Fig. 8.20 Measure system of deployment process

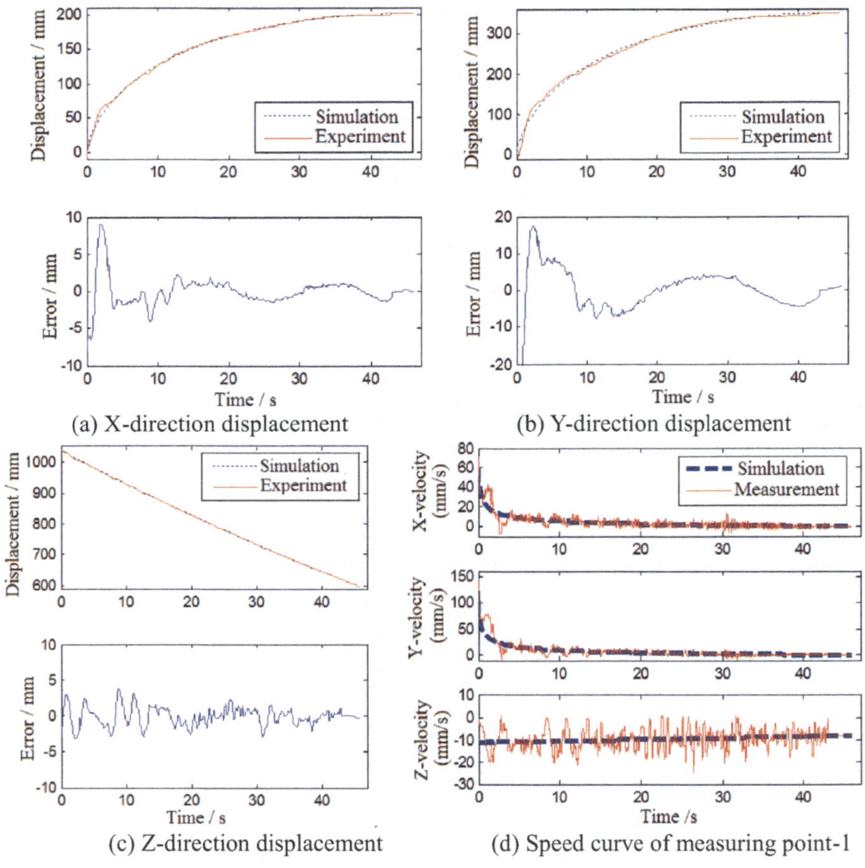

Fig. 8.21 Displacement and speed curves of measuring point-1

due to the influences of structural clearance and flutter in the gravity-compensating device, the actual deployment speed curves have more burrs.

8.3.4 Deployment Reliability Experiment

For an arbitrary parallelogram unit, it is a necessary condition for the smooth deployment of the deployable antenna that the oblique telescopic rods can slide smoothly without jamming. Since the movements of the telescopic rods are easy to implement, a set of experimental devices can be designed and manufactured to carry out the motion test of the telescopic rods during deployment to obtain the reliability of the telescopic rods sliding without jamming. The experimental device is shown in Figs. 7.8 and 7.10. The test methods and procedures are shown in Fig. 7.9. The tensile data of the test are shown in Fig. 8.22.

The curve shown in Fig. 8.22 is obtained by taking three experimental data points from a set of experiments (100,000) as samples. The ordinate is the tensile force F, and the abscissa is the corresponding time t. In this experiment, the set tension value is 4.98 N; that is, when the value displayed by the digital display force gauge is greater than 4.98 N, it is considered to be invalid. There is an experiment period from a to c and from c to e and each experimental period includes a sliding inner rod rising stage (segments ab, cd, and ef) and descending stage (segments bc and de).

(1) It can be seen from the experimental data that a sharp decrease in the pulling force occurs in going from the end of each experimental cycle to the beginning of the next experimental cycle (the three points a, c, and e in Fig. 8.22). The main reason is that when the sliding inner sleeve is lowered to the lowest end and is in contact with the lower limit switch, the limit switch has an upward supporting force on the sliding inner sleeve. The three decreases at a, c and e are mainly caused by collisions between the sliding inner and outer jacket walls.
(2) It can also be seen from the experimental data that there will be a situation in which the pulling force first increases and then suddenly decreases near the three points b, d and f. The reasons for the increase in tensile force are analyzed. The three points b, d and f are the positions where shifts from the last part of the ascending stage to the rising stage of the experiment. Here, the sudden increase in the tension is mainly due to the impact of the top end of the sliding inner

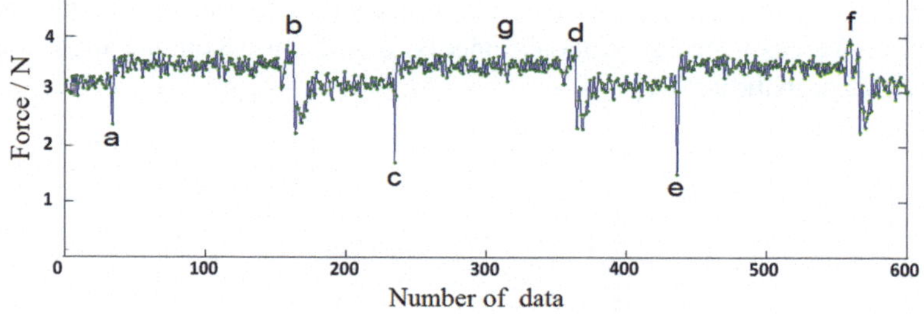

Fig. 8.22 Tensile data for three experiments

sleeve on the upper end of the limit switch. The reason for the sudden drop afterwards is mainly the sudden relaxation of the motor reverse rope, which caused the pulling force to decrease.

(3) It can also be noted from the data that a small protrusion appeared at point g, and the corresponding stage in the other two experiments did not appear. The reason for the protrusion is that the head of the sliding inner sleeve is adhered to the sensing device, which causes the outer diameter of the sliding rod to be increased slightly. In the process of the inner sleeve rod entering the outer sleeve rod, after the inner diameter of the outer sleeve contacts the sensor, the tension value increases, corresponding to the position of point g in the figure. The outer diameter of the inner sleeve rod is 12 mm, and the inner diameter of the outer sleeve rod is 13 mm. Due to the clearance between the inner sleeve rod and the outer sleeve rod and the rotation and left and right swing during the ascent of the inner sleeve rod, the solid glue colliding with the outer sleeve rod is also random.

Through 100,000 experiments conducted by the experimental device, a total of 2,000,000 experimental data points were obtained. Based on the knowledge of probability theory, the measured experimental data were used to find the probability distribution and distribution parameters of the random variables. Then, the reliability could be obtained according to the probability distribution.

According to the law of large numbers and the central limit theorem, when the number of tests n is large enough, the random variable (maximum tensile force in each test) obeys a normal distribution with parameters of μ and σ; that is, $X \sim N(\mu, \sigma)$, where X is the maximum tensile force, so only the parameters μ and σ need to be obtained. According to the method of moment estimation in the method of point estimation, the unknown parameters μ and σ can be estimated by means of the experimentally measured sample data. The front moment k of the overall force X can be written as

$$\mu_1 = E(X^1) = \int_{-\infty}^{\infty} x^1 f(x;\, \theta_1,\, \theta_2,\, \ldots,\, \theta_k)dx \tag{8.1}$$

The sample moment $A_i = \frac{1}{n}\sum_{i=1}^{n} X_i^l$ converges to the corresponding total moment $\mu_i(l = 1, 2, \ldots, k)$ based on probability. The continuous function of the sample moment converges to the continuous function of the corresponding total moment according to probability.

Assume that

$$\begin{cases} \mu_1 = \mu_1(\theta_1\theta_2, \ldots, \theta_k) \\ \mu_2 = \mu_2(\theta_1\theta_2, \ldots, \theta_k) \\ \vdots \\ \mu_k = \mu_k(\theta_1\theta_2, \ldots, \theta_k) \end{cases} \tag{8.2}$$

This is a simultaneous equation system with k unknown parameters $\theta_1, \theta_2, \ldots, \theta_k$, from which we can obtain the following equation:

$$\begin{cases} \theta_1=\theta_1(\mu_1\mu_2, \ldots, \mu_k) \\ \theta_2=\theta_2(\mu_1\mu_2, \ldots, \mu_k) \\ \vdots \\ \theta_k=\theta_k(\mu_1\mu_2, \ldots, \mu_k) \end{cases} \tag{8.3}$$

Substituting A_i for $\mu_i(l = 1, 2, \ldots, k)$ into the above formulas, we can use $\hat{\theta} = \theta_i(A_1, A_2, \ldots, A_k)$ as the estimate of θ_i $(i = 1, 2, \ldots, k)$.

From the experimental data, the values μ and σ of the random variable tensile force are calculated as $\mu = 4.16$ and $\sigma = 0.18$. We make a plot of the probability distribution of the data (Fig. 8.23) and compare it with the cumulative probability distribution of the normal distribution with the parameters $\mu = 4.16$ and $\sigma = 0.18$. The two curves are basically consistent, which further confirms that the random variables obey the above normal distribution.

The normal distribution is used to estimate the correspondence between the threshold and reliability and is compared with the experimental value of the threshold and reliability, as shown in Table 8.8. It can be seen that, depending on the number of tests, when F is greater than 4.9 N, there is no failure within a limited number of trials, and the reliability is 1. In fact, as the number of trials increases, it is still possible to fail, but the probability of failure decreases. In addition, the threshold corresponding to the required reliability can be obtained from the table. In fact, the normal distribution curve can be used to obtain the reliability and failure probability corresponding to the threshold value and the corresponding threshold when the predetermined reliability is reached. Since the threshold value in this experiment is

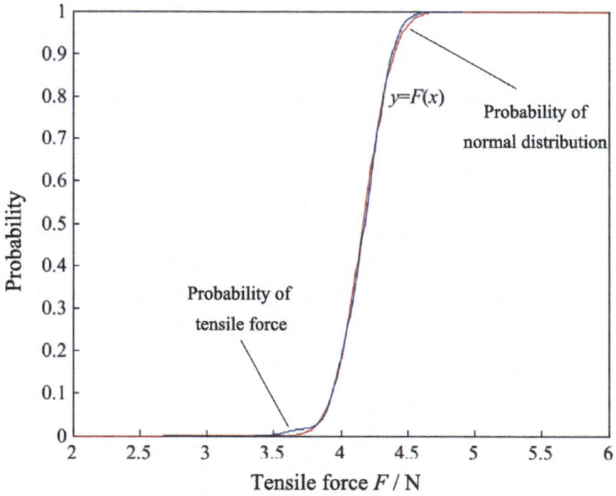

Fig. 8.23 Comparison of probability distribution and normal distribution of tensile force

Table 8.8 Correspondence between threshold and reliability obtained by two approaches

Threshold/(F/N)	Acquired by distribution curve		Acquired by experimental statistic	
	Failure probability	Reliability	Failure probability	Reliability
5.10	0.000000033	0.999999967	0.000000000	1.000000000
5.05	0.000000161	0.999999839	0.000000000	1.000000000
5.00	0.000000733	0.999999267	0.000000000	1.000000000
4.95	0.000003060	0.999996940	0.000000000	1.000000000
4.90	0.000011758	0.999988242	0.000016667	0.999983333
4.85	0.000041610	0.999958390	0.000016667	0.999983333
4.80	0.000135644	0.999864356	0.000050000	0.999950000
4.70	0.001129152	0.998870848	0.000416667	0.999583333
4.60	0.006818749	0.993181251	0.003183333	0.996816667
4.50	0.030097021	0.969902979	0.018800000	0.981200000

4.98 N, it can be concluded from Table 8.8 that the probability of the sliding failure of the telescopic rod is $[7.33 \times 10^{-7}, 3.06 \times 10^{-6}]$, and the reliability is [0.999996940, 0.999999267].

References

1. Zhang Y (2013) The integrated design of structure and control for flexible space deployable antennas. Xidian University, Xi'an
2. Yang D, Ning X, Cao J, Yang L, Duan B (2016) A cable cutting machine for mesh reflector, China, CN201410563241.7. Accessed 7 Sept 2016

Chapter 9
Spaceborne Deployable Antenna Design Software (SDADS) and Its Application

9.1 Introduction

Spaceborne deployable antennas have been widely used in satellite communications, deep space exploration and electronic reconnaissance [1, 2]. Designing an antenna quickly and effectively with a reasonable structure and performance has become one of the most important issues. The first reflector antenna was made and applied in astronomical observations by Grote Reber (Fig. 9.1) in his backyard in 1837. At that time, the design of such antennas relied mainly on simple numerical calculations and engineering experience. The method of predicting its performance is by physical prototype measurement [3]. If the results are not satisfactory, modified or compensating technologies are necessary. If the modification or compensation still does not work, the only remedy is to completely redesign and remanufacture. If this procedure were used today, it would obviously be unacceptable. Therefore, developing a comprehensive tool of analysis and design for spaceborne deployable antennas is urgently needed. With this tool, the designer can quickly, easily and efficiently convert design ideas into digital models to perform comprehensive analysis, simulation and evaluation of the performance of spaceborne antennas.

To develop a comprehensive design platform for satellite deployable antennas, two problems must be solved. One is how to reasonably analyze and abstract the design process and method, refine the main parameters, establish a parametric model, and generate a digital model according to the parametric model. This process can transform the abstract ideas of designer's mind into concrete models. The second problem is that in order for designers to implement a variety of performance analyses under complicated space environments, such as structural analysis, thermal analysis, deployment dynamics analysis and control, electronic performance analysis, and reliability analysis, the secondary development of these mature or submature softwares is necessary. To avoid duplication of resources and enable effective information transmission and communication among different systems, an efficient and correct data transfer process and the seamless connection of CAD/CAE are badly needed. This

B. Duan et al., *Large Deployable Satellite Antennas*,
Springer Tracts in Mechanical Engineering,
https://doi.org/10.1007/978-981-15-6033-0_9

Fig. 9.1 Grote Reber,1911—2002, American astronomer, one of the pioneers of radio astronomy. The first radio frequency survey was conducted using the parabolic reflector antenna for the first person in radio astronomy

requires solving the problem of the coordination and integration of heterogeneous systems.

Since SDADS (Spaceborne Deployable Antenna Design Software) is a large and complex software system based on the rule of easy-to-difficult and partial-to-whole, it starts with the widely used spaceborne mesh reflector antenna, and then gradually deepens. For this reason, the following works will be described, including digital overall modeling, databases and so on. The software can not only complete the digital modeling of a whole structure quickly easily and effectively, but also solve the CAD/CAE integrated modeling problem of the spaceborne deployable antenna and provide data protection for various analysis processes.

9.2 Basic Framework and Process of SDADS

9.2.1 Overall Framework

To meet the needs of integrated design and auxiliary innovation for spaceborne deployable antennas, the SDADS tool needs to meet the following requirements:

(1) It is convenient to carry out the layer design of the system and reasonably divide the functions of each layer so that they are logically relatively independent, the logical structure of the whole system is clear and the maintainability, upgradeability and scalability of the system and software are improved.
(2) Each module in the SDADS tool can be developed in parallel, and the overall progress is not affected by temporary loss of function in a single module, which would result in reduced R&D efficiency. At the same time, the SDADS can

Table 9.1 Three levels of SDADS

Application layer	Digital modeling	Structural analysis	Structural optimization	Deployment analysis	Thermal design	…
Information layer	The information library of integrated design system of deployable antenna					
Supporting layer	The support platform (CAD/CAE/PDM/DBMS)					

reduce the difficulty of developing and maintaining the processing logic of each layer.

(3) The user-open interface can be effectively isolated from the stored data to prevent unauthorized users from illegally obtaining relevant data, which ensures the data security of the platform.

Based on the above considerations, the SDADS is mainly constructed from three levels, namely, the application layer, the information base and the support platform. The layers are relatively independent and intrinsically connected. The application layer, database and underlying supporting software are shown in Table 9.1.

We now describe the three levels.

(1) Application layer

The application layer provides designers with human-computer interfaces, including the system's graphical user interface, windows and so on. It provides an input-output interface for sub-modules such as digital modeling, structural performance analysis, deployment process analysis and control, mesh surface precision control, structural optimization, thermal design, reliability analysis, and electronic performance analysis, including analysis of satellite antennas through different system windows.

(2) Information layer

The information layer is the base of the SDADS for the entire spaceborne deployable antenna integrated design, which provides data storage and sharing capabilities. Among these, the accuracy and rationality of the information layer design is one of the keys to determining the success of the system design.

The parameter model, assembly model, optimization model, finite element model, dynamic analysis model, basic system data, and project management information are stored in the information layer.

(3) Supporting layer

To simulate the complex environment of space, the SDADS of the spaceborne antenna needs to provide analysis functions for multiple performances of the antenna. Therefore, from the perspective of saving platform development time, commercial software or modules are employed as the basic supporting application layer, and secondary development is carried out to realize the corresponding functions.

9.2.2 Process and Analysis

According to the structural characteristics of spaceborne deployable antennas and the functional requirements of the SDADS, the connection among the modules are determined through the refinement of the technical approach, and the flow of the software shown in Fig. 9.2 is established.

(1) Parametric modeling. After the antenna designer inputs the structural design parameters, the tool can automatically generate the parametric model of the onboard antenna and then automatically convert it into a three-dimensional geometric model by a parametric module.
(2) Form-finding analysis. The CAD/CAE integrated modeling interface is used to read the parameter model, and the geometric model is automatically converted into a finite element analysis model by the form-finding (shape- and state-finding) analysis module so that the automatic form finding of the cable net and truss combined structure can be found. It can also perform the performance analysis of both the stowed and deployed states.
(3) Optimization design. According to the two-state(stowed and deployed states) performance index of the satellite antenna, three main factors of optimization, namely the design variable, objective functions and constraints, need to be determined to establish the two-state dynamic optimization model. It should be pointed out that the analysis and optimization module and digital modeling can be seamlessly connected.
(4) Integrated model of mechanical (or structural), electronic and thermal analysis. After reading the antenna analysis model from the shared database, by dynamics analysis, reliability analysis, thermal analysis and electronic performance analysis, weak links can be found to guide modification of the geometry and physical parameters of the antenna.
(5) Adjustment of surface precision. Adjustment and measurement are a whole organic unit. The measurement can be performed by high-precision photographic technology. By comparing the measured values of the mesh reflector with the design requirements, the platform can automatically adjust the appropriate value (cable force or cable length) of the adjustable cable to guide engineering technicians to adjust the reflector surface precision.

9.2.3 Combination of Digital Modeling

As a sub-module of the tool and the recipient of the initial structural information, digital modeling plays a vital role in the SDADS. The study of its overall architecture scheme not only ensures the rapid generation of the antenna geometry model but also lays the foundation for the accurate modeling of subsequent structural analysis. Therefore, from the overall design scheme, combined with the structural characteristics of spaceborne antennas, the digital design concept (Fig. 9.3)

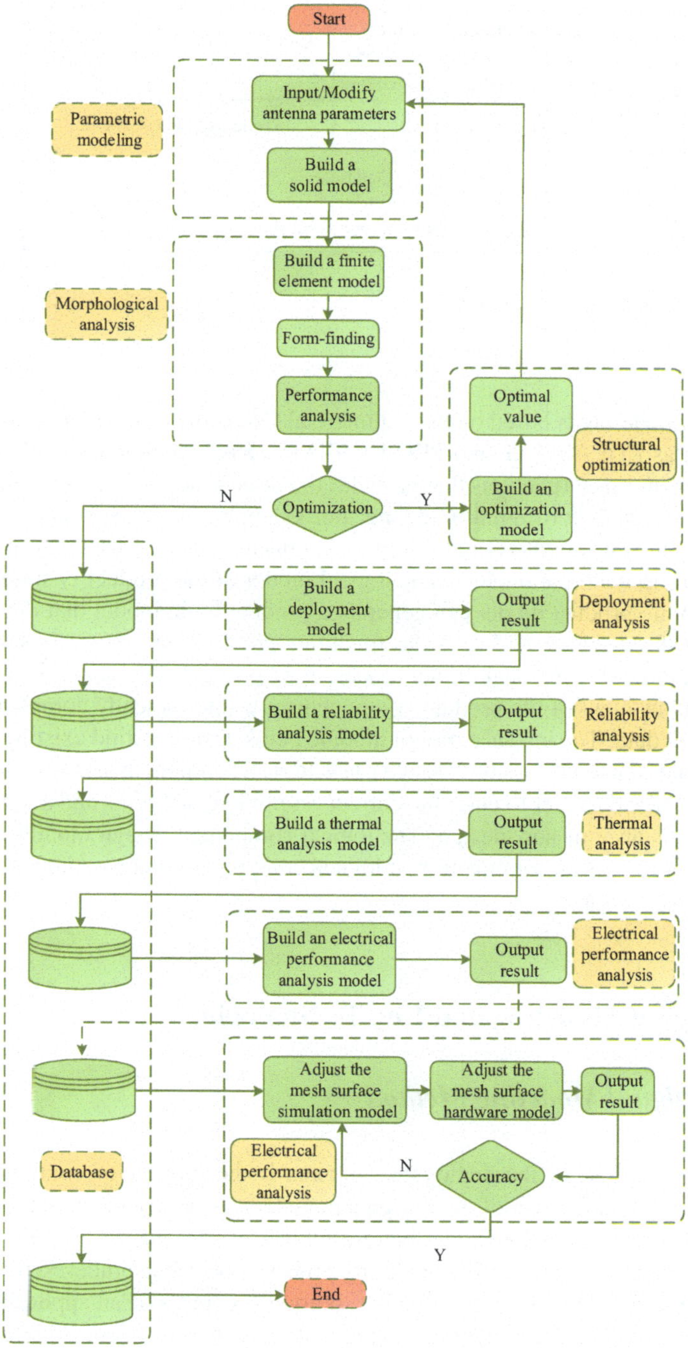

Fig. 9.2 Flow chart of SDADS

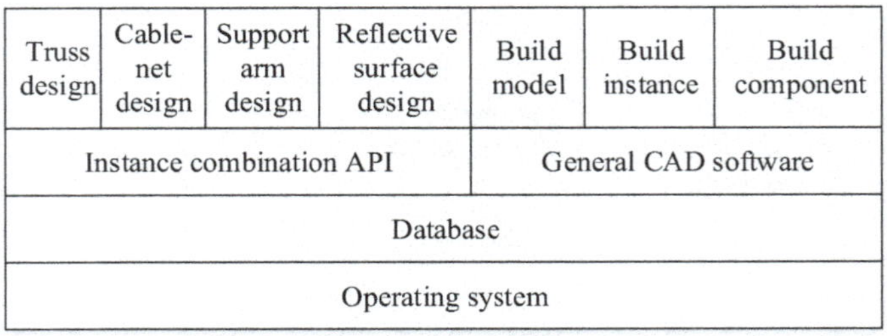

Fig. 9.3 Digital design concept map

is proposed.

Digital modeling is based on the commercial CAD software with digital capabilities (such as I-DEAS, UG, and PRO/E). The instance combination API is a set of application interface functions that provides the basic functions of combining corresponding instances for digital modeling. For example, the truss, support cable net and mesh are user-oriented digital logic components, and the user can modify the corresponding main parameters. The logical model of the product is automatically generated; the product model can generate a product logic model that the program reads into the database and uses the functions provided by the CAD software to generate a three-dimensional digital model of the product. The selection, assembly, and size adjustment of the product components are automatically completed based on the input data. The instance design module helps designers find existing instance libraries. If the designer wants to design a new logical component instance, the system will prompt the designer to enter the correct assembly constraints and automatically record an instance of new design. The logical component design module will help designers find existing combination of logical components and automatically record new product designs.

9.3 Digital Modeling and Key Technologies

9.3.1 Digital Modeling Method

The design process of the deployable antenna is the process of continuous optimization and modification of the design parameters. To verify the structural performance, the digital model needs to be re-established and verified after the parameters are changed. This repetitive modeling process is cumbersome, resulting in low design efficiency. Therefore, an effective solution is to propose an applicable digital modeling method for the structure of deployable antennas to generate digital antenna models quickly and accurately. At the same time, additional interfaces of the components are reserved to support the design of possible new antenna structures in the future.

By abstracting and classifying the design tasks of spaceborne deployable antennas, it can be seen that the structural design of antennas is composed of regular and irregular tasks. The development and assembly of parts are regular tasks that can be automatically completed by a computer. In contrast, the generation of parameter models, topological relationships and assembly information is an irregular task that needs to be combined with specific design parameters and relevant algorithms.

Regarding regular tasks, the main components of the structure, such as cables, rods, joints and common parts, can be obtained through the overall analysis of spaceborne deployable antennas. Examining the details shows that although parts of the same kind have the same function, they may have different ways of implementation (such as with different shape or size). If the practical properties of these parts are ignored and the components that implement the same functionality are abstracted into one class (called the logical components [4]), it can be abstractly concluded that the spaceborne deployable antenna structure is composed of these logical components. At the same time, it could be concluded that the logical components have multiple instance forms. For example, a hollow pipe, solid pipe or I-steel are examples of the logical component of the bar.

The generation of logical components has become a regular task. For a logical component, a plurality of instance forms can be established, and a basic model library of the antenna is constructed. When the model is used to query the database and retrieve the corresponding instance, if there is no corresponding instance, the model can be added to the model library, which can save a large amount of time on repetitive modeling and ensure support for new structure design.

A logical component should be designed for the specific structure of the deployable antenna, and the minimum functional part inside the antenna should be selected as the unit. Taking the AstroMesh deployable antenna as an example, it includes a deployable truss, support cable net, mesh and arms between the antenna and satellite. Among these, the support cable net forms the required reflector shape under the action of pre-tension, and the mesh, as the reflector, is attached to the front support cable net, as shown in Fig. 3.11.

Through this division method, each part of the antenna structure of the AstroMesh deployable antenna can be further divided in detail, where the selected logical components are the basic elements of the structure of the upper layer. This can be achieved through parameter control (Fig. 9.4).

On the one hand, the extraction of the specific parameters of logical components can facilitate information exchange among different tasks to conveniently describe antenna parameters. On the other hand, the corresponding parts library can be established for these logical components to lay the foundation for digital modeling.

After dividing the logical components of the deployable antenna, the corresponding part library can be established. Each logical component corresponds to a component library formed by multiple instances. The establishment of the parts uses the digital design idea; that is, the size of the digital model corresponds to a functional relationship. By changing the parameters, all the relevant sizes can be changed automatically.

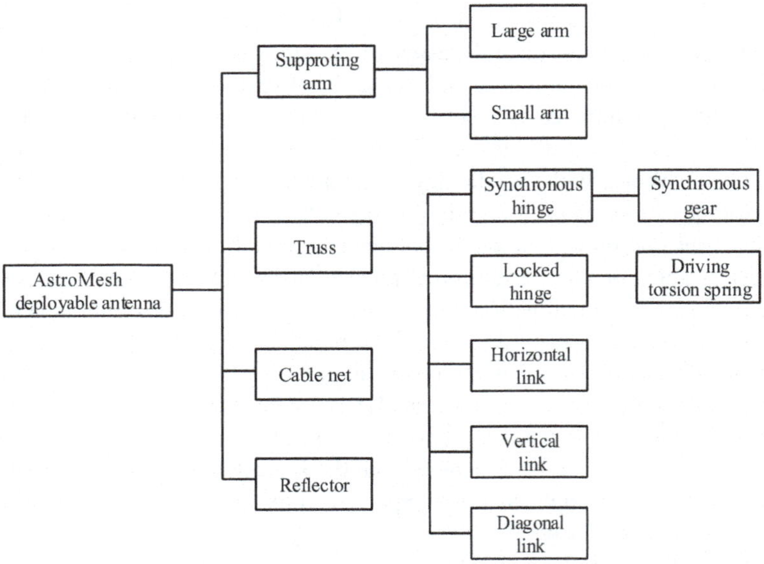

Fig. 9.4 Structure devision diagram

With the part library, there is material for the digital modeling of the antenna. At this time, irregular work needs to be studied, which mainly involves two tasks: the generation of a parameter model and the assembly of parts. The generation of the parameter model and topological relationships is related to the specific design method of the spaceborne antenna. For example, considering the support cable net, different meshing methods correspond to different structural topological relations. The assembly of parts requires the precise calculation of the positional relationships among the corresponding logical components to ensure that the assembly does not interfere. These tasks will be performed by combining the corresponding design quasi-test with the secondary development of I-EDAS. First, the parameter calculation method shall be programmed, and then the secondary development of I-EDAS is performed. The specific design algorithm can be found in the relevant literature; here for simplicity, only the secondary development of I-EDAS is briefly described.

There are two choices for secondary development based on I-DEAS. One is secondary development based on Open I-EDAS, which is fast and stable in performance but requires complicated programming work. The other is secondary development based on the program file. This method has strong pertinence and convenient implementation and is suitable for a series of tasks requiring frequent execution of the same or similar operations. In view of the structural characteristics of spaceborne antennas, the latter is suitable because there are more regular tasks.

Program files are mainly sequenced collections of a series of existing instructions in the software. By reading and running the program files, I-EDAS can automatically perform specific functions specified by users. A program file contains multiple lines of statements, each of which contains a header and program code. The header indicates the type of program code in that line, which usually consists of letters and a colon. For example, "C:" represents the content of the behavior annotation, "P:" indicates that the program file will be temporarily suspended when it runs to the line, and "E:" means the end of the program file. Program files can be generated either in I-EDAS software or a text editor.

In application, the program files required to generate the model at one time are error-prone and less efficient. To address this, the platform for regular tasks and its functions are packaged into a program file, that is, a specific program file to achieve a specific function; when it is needed, the program file can be called. This not only greatly reduces the time required to generate files but also improves accuracy.

In the modeling process, the more commonly used functions mainly include checking the instance parts in and out, modifying the parameters of instance parts, duplicating instance parts, generating the assembly parts, adding the parts in the assembly, and moving and rotating the parts. All the program files form a library of program files. When the antenna is modeled, the parameterized model antenna can be automatically generated by calling files in the program file library. For example, for the key parameters of the deployable antenna, the correlation between parameters is established, and the parameters in the corresponding program file are modified to perform the function of generating and assembling specific antenna parts and eventually automatically generating the overall model, as shown in Fig. 9.5.

Thus far, the digital modeling of the deployable antenna includes three main steps. First, the basic logical components are extracted according to the antenna structure, and the corresponding instance parts are built and stored in the part library as the basis of parameterization. Second, according to the specific structure of the antenna, the corresponding program file library is constructed for the model generation process.

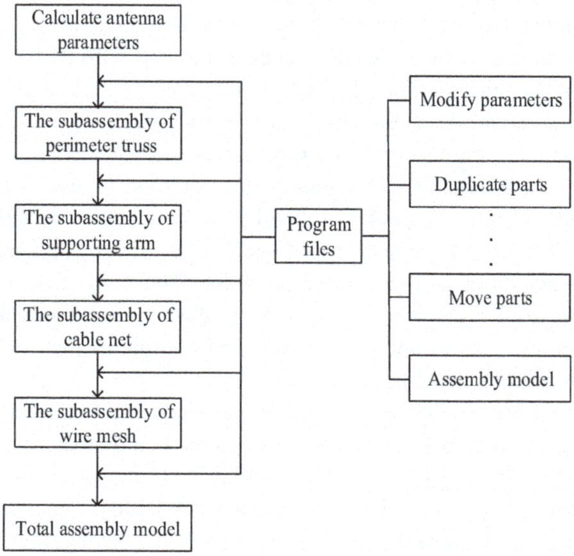

Fig. 9.5 Modeling processs

Third, the determination of the deployable antenna parameters is programmed to realize the real-time generation of the parametric model.

9.3.2 CAD/CAE Integration Technology

Based on the digital modeling of antennas, the integrated design platform can comprehensively evaluate the design scheme of spaceborne deployable antennas and propose suggestions for improvement. This process involves multiple sub-modules. Although different modules complete the specified functions independently and work independently of each other, the objects they analyze have consistency and inheritance of data. Therefore, it is important to design interfaces, such as communication between modules and modules, and between modules and systems, which need to be solved by integration of CAD/CAE.

At present, there are three ways to integrate CAD/CAE systems. One is to achieve CAD/CAE integration on the CAE platform. This method has a high efficiency, but its CAD modeling ability is lower than that of professional CAD systems. Second, CAD/CAE integration can be realized on the CAD platform. This method is faced with the problem of integrating the finite element model and geometric modeling, so it is difficult to simplify the CAD model. The third method is to achieve CAD/CAE integration by using a third-party platform, which can make use of professional platforms and avoid data loss and incompatibility problems caused by direct data file transfer. However, the corresponding data transmission mode and storage structure must be developed.

In the SDADS tool, the third scheme is employed, which can make use of the advantages of mature analysis software and use secondary development means to achieve more consistent function of antenna analysis. The tool involves a variety of analysis software, so this scheme fits well with it. To compensate for the deficiency of the third method, the SQL database is specifically used to perform CAD/CAE data exchange. As a third-party data management system, the SQL database can store a large amount of data and has high security.

In application, secondary development is used in the CAD system to extract the feature parameters related to finite element modeling from geometric models, including shape size, topological relationships and related attribute data, and then send them to the central database. The CAE system reads relevant information from the database, processes it through corresponding modules, generates the finite element analysis model, and completes the analysis according to the design requirements (Fig. 9.6).

Accordingly, it is necessary to study the data storage structure suitable for the characteristics of spaceborne deployable antennas. Reasonable design of the storage structure can guarantee the standardization of data and the effectiveness of the transmission process. Take the AstroMesh deployable antenna as an example. First, the structural characteristics are analyzed, and the parameters of the same topic are stored in one table, which maintains the information of each topic independently of other

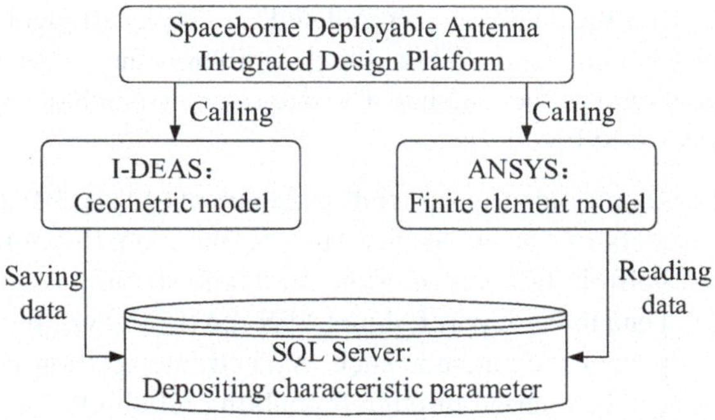

Fig. 9.6 CAD/CAE flow chart

topics; the structure is clear, so as to avoid data confusion. Antenna information can be divided into truss information, support cable net information, support arm information, etc. At the same time, each category can be further subdivided into structural shape characteristic parameters and related attribute parameters. Second, the extraction of the main driving parameters [5] can directly control the structural shape characteristics of the model parameters, make full use of mathematical relations to reflect the geometric relationship among objects, and minimize the number of driving parameters (main parameters). For example, diameter is undoubtedly one of the main parameters, and the diameter parameters can be used to determine the length of the bar and other information. Furthermore, the design rules, namely, the rules of the variable and the parameter table, should be established to intuitively and concisely reflect the parameter information, such as the truss parameter information of ATN_TRUSS_PARA corresponding to the antenna.

Based on this, a variety of parameter data tables are designed in the SQL database to store data. Each table contains a class of related main parameters, and the antenna version number is used to correlate the tables. The table contents are roughly as follows:

(1) Data table of antenna truss (ATN_TRUSS_PARA): records the antenna version number, antenna diameter, height (deployed and stowed state), grid type, element number, truss rod size and material information, etc.
(2) Data table of antenna support cable net (ATN_NET_PARA): records the antenna version number, cable net focal length, grid type, cable cross-sectional size, cable material information, etc.
(3) Data table of antenna support arm (ATN_ARM_PARA): records the antenna version number, relevant size of the support arm, angle of the support arm, material information of the support arm, etc.
(4) Information table of surface cable net (ATN_STICK_LINKS): records the antenna version number, number of ropes of the cable net surface, number of start and end nodes of the cable net surface rope, etc.
(5) Information table of node (ATN_STICK_NODES): records the antenna version number, node number and node coordinate information, etc.

After the database is completed, a complete logical model of spaceborne deployable antenna is available, which facilitates the structural performance analysis and evaluation of the antenna. The database also includes the following specialized data tables (see Fig. 9.7):

(1) Data table of space orbit environment (ATN_ORBIT_INFO): records the parameters of various orbits, such as the planet's orbiting period, half-length axis, half-circumference of the orbit, and so on.
(2) Data table of part failure (ATN_FAIL_INFO): records the failure probability of part of the antenna, such as electronic ignition failure and gear failure, which can be queried through probability reliability.

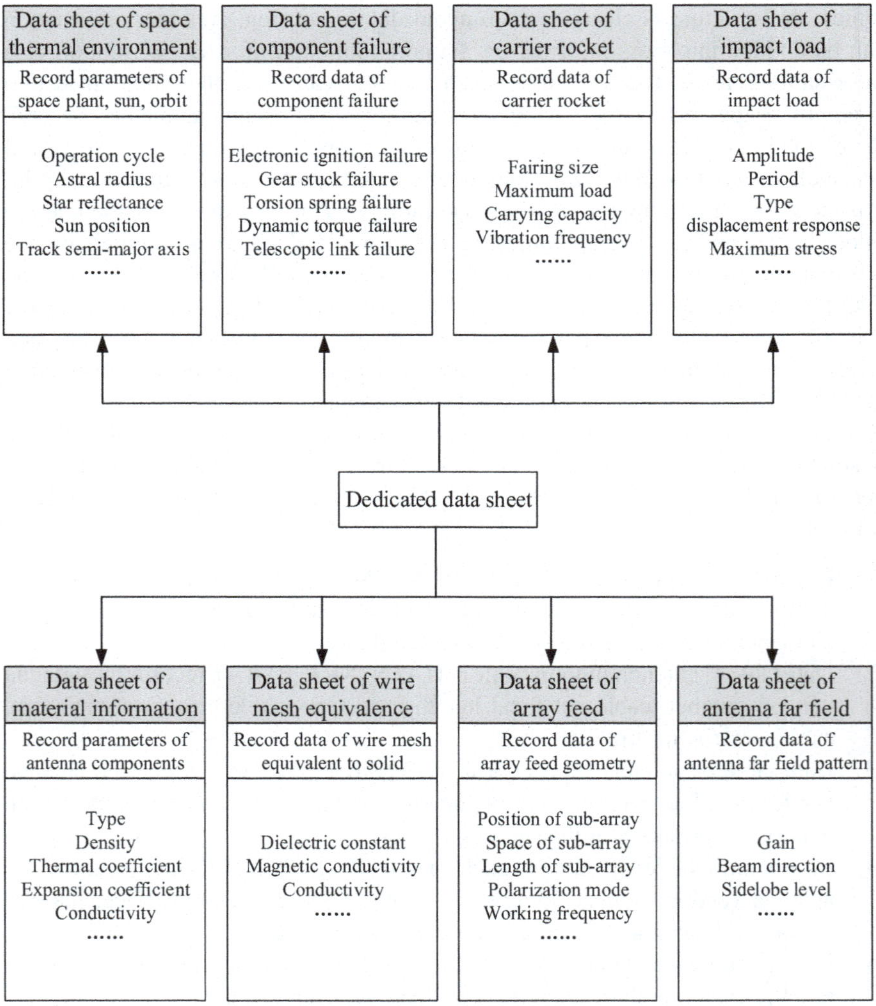

Fig. 9.7 Schematic diagram of dedicated data table

(3) Data table of carrier rocket (ATN_ROCK_INFO): records the relevant parameters of the carrier rocket, such as the fairing size, bearing capacity, maximum overload, vibration frequency and other parameters. These parameters restrict the design elements such as the furled size, weight and fundamental frequency of the deployable antenna.
(4) Data table of impact load (ATN_SHOCK_INFO): records the relevant parameters of various impact spectra, such as type, period, amplitude, etc.
(5) Data table of material information (ATN_MAT_INFO): records common parameters including a variety of deployable antenna components as well as their physical and electrical performance parameters, etc.
(6) Wire mesh equivalent electromagnetic parameters table (ATN_EM_INFO): records the preparation, dielectric constant, permeability, etc., of the reflector wire mesh.
(7) Array feed parameter table (ATN_FEEDARRAY_INFO): records the position, length, spacing, polarization mode and frequency of the array.
(8) Data table of antenna far-field pattern (ATN_PATTERN_INFO): records the gain, sidelobe level, half-power beam width, etc., of the far-field pattern.

9.4 Main Functions and Modules of SDADS

9.4.1 The Composition of Digital Modeling

Based on the above description, the deployable antenna digital modeling module is developed with the integration of the I-DEAS and C++ Builder program technologies. Since the SDADS includes this module, the user can conveniently and efficiently perform digital modeling of antenna parameterization by visual human-computer interaction [6].

This module consists of three parts: the menu bar, parameter control area and window display area. The parts have the following functions:

(1) Menu bar: This can create, open and save the project and has the functions of engineering maintenance, material library maintenance and model display.
(2) Parameter control area: In addition to completing the parameter input and modification of the perimeter truss, hinge rib and supporting net structure, the corresponding digital model can also be generated.
(3) Graphic display area: The antenna digital model can be displayed from multiple angles.

As shown in the data flow of the digital design (Fig. 9.8), first, the relevant size parameters are input, including the parameters of the perimeter truss, mesh and support arm, the parameter model of each component is obtained, and the relevant size information is stored in the database. The empirical net formula and expert knowledge are used to derive the cable net topology data, that is, the cable net node and the node information of the cable element. Finally, based on the program file,

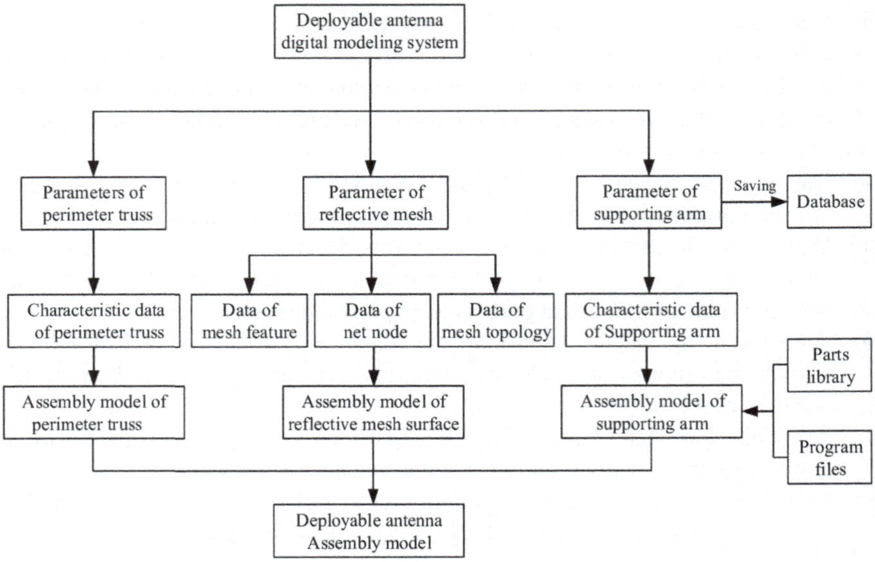

Fig. 9.8 Data flow chart of digital modeling

the model in the part library is called, and the digital model of the whole entity is generated in I-DEAS.

9.4.2 Main Functions of SDADS

The digital modeling module of a large spaceborne deployable antenna has been integrated into the integrated design platform. The main menu is shown in Fig. 9.9. The SDADS tool includes eight modules: parameterized modeling, form-finding and structural analysis, two state dynamic optimization, flexible dynamic analysis and control of deployment, deployment reliability analysis, thermal analysis, electronic performance analysis and shape precision adjustment. Aided by the SDADS, designers can implement systematic simulation analysis, innovative design and exploration of new solutions. As a result, it is able to achieve the purpose of shortening the design cycle, reducing the design cost, and improving the design performance. The functions of each part of the platform are as follows:

(1) Parameterized digital modeling

In this module, the parameterized modeling of the spaceborne deployable antenna can be developed automatically after determining, from the professional database, the structural components with the required features, assembly relationship, and material factors. In the modeling, the required information of nodes, elements and materials used in the subdisciplines can be related automatically. For popular antennas such as

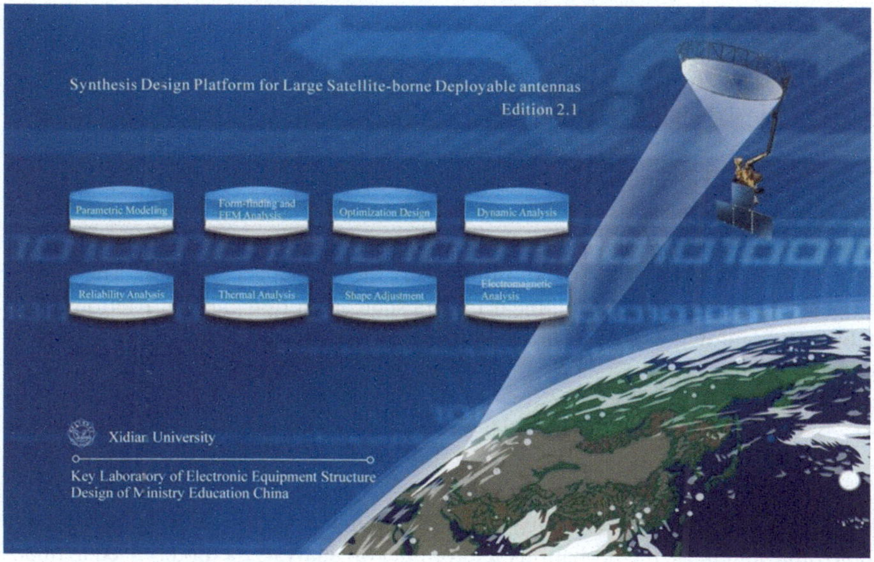

Fig. 9.9 Platform main interface

AstroMesh, module and radial rib antennas, the corresponding digital model can be easily established by inputting a few factors of the antenna's geometry and material.

(2) Form-finding and structural analysis

In this module, numerical analysis of combined structures with a cable net and truss (frame) can be implemented for AstroMesh, module and radial rib antennas. To do this, a finite element analysis model is first established by CAD/CAE integration. Then, structural analysis of the antenna is carried out. The appropriate state, that not only satisfies the reflector precision requirement but also makes the cable tensions as uniform as possible, can be found. This structural state, with the eigenfrequency, eigenmodes of vibration, deformation and material strength, and the reflector performance, can be considered either the input of optimization or the final output.

(3) Two-state dynamic optimization

This can be used in the optimization design of the deployed and furled states. First, the nonlinear programming models are established with the lightest weight or optimal electronic performance (such as high gain and low sidelobe level) as the target and cable tension, cable cross-sectional dimension and truss link element cross-sectional area as design variables; the material strength, the structural natural frequency and the antenna electronic performance (such as beam forming and beam pointing accuracy) are considered as constraints. The corresponding solving methods are selected according to the prompts. Second, functions of process and control visualization are available, such as flexible modeling; that is, the target and constraint functions can be exchanged and altered during the optimization process, and the design variables can also be adjusted according to needs during the iterative process.

(4) Flexible dynamic analysis and control of deployment

The module includes the following functions: First, multi-flexible body dynamic analysis can be achieved, which includes rigid bodies, flexible bodies and driving cables, etc. The second function is the speed and acceleration planning of the deployment process, with the principle of minimizing the impact acceleration. The third is to consider the influence of nonlinear factors such as motion gap and friction on the stability of deployment. The fourth is to consider the impact of the support cable net on deployment.

(5) Reliability analysis of deployment

This module has a reliability analysis function for deployment. By deeply analyzing the factors affecting the reliability of the deployment process of large spaceborne deployable antennas, the probability and/or interval analysis method is used to quantitatively describe the reliability of these factors, and then two types of deployment process failures with probability and/or interval parameters are established. Furthermore, reliability modeling and analysis of the important events affecting the antenna deployment reliability are carried out, and the reliability of the antenna in orbit is predicted.

(6) On-orbit thermal analysis

The module can automatically generate the temperature field and the antenna thermal load along the orbit according to the running orbit parameters, especially the thermal load function of the antenna in the process of entering and exiting the shadow zone, and then analyze the performance of the spaceborne antenna under the thermal load, including the static structure performance, dynamic performance, antenna electromagnetic performance, etc.

(7) Electronic performance analysis

The electronic performance analysis of the antenna can be performed, including a simplified calculation formula between the mesh and antenna electronic performance, taking into account the field coupling theory model for the electromagnetic field, the structural displacement field, and the temperature field. In addition, according to the simulation results of electronic performance, the designer can easily find the structural deficiencies and determine how the structure should be modified.

(8) Shape precision adjustment

The shape precision adjustment module consists of three parts. First, the simple and accurate measurements of surface precision and the vertical cable tension are carried out. Second, the modification of the finite element analysis model can be implemented to a certain extent. Third, the adjustment of the support cable net surface precision can be implemented under the guidance of theory and methodology.

9.5 Case Demonstration

To confirm the correctness and effectiveness of the integrated SDADS design tool, three typical mesh reflector antennas are applied. Considering limitations of space, only the main results are given below.

These three types of antennas are all large deployable cable net and truss combined reflector antennas. They differ in the type of truss and support cable net used, such as the AstroMesh antenna (Fig. 3.11), modular antenna (Fig. 3.37) and radial rib antenna (Fig. 3.43).

To quantitatively compare the computational efficiency of the SDADS tool, three types of antennas with deployed diameters of 16, 5 and 5 m are considered. The time consumption for each type of antenna is quantitatively compared in two cases, that is, when the SDADS tool is used and when it is not, to verify the computational efficiency advantage of the platform. The following criteria are listed for illustration.

(1) Digital modeling, which includes model building of the parts, model pre-processing and model assembly. Table 9.2 gives the contrast time, which shows a visible efficiency increment of more than 70%.
(2) Surface precision adjustment, including testing (photography), calculating adjustment point error and performing specific adjustments. The adjustment efficiency is investigated through the physical principle prototype of the 2 m caliber shown in Fig. 4.2.
    The actual test and adjustment also tell us that the measured results meet the design requirements (better than 1 mm) and fall within the prediction range of the design platform, which verifies the accuracy of the SDADS tool for antenna reflector surface precision design, manufacturing error impact mechanism analysis and other functions.
    Table 9.3 gives the time consumption comparison between the SDADS tool and the traditional adjustment. The reduction is approximately 2/3. Specifically, the traditional method needs 12 adjustments to reduce the RMS error from 1.32 to 0.99 mm. With the SDADS, the RMS error is reduced from 1.35 to 0.73 mm with just 4 adjustments.
(3) Dynamic analysis and deployment control are carried out. As mentioned previously, the antenna in the furled state should be deployed automatically after it is put into the predetermined orbit by the rocket; otherwise the whole satellite will be abandoned and the loss is great. To verify the correctness and effectiveness of the dynamic analysis and deployment process control module of the flexible multi-body of the SDADS, a variety of experiments with the 2 m diameter physical model are carried out to verify the effectiveness of the SDADS. Table 9.4

Table 9.2 Comparison of digital modeling efficiency of three types of antennas

Antenna type	Modeling stage	Manual modeling/h	Parametric modeling/h	Efficiency improvement (%)
AstroMesh antenna (16 m)	Building part model	9.5	0.5	**71.3**
	Model pretreatment	0	9.2	
	Model assembly	33	2.5	
	Total modeling time	42.5	12.2	
Radial rib antenna (5 m)	Building part model	8	0.5	**85.7**
	Model pretreatment	0	1.5	
	Model assembly	9.5	0.5	
	Total modeling time	17.5	2.5	
Modular antenna (5 m)	Building part model	10	0.5	**86.5**
	Model pretreatment	0	2	
	Model assembly	16	1	
	Total modeling time	26	3.5	

Table 9.3 2m-diameter prototype shape precision comparison

	Time consuming for traditional method/h	Time consuming for integrated platform/h	Reduction rate
Testing time	12	4	2/3
Computation error time	6.0	2	2/3
Adjustment time	12	4	2/3
Total time	30	10	2/3

gives one of the experimental results and shows the tension of cable elements 7, 80 and 81 at 79°, 82° and 90°, respectively. The results calculated by the SDADS are close to the measured results, which can be used as an important reference for engineering design.

Table 9.4 2m-diameter prototype multi-flexible body analysis and calculation and measurement results of control cable tension

Cable unit number		7	80	81
Tension at 79°/N	Calculation	30.08	11.75	89.94
	Actual measurement	29.36	13.25	100.3
	Error	2.40%	8.50%	10.30%
Tension at 82°/N	Calculation	49.9	26.31	146.8
	Actual measurement	57.37	24.52	171.2
	Error	13.00%	6.80%	14.30%
Tension at 90°/N	Calculation	15.01	33.07	33.07
	Actual measurement	17.11	37.96	38.62
	Error	12.30%	12.90%	14.40%

References

1. Gunnar T (2002) Deployable tensegrity structures for space applications. Ph.D. Dissertation, Royal Institute of Technology, Department of Mechanics, Sweden
2. Thomson MW (1998) The AstroMesh deployable reflector. In: IUTAM-IASS Symposium on deployable structures: theory and applications. Cambridge, UK, September 6–9
3. Duan B (2005) Structural analysis, optimization and precision control of flexible antennas. Science Press, Beijing
4. Jing Zhao, Jianyuan Jia, Xiaodong Shao (2003) Digital design and research of virtual products. Electromechan Eng 19(2):23–27
5. Weiwei Wu, Renzhong Tang, Liang Hou et al (2005) Research and Implement of the parameter-based mechanical product size and deformed design. China Mech Eng 16(3):215–222
6. Hao J, Baoyan D, Li T, et.al (2015) Synthesized design platform and digital modeling of large spaceborne combined cable net and truss deployed antennas. Space Electron Tech 12(3):35–42

Chapter 10
Electrostatic Forming Membrane Reflector Antenna

10.1 Introduction

The principle of the membrane antenna is to coat the membrane with a metal reflective medium to form a specified shape under a certain external force to reflect electromagnetic waves. The forming methods of membrane reflectors mainly include inflation forming [1], cable tensioning [2, 3], PVDF control [4, 5] and electrostatic stretching forming [6]. Compared with other types of membrane antennas, electrostatic forming membrane reflectors have attracted much attention because of their low areal density and the ability of performing real-time surface control by electrostatic adjustment.

To date, research on electrostatic forming of membrane reflectors has mainly concentrated on membrane mirror and membrane reflector antennas. Although the electrostatic forming principles of the mirror and reflector antennas are similar, the analysis method and the fabrication of the membrane reflector are obviously different because of the difference in the diameter and curvature of the surfaces. The diameter and curvature of the antenna reflector are much larger than those of the mirror. Therefore, the analysis theory and fabrication process of the antenna reflector are much more complicated than those of the mirror. The shape of the electrostatic forming membrane reflector is stretched and maintained with electrostatic force, which can be associated with the homologous design concept for reflector antennas design proposed by Sebastian von Herner (Fig. 10.1).

NASA studied the reflector of flexible materials in the mid-20th century. A 4.88 m polyimide membrane mirror with multi-ring electrodes was developed by NASA as early as 1979. However, due to the limited preparation technology for large-area membranes, research has gradually shifted to focus on small-diameter membrane mirrors [6, 7]. In 2000, on the basis of electrostatic forming membrane mirrors, the University of Arizona in the United States has employed the micro-control of piezoelectric ceramics to effectively improve imaging quality and developed a 6 inches electrostatic forming membrane mirrors [8]. In 2005, the US SRS Company and

B. Duan et al., *Large Deployable Satellite Antennas*,
Springer Tracts in Mechanical Engineering,
https://doi.org/10.1007/978-981-15-6033-0_10

Fig. 10.1 Sebastian von Herner, 1919–2003, German astrophysicist and radio astronomer. He designed the Effelsberg 100-m fully steerable paraboloid radio telescope in Germany and proposed the homologous design concept for large reflector antennas

the Air Force Research Laboratory (AFRL) cooperated to develop a 0.25 m electrostatic forming membrane mirror and then developed a 1 m diameter electrostatic forming membrane mirror [9]. In 2006, the SRS Company successfully developed a 0.7 m membrane mirror by means of a hybrid control method of central inflation and marginal electrostatic stretching, and the central profile precision of the mirror was greatly improved [10].

The research on electrostatic forming membrane mirrors in China began in the early 21st century. Since 2004, theoretical and experimental studies of electrostatic forming membrane mirrors have been carried out at Suzhou University [11, 12]. An experimental prototype with a 200 mm diameter and 17.7 m curvature radius has been manufactured. Finite element analysis and experimental study of the electrostatic forming method of membrane mirrors were carried out by Changchun Institute of Optics and Machinery [13, 14], and a good mirror surface was obtained.

In 2004, based on the in-depth study of electrostatic forming membrane reflectors by SRS and Northrop Grumman Corporation, the first 5 m diameter electrostatic forming membrane antenna was fabricated [15]. It is constructed on the base of the AstroMesh deployable antenna. A number of electrodes are laid on the upper support cable net, and a suspended membrane reflector is installed above the electrodes. The membrane reflector is formed by the attraction of the electrostatic force. Experiments on the antenna fully demonstrate the feasibility of the electrostatic forming membrane reflector. Xidian University in China has also conducted an in-depth study and has developed 0.5, 2 and 5 m prototypes.

10.2 Fundamental and Multifield Coupling Theory Model

The process of forming the electrostatic forming membrane reflector antenna is complicated and is composed of the following steps: the change in electrostatic force caused by a voltage change leads to a change in the membrane surface, and the change in the membrane surface in turn causes a change in the electrostatic force acting on the membrane. In the whole forming process, the electrostatic force and membrane surface iteratively change until the electrostatic force and internal stress of the membrane gradually reach equilibrium, and then the whole structure tends to be stable. It can be seen that the forming process of the membrane is a two-field coupling problem where the electrostatic field and the displacement field of the membrane structure interact with and depend on each other.

In the electrostatic forming analysis of the membrane reflector, the mechanical analysis of the membrane is carried out first to obtain the electrostatic force needed for forming, and then the corresponding electrode voltage is obtained according to the approximate formula of plate capacitance. The problem can be simplified by the separation of the electrostatic field and the membrane displacement field, but there are two obvious shortcomings. (1) There is a certain error in the electrostatic force determined according to the plate capacitance formula because the application of the plate capacitance formula has a strict restriction: the distance between the membrane and the electrode is required to be uniform, the curvature of the membrane is small, and the deformation of the membrane might be ignored with respect to the distance from the electrode. However, the characteristics of the electrostatic forming membrane reflector antenna are that the forming curvature of the membrane is large and that the electrode is composed of many small planar electrodes. There is of a big error existed in principle. (2) A change in the surface of the membrane will lead to a change in the amplitude and direction of the electrostatic force in the process of electrostatic forming. The formula of plate capacitance cannot precisely describe this process.

It is difficult to improve the surface precision of electrostatic forming membrane reflectors effectively if the design idea of field separation is still used. For this reason, the electrostatic forming process of the membrane is studied based on the two-field coupling theory.

10.2.1 Fundamental Theory of Electrostatic Forming Membrane Reflector Antennas

The forming process of the electrostatic forming membrane reflector antenna is realized by controlling the magnitude and distribution of the potential between the electrode and membrane reflector. Electrostatic stretching is a dynamic process in forming a membrane reflector antenna, which is accomplished by the work of the electrostatic force. Under working conditions, the shape of the membrane reflector is maintained mainly by the attraction of the electrostatic force, while the electrostatic potential puts a certain constraint on the charge distribution of the membrane surface and base electrodes so that the electrostatic force is stable.

The electrostatic forming of the membrane reflector includes two main parts. One part is the initial paraboloid configuration of the membrane reflector. One method for the initial forming of the membrane is direct forming using curved membrane

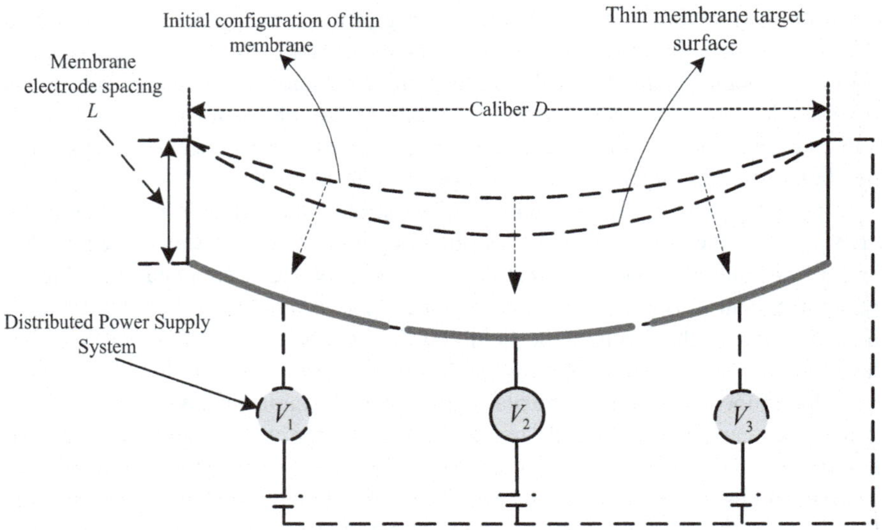

Fig. 10.2 Diagram of electrostatic forming membrane reflector

processing technology, and the other is planar membrane splicing based on curved membrane flattening theory. The latter is easier to achieve and more widely used. The other part is shape adjustment by altering the electrostatic force. The membrane reflector is shifted from the initial configuration to the target paraboloid by changing the voltage applied to the base electrodes. The electrostatic forming mechanism of the reflector antenna is shown in Fig. 10.2.

10.2.2 Two-Field Coupling Model

The electrostatic forming membrane reflector consists of a polymer membrane coated with a conductive layer, electrodes and a distributed power supply system, as shown in Fig. 10.3. Electrodes and membrane reflectors serve as the boundary of the elec-

Fig. 10.3 Diagram of electrostatic field analysis

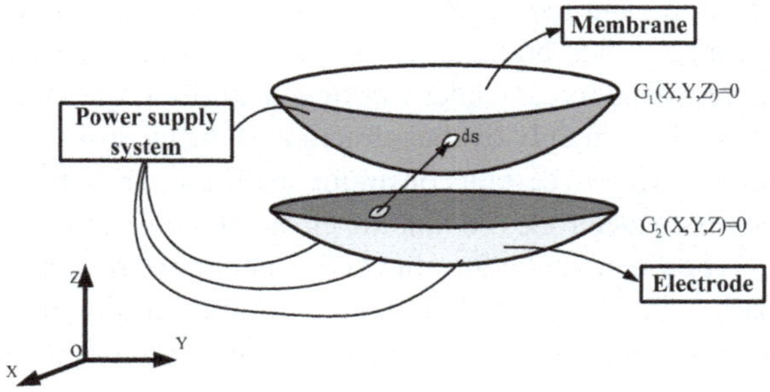

trostatic field. The displacement field of the membrane reflector and the electrostatic field interact with the electrostatic force on the reflector and the surface of the reflector formed under this force. Therefore, the electrostatic force acting on the membrane surface is first determined based on the equation of the electrostatic field, and then the two-field coupling model of the membrane displacement field and the electrostatic field is developed using the moment-free theory of the membrane.

1. Electrostatic force acting on membrane surface

In Fig. 10.3, $G_1(X, Y, Z) = 0$ is the surface equation of the membrane, $G_2(X, Y, Z) = 0$ is the surface equation of the electrodes, and ds is the differential element of any point P on the surface of the membrane. The electric field of the electrostatic forming membrane reflector is a three-dimensional passive electrostatic field. The potential $\varphi(X, Y, Z)$ in the Cartesian coordinate system *OXYZ* satisfies the Laplace equation [16].

$$\nabla^2\varphi = 0 \tag{10.1}$$

The Laplace equation in electrostatic forming is the first boundary-value problem of partial differential equations. The potential of the membrane is zero since it is grounded, and the electrode potential is denoted by φ_2. As a result, the boundary condition can be written as

$$\varphi(X, Y, Z)_\Gamma = \begin{cases} 0, \ \Gamma_t = G_1 \\ \varphi_2, \ \Gamma_2 = G_2 \end{cases} \tag{10.2}$$

The electric field intensity ***E*** of an arbitrary point Q (X, Y, Z) in the electric field can be written as

$$\mathbf{E} = \begin{Bmatrix} E_X \\ E_Y \\ E_Z \end{Bmatrix} = \begin{Bmatrix} \frac{\partial\varphi}{\partial X} \\ \frac{\partial\varphi}{\partial Y} \\ \frac{\partial\varphi}{\partial Z} \end{Bmatrix} \tag{10.3}$$

Based on the Maxwell stress tensor, the electrostatic force of any point P on the membrane surface is

$$\mathbf{T}_e = \mathbf{E}\mathbf{D}^{\mathrm{T}} - \frac{1}{2}\mathbf{E}^{\mathrm{T}}\mathbf{D}\boldsymbol{I} \tag{10.4}$$

where $\boldsymbol{I}$ is the unit matrix and $\boldsymbol{E}$ and $\boldsymbol{D}$ are the electric field strength and displacement vectors at P, respectively, which satisfy the following equation

$$\mathbf{D} = \varepsilon\mathbf{E} \tag{10.5}$$

where $\varepsilon = 8.85 \times 10^{-10}$ F/m is the permittivity of the vacuum.

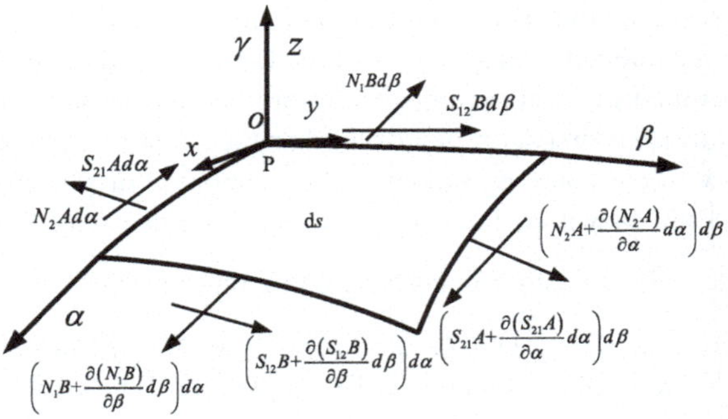

Fig. 10.4 Differential element analysis diagram in curvilinear coordinate system of membrane

Therefore, the electrostatic force of the differential element *ds* at point *P* is

$$\mathbf{F} = \mathbf{P}_e \mathrm{d}s \tag{10.6}$$

where $\mathbf{P}_e = \mathbf{T}_e\mathbf{n}$ and $\mathbf{n}$ is the unit normal vector at point *P*.

2. Coupling equation between membrane displacement and electrostatic fields

To obtain the two-field-coupling equation in the general sense, the curvilinear coordinates are introduced to describe the shape of the membrane surface. The coordinate lines can be placed on the curvature line and normal line of the middle surface to obtain the simplest and most universal equation. The orthogonal curvilinear coordinate system of the membrane surface is established with the axes in the direction of the principal curvature lines α, β and normal γ of the middle surface of the membrane. The Cartesian coordinate system *Oxyz* is established at the differential element *ds* in Fig. 10.3, where the *z* axis is in the direction of γ, and the *x* and *y* axes are tangent to the curvature lines α and β, respectively, as shown in Fig. 10.4. Since the membrane is very thin, accordingly the *oz* axis is along the normal direction of the membrane surface. The three components of the electrostatic force in the local coordinate system *oxyz* of the differential element *ds* are as follows:

$$P_x = 0,\ P_y = 0,\ P_z(\varphi, X, Y, Z) = ||\mathbf{P}_e|| \tag{10.7}$$

According to the moment-free theory of the membrane [17], the equilibrium equation of the differential element *ds* at an arbitrary point *P* on the membrane surface can be established in the local coordinate system *oxyz*; that is, the coupling equation between the structural displacement field and the electrostatic field becomes

$$\begin{cases} \dfrac{\partial(N_1\mathrm{B})}{\partial\alpha} - N_2\dfrac{\partial\mathrm{B}}{\partial\alpha} + S_{12}\dfrac{\partial\mathrm{A}}{\partial\beta} + \dfrac{\partial(S_{21}\mathrm{A})}{\partial\beta} = 0 \\ \dfrac{\partial(N_2\mathrm{A})}{\partial\beta} - N_1\dfrac{\partial\mathrm{A}}{\partial\beta} + S_{21}\dfrac{\partial\mathrm{B}}{\partial\alpha} + \dfrac{\partial(S_{12}\mathrm{B})}{\partial\alpha} = 0 \\ K_1N_1 + K_2N_2 + P_z(\varphi, \mathrm{X,Y,Z}) = 0 \end{cases} \tag{10.8}$$

where A and B represent the Lame coefficient of the curvilinear coordinate system, K_1 and K_2 are the principal curvatures along the curvature lines α and β, respectively, N_1 and N_2 are the resultant internal forces corresponding to the normal stress σ_1 and σ_2 in the curvature directions, and S_{12} and S_{21} are the resultant internal forces of the shear stress τ_{12} and τ_{21}. The expression of the forces is

$$\begin{cases} N_1 = \dfrac{Eh}{1-\nu^2}(\varepsilon_1 + \nu\varepsilon_2) \\ N_2 = \dfrac{Eh}{1-\nu^2}(\varepsilon_2 + \nu\varepsilon_1) \\ S_{21} = S_{12} = \dfrac{Eh}{1+\nu}\varepsilon_{12} \end{cases} \tag{10.9}$$

where E, h and ν are the elastic modulus, thickness, and Poisson's ratio of the membrane, respectively, and ε_1, ε_2 and ε_{12} are the normal strains and shear strain along the principal curvature lines α and β at point P, respectively.

Since the membrane has a small strain and large deformation under the action of the electrostatic force, and considering the geometric nonlinearity of the membrane structure, the geometric equation can be written as

$$\begin{cases} \varepsilon_1 = \dfrac{1}{\mathrm{A}}\dfrac{\partial u}{\partial \alpha} + \dfrac{v}{\mathrm{AB}}\dfrac{\partial \mathrm{A}}{\partial \beta} - \mathrm{K}_1 w \\ \varepsilon_2 = \dfrac{1}{\mathrm{A}}\dfrac{\partial v}{\partial \beta} + \dfrac{u}{\mathrm{AB}}\dfrac{\partial \mathrm{B}}{\partial \alpha} - \mathrm{K}_2 w \\ \varepsilon_{12} = \dfrac{\mathrm{B}}{\mathrm{A}}\dfrac{\partial}{\partial \alpha}\left(\dfrac{v}{\mathrm{A}}\right) + \dfrac{\mathrm{A}}{\mathrm{B}}\dfrac{\partial}{\partial \beta}\left(\dfrac{u}{\mathrm{B}}\right) \end{cases} \tag{10.10}$$

where u, v and w are the displacement components in the α, β, and γ directions, respectively.

The boundary condition for a membrane with a complete fixed boundary is

$$(u, v)\big|_{\alpha=\alpha_0} = 0 \tag{10.11}$$

According to the uniqueness theorem for the electrostatic field solution, the stress distribution of the membrane structure can be obtained by the coupling equation, and the corresponding displacement field can be obtained with the given boundary condition. The electrostatic force can be derived from the displacement field of the membrane, and the potential distribution can be obtained with the boundary condition of the membrane surface potential being zero. The coupling model of the electrostatic and displacement fields of the membrane provides a theoretical description of the electrostatic forming of the membrane reflector. However, because the two-field coupling equations are a set of high-order nonlinear partial differential equations, it is possible to obtain a complete analytical solution for only a few cases, such as a rotationally symmetric membrane surface and continuous distribution of the potential. This is too strict for practical engineering problems, such as forming analysis of electrostatic forming membrane reflectors. To apply the two-field coupling model to the analysis of electrostatic forming, a key problem is to find a suitable numerical solution method, and the finite element method is a feasible choice.

The finite element method is a numerical method based on the variation principle and approximate interpolation discretization. It has good applicability for solving the electrostatic field and analyzing the structural displacement field [18, 19].

The boundary value problem of two fields can be transformed into a functional extremum problem using the variation principle for the electrostatic field and deformation field of the membrane structure. Then, the problem of two two fields is transformed into two algebraic equations by discretization of the field domain, and the basic equations of two-field coupling analysis can be obtained as a set of simultaneous equations.

$$\begin{bmatrix} \mathbf{K}^S(\delta, \varphi) & \mathbf{0} \\ \mathbf{0} & \mathbf{K}^{\mathrm{E}}(\delta) \end{bmatrix} \begin{Bmatrix} \delta \\ \varphi \end{Bmatrix} = \begin{Bmatrix} \mathbf{F}^{\mathrm{E}}(\delta, \varphi) - \mathbf{F} \\ \mathbf{0} \end{Bmatrix} \tag{10.12}$$

where $\mathbf{K}^S(\delta, \varphi) = \mathbf{K}_{\mathrm{L}} + \mathbf{K}_{\mathrm{NL}}$ is the stiffness matrix for geometric nonlinear analysis of the membrane structure, $\mathbf{K}_L$ and $\mathbf{K}_{NL}$ represent the linear and nonlinear stiffness matrices, respectively, and $\mathbf{K}^{\mathrm{E}}(\delta)$ is the stiffness matrix for electrostatic field analysis. $\mathbf{F} = \int_V \mathbf{B}_L^T \boldsymbol{\sigma} dV$ is the equivalent nodal force vector due to membrane stress. $\mathbf{F}^{\mathrm{E}}(\delta, \varphi) = \int_V \mathbf{B}_L^T \boldsymbol{\sigma}_{\mathrm{E}} dV$ is the equivalent nodal load vector corresponding to the electrostatic force. $\boldsymbol{\sigma}_{\mathrm{E}} = \{\sigma_{\mathrm{XX}}, \sigma_{\mathrm{YY}}, \sigma_{\mathrm{ZZ}}, \sigma_{\mathrm{XY}}, \sigma_{\mathrm{YZ}}, \sigma_{\mathrm{XZ}}\}^{\mathrm{T}}$ is the vector form of the Maxwell stress tensor of the electrostatic field.

The above equations are a set of nonlinear equations that can be solved by a direct method or iterative method. The iterative method is more commonly used in the analysis of general coupling problems because it can use the single-field analysis function of finite element analysis software to a large extent. The basic principle of the iterative method is to find the nodal displacement and electrostatic force of the membrane according to the single field analysis of the electrostatic field and the deformation field of the membrane structure, substitute the analysis result of one field into the analysis of the other, and iterate repeatedly until the two fields are balanced.

10.3 Optimization of Voltage and Electrode Layout

The structure of the electrostatic forming membrane reflector antenna is shown in Fig. 10.2. The main part of the antenna consists of a membrane reflector and a cable net serving as the base supporting structure. The base electrodes are laid on the front support cable net. When applying a voltage, an electrostatic field is generated between the membrane reflector and the electrodes. The voltage of each electrode can be controlled with an independent power supply channel. By means of an effective voltage optimization method, the membrane reflector can achieve high precision under the working voltage of each base electrode. At the same time, the stress distribution on the surface should be as uniform as possible to resist unfavorable interference from outer space, for instance, dramatic temperature changes and space radiation. For these reasons, an effective electrode layout and voltage optimization method is proposed so that the membrane reflector antenna can meet the predetermined precision and stress requirements.

10.3.1 Basic Electrode Configuration of Electrostatic Forming Membrane Reflector Antenna

For electrostatic forming membrane reflector antennas, the influence of the basic electrode configuration on the accuracy of the reflector surface is obvious. The ring electrode and the refined ring electrode are usually used. The membrane mirror developed in the United States in the 1980s adopted a 5-ring electrode [6]. Suzhou University in China used 2-ring and 4-ring electrodes [12], as shown in Fig. 10.5.

The refinement of the ring electrode subdivides electrodes along the ring. The advantage is that multiple regions of the electrostatic field distribution between the base electrode and the membrane reflector can be finely adjusted so that the membrane reflector can achieve high profile precision. The disadvantage is that too many electrodes will give rise to complicated voltage control systems. In addition, an increasing number of electrodes will increase the possibility of discharge between the base electrodes. As a result, the basic electrode layout has to be designed with a reasonable method. Figures 10.6, 10.7, 10.8 and 10.9 show the refined ring electrodes used by the University of Arizona, the Lincoln Laboratory of the Massachusetts Institute of Technology, the US Air Force Research Laboratory, and NASA [5–7].

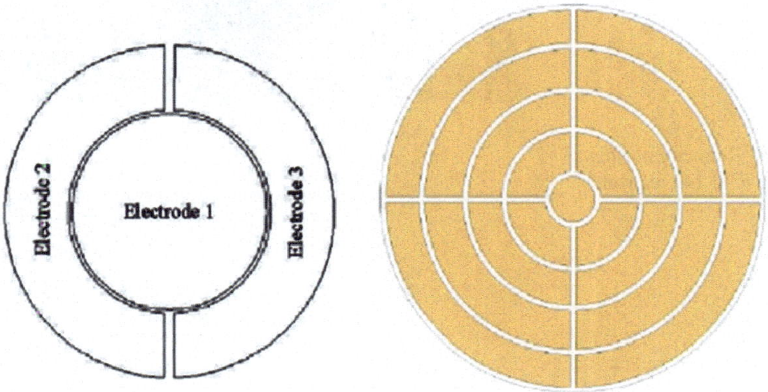

Fig. 10.5 2-ring and 4-ring electrodes

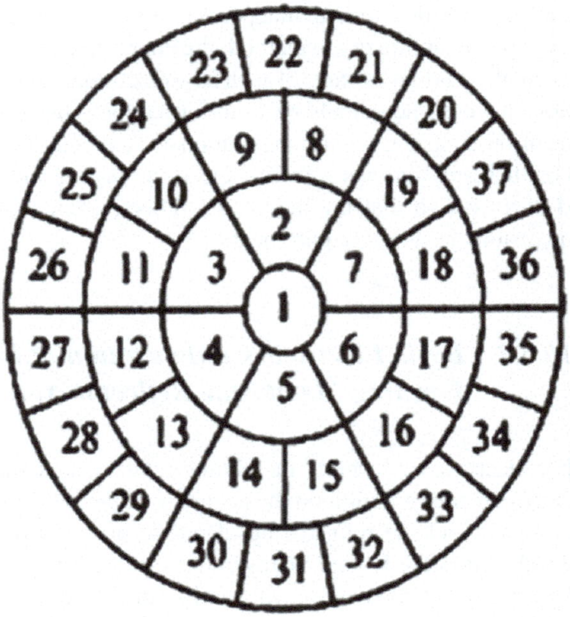

Fig. 10.6 Refinement ring electrode with 37 electrodes

Fig. 10.7 Refinement ring electrode with 19 electrodes

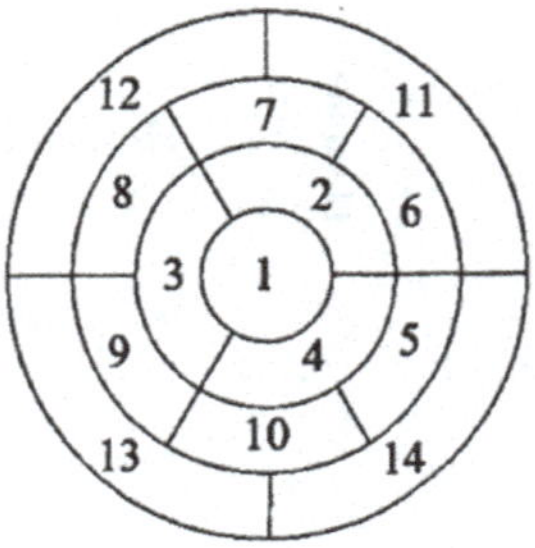

Fig. 10.8 Refinement ring electrode with 14 electrodes

Fig. 10.9 Refinement ring electrode with 29 electrodes

10.3.2 Optimization of Electrode Voltage

1. Electrode voltage optimization model

(1) Optimization design variables

The shape adjustment of the electrostatic forming membrane reflector is performed by changing the voltage of the base electrodes. Each base electrode is connected to a voltage channel of the power supply system so that the electrode voltage changes with the channel voltage. The electrostatic force on the reflector also changes accordingly so that the membrane reflector exhibits a specific shape. The voltage of each channel is taken as the design variable and is denoted as

$$\mathbf{U} = [U_1, U_2, \cdots U_n]^{\mathrm{T}} \tag{10.13}$$

where U_i is the voltage of the ith ($i = 1, 2,\ldots, n$) channel and n is the total number of voltage channels.

(2) Objective function of optimization

The nodal displacement of the reflector after adjustment of the electrostatic force can be derived from the stiffness equation as follows:

$$\mathbf{K}^S\mathbf{X} = \mathbf{F}_E \tag{10.14}$$

where $\mathbf{F}_E = \mathbf{B}_u\mathbf{U}^2$ is the electrostatic force on the membrane reflector, $\mathbf{B}_u$ is the coefficient matrix, and $\mathbf{X}$ is the nodal displacement caused by the electrostatic force.

Generally, the effective area illuminated by the antenna feed does not need to be the entire reflector, and the nodal displacement obtained by Eq. (10.14) is composed of two parts, namely, the nodal displacement in the effective illuminated area and outside of the effective illuminated area. When an electrostatic force is applied, the nodal displacement in the effective illuminated area is denoted by $\mathbf{X}_E$, the nodal coordinates in the initial configuration are denoted by $\mathbf{X}_0$, and the nodal coordinates of the target position are denoted by $\mathbf{X}_1$. Then the residual position deviation of each node of the reflector is

$$\boldsymbol{\delta} = \mathbf{X}_0 + \mathbf{X}_E - \mathbf{X}_1 \tag{10.15}$$

When the membrane has exactly of the desired shape, the residual deviation of each node should be zero, so the sum of the squares of the residual deviation of the nodes in the illuminated area.

$$J(\mathbf{U}) = \boldsymbol{\delta}^{\mathrm{T}}\boldsymbol{\delta} \tag{10.16}$$

is the objective function.

(3) Mathematical model of optimization

In summary, the mathematical model of voltage optimization can be described as

$$\begin{cases} \text{find } \mathbf{U} = [U_1, U_2, \cdots U_n]^{\mathrm{T}} \\ \min J(\mathbf{U}) = \boldsymbol{\delta}^{\mathrm{T}}\boldsymbol{\delta} \\ \text{s. t. } \mathbf{K}^S\mathbf{X} = \mathbf{F}_E \\ 0 < U_i < U^{up},\ (i = 1, 2, \cdots n) \end{cases} \tag{10.17}$$

where $\mathbf{U}$ is the voltage vector of the voltage channel, U^{up} is the upper limit of the electrostatic voltage in the space environment, and n is the total number of voltage channels.

2. Solution of Electrode Voltage Optimization Model

The relative displacement of the membrane reflector under the electrostatic force is minute, and the membrane displacement caused by the change of the electrostatic force corresponding to each electrode is also minute. The deformation of the reflector is within the linear elastic range and conforms to the deformation superposition principle for small deformations. In view of this, the problem can be solved using the least-squares optimization method.

If a unit adjustment is applied to the jth voltage, and the residual position deviation of the ith node relative to the target surface is denoted by A_{ij}, all the A_{ij} can be arranged into a matrix form, namely, the sensitivity matrix $\mathbf{A}$ of the residual position deviation with respect to the electrode voltage. We denote the voltage adjustment by $\Delta\mathbf{U}$, and the residual position deviation of the reflector node with respect to the target surface by $\boldsymbol{\delta}^0$ when the voltage adjustment is applied. According to the principle of deformation superposition, the sensitivity matrix, voltage adjustment and residual deviation satisfy the following relation:

$$\mathbf{A}\Delta\mathbf{U} = \boldsymbol{\delta}^0 \tag{10.18}$$

Using the above relation, the adjustment of the electrode voltage can be effectively determined. Accordingly, the following iterative procedure for optimizing the reflector shape is designed:

(1) According to Eq. (10.15), compute the residual position deviation of the membrane reflector under the present electrostatic force,

$$\boldsymbol{\delta}^0 = [\delta_1^0, \delta_2^0, \cdots, \delta_m^0]^{\mathrm{T}} \tag{10.19}$$

where m is the total number of nodes of the membrane reflector.

(2) Apply a unit adjustment to the first voltage and compute the residual deviation of the reflector nodes with respect to the target surface by finite element analysis. Thus, the first column of the sensitivity matrix is obtained, and the other columns can be obtained in the same way.
(3) According to $\mathbf{A}\Delta\mathbf{U} = \boldsymbol{\delta}^0$, the adjustment of the electrode voltage is obtained using the least-square method,

$$\Delta\mathbf{U} = (\mathbf{A}^{\mathrm{T}}\mathbf{A})^{-1}\mathbf{A}\boldsymbol{\delta}^0 \tag{10.20}$$

(4) Apply the obtained voltage adjustment in the finite element model for analysis. If the objective function $J(\mathbf{U}) = \boldsymbol{\delta}^{\mathrm{T}}\boldsymbol{\delta}$ satisfies the convergence criterion, stop the iteration; otherwise return to step 1).

10.3.3 Mathematical Model and Solution of Electrode Layout Optimization

1. Mathematical model of electrode layout optimization

On the basis of nonlinear finite element analysis, electrode layout optimization is mainly carried out by grouping optimization of the basic electrodes laid on the front support cable net, connecting the electrodes in each group to the same voltage channel, and then adjusting the voltages of the channels to meet the shape precision requirement. Thus simultaneous optimization of the electrode layout and voltage is required.

1) Design variables

The purpose of electrode layout optimization is to connect a large number of base electrodes to a limited number of voltage channels so that the membrane reflector can meet the requirements of a given surface accuracy. In this way, each voltage channel controls several base electrodes at the same time, and the determination of the corresponding relationships between the base electrodes and the voltage channel is an optimization problem. At the same time, under the condition of satisfying the precision requirement of the reflector, the number of base electrodes and voltage channels should to be as low as possible to reduce the structural weight and system complexity as much as possible. Regarding electrode layout optimization, there are two stages. The first is to determine the connection between the base electrodes and voltage channels for the given number of base electrodes and voltage channels. In this process, the voltage of each voltage channel is determined by the preceding electrode voltage optimization method. The second stage is to minimize the number of base electrodes and voltage channels.

Thus, the design variables of optimization include the number of base electrodes, the number of voltage channels, and the voltage of each voltage channel; namely,

$$a\ ;\ n;\ \mathbf{w} = [w_1, w_2, \cdots, w_a]^{\mathrm{T}}\ ;\ \ \mathbf{U} = [U_1, U_2, \cdots, U_n]^{\mathrm{T}} \tag{10.21}$$

where *a and n* are the total numbers of base electrodes and voltage channels, separately, and $n << a$. $\boldsymbol{w}$ is the voltage channel number for each base electrode, $w_i \in \{1, 2, \cdots n\}$, and $\mathbf{U}$ is the voltage of each voltage channel.

2) Objective function

The objective is to improve the precision of the antenna reflector and to minimize the number of base electrodes and voltage channels, so the objective functions can be selected as

$$\begin{cases} f_1 = a \\ f_2 = n \\ f_3 = J = \boldsymbol{\delta}^{\mathrm{T}}\boldsymbol{\delta} \end{cases} \tag{10.22}$$

Clearly, this is a multi-objective optimization problem. The weighted coefficient method can be employed to turn the multiple objective functions into a single objective function so that a single objective optimization method can be used to solve the problem:

$$f = \alpha_1 f_1 + \alpha_2 f_2 + \alpha_3 f_3 \tag{10.23}$$

where α_1, α_2 and α_3 are the weighted coefficients of the three objective functions.

3) Constraints

To improve the ability of the reflector to resist external disturbances and prevent wrinkles, the stress of the membrane reflector should be greater than zero, and the stress distribution should be as uniform as possible. The corresponding constraints can be written as

$$\min \boldsymbol{\sigma} > 0 \tag{10.24}$$

$$\max \boldsymbol{\sigma} / \min \boldsymbol{\sigma} < \upsilon \tag{10.25}$$

where $\min \boldsymbol{\sigma}$ and $\max \boldsymbol{\sigma}$ are the minimum and maximum stresses of the reflector and υ is the upper limit of the ratio of the maximum stress to the minimum stress.

4) Mathematical model

At this point, the electrode layout optimization can be stated mathematically in the following form,

$$\begin{cases} \text{find}\, a, n, \mathbf{w}, \mathbf{U} \\ \min f(a, n, \mathbf{w}, \mathbf{U}) = \alpha_1 f_1 + \alpha_2 f_2 + \alpha_3 f_3 \\ \text{s.t.}\, \mathbf{K}^S \mathbf{X} = \mathbf{F}_E \\ \quad \min \sigma > 0 \\ \quad \max\sigma / \min \sigma < \upsilon \\ \quad a \in \{1, 2, \cdots a^{up}\} \\ \quad n \in \{1, 2, \cdots n^{up}\} \\ \quad w_i \in \{1, 2, \cdots n\} (i = 1, 2, \cdots, a) \\ \quad 0 < U_i < U^{up} (i = 1, 2, \cdots n) \end{cases} \tag{10.26}$$

where a^{up},b^{up} and U^{up} are the upper limits of the total numbers of base electrodes, the total number of voltage channels and the channel voltage, respectively.

2. Solution method

Electrode layout optimization is a highly nonlinear problem with discrete and continuous variables. Because the objective function depends on the nodal coordinates, the nodal coordinates depend on the electrostatic force acting on the membrane reflector, and the electrostatic force depends on the layout of the basic electrodes and voltage, the objective function is a highly nonlinear implicit function of the design variables. Similarly, the constraint functions are implicit functions of the design variables. In addition, the membrane structure is highly geometrically nonlinear, making the problem difficult to solve.

Considering the above characteristics of nonlinear programming, the genetic algorithm [20] is employed because it can conveniently deal with discrete, continuous and mixed variables. In optimization, no auxiliary information, such as sensitivity information, is needed, and there is no special requirement for the objective function or constraint function. A group of points are searched in the solution space at the same time and an evolutionary population sequence is formed. This algorithm has a high probability of obtaining the global optimal solution and has good robustness, likely global convergence and wide adaptability. Based on the above advantages, a genetic algorithm combined with the least-squares method is employed to solve the optimization problem.

Electrode layout optimization is a constrained optimization problem that cannot be solved directly by a genetic algorithm. Therefore, it is necessary to transform the constrained optimization problem into an unconstrained problem and then solve it with a genetic algorithm. First, we transform the constraint functions of Eqs. (10.24) and (10.25) into the form of a unilateral inequality,

$$\begin{cases} g_1 = -\min\sigma < 0 \\ g_2 = \max\sigma/\min\sigma - \upsilon < 0 \end{cases} \tag{10.27}$$

Then, combined with the penalty function method, the constrained problem is transformed into an unconstrained problem; that is, the objective function is treated as the following [20]:

$$f(a, n, \mathbf{w}, \mathbf{U}) = \alpha_1 f_1 + \alpha_2 f_2 + \alpha_3 f_3 + C \sum_{i=1}^{2} \max\{0, g_i(\mathbf{w})\} \tag{10.28}$$

where C is the constraint penalty factor.

When the objective function is determined, a new generation of the population can be obtained by selecting, crossing and mutating the individuals. Repeating the above process, the fitness of the optimal individuals will continue to improve until stability is reached, that is, the optimal solution is obtained.

10.4 Physical Model Development and Experiments

To systematically investigate and verify the two-field coupling model and the optimization method of voltage and electrode layout during the electrostatic forming process, a 2 m-diameter prototype of an electrostatic forming membrane reflector is developed. For the prototype, the cable net reflector of the AstroMesh antenna is employed as the basic supporting structure, triangular membrane electrodes are laid on the supporting cable net, each group of electrodes is connected to a different voltage channel of the high-voltage power supply system, and a membrane reflector is suspended above the basic electrode. The fabrication of the membrane reflector is a key step in prototype development.

10.4.1 Fabrication of Base Electrode

Triangular membrane electrodes are laid on the front support cable net of the AstroMesh antenna, which are connected to the 9 channels of the high-voltage power supply in the 2 m-diameter prototype of the electrostatic forming reflector. The support cable net, triangular electrodes and corresponding electrode group are shown in Fig. 10.10. The triangular membrane electrodes are cut with specially designed templates and pasted onto the support cable net at the corresponding position with polyimide tape. Considering the connection and insulation problems between electrodes, the triangular membrane electrodes linked to the same electrode channel are connected with a conductive cloth, and the electrodes linked to different channels should be separated by a certain distance.

Fig. 10.10 Electrode layout

If the electrodes are not properly handled, discharge may occur at the juncture of different channel electrodes and the conductive connection between electrodes in the same channel. There are two main reasons for discharge, i.e., sufficient insulation between different channel electrodes and tip discharge. To investigate the discharge breakdown between electrodes, an electrical breakdown test is carried out. Two membrane electrodes are placed in parallel and spaced a certain distance apart, as shown in Fig. 10.11, and then the voltage between the two electrodes is continuously increased until discharge breakdown occurs. The insulation distance and critical breakdown voltage obtained from the test are shown in Table 10.1. To avoid discharge in the 2 m-diameter prototype, the insulation distance between different channel electrodes is set to 11 mm, corresponding to a breakdown voltage of 12.5 kV.

Fig. 10.11 Discharge experiment

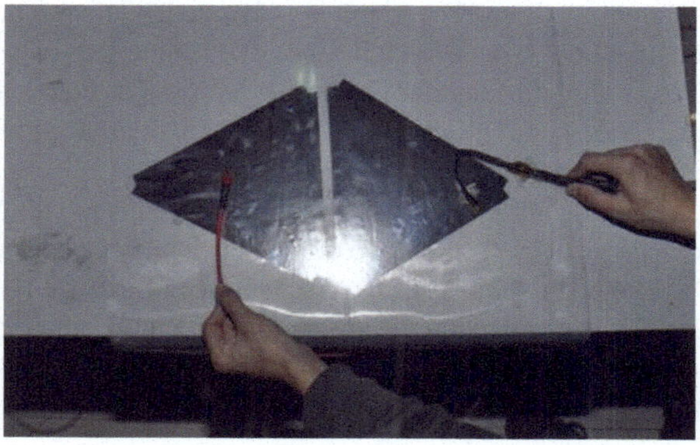

Table 10.1 Discharge test results

Insulation distance between electrodes/mm	Critical breakdown voltage/V
5	6000
8	9000
11	12500

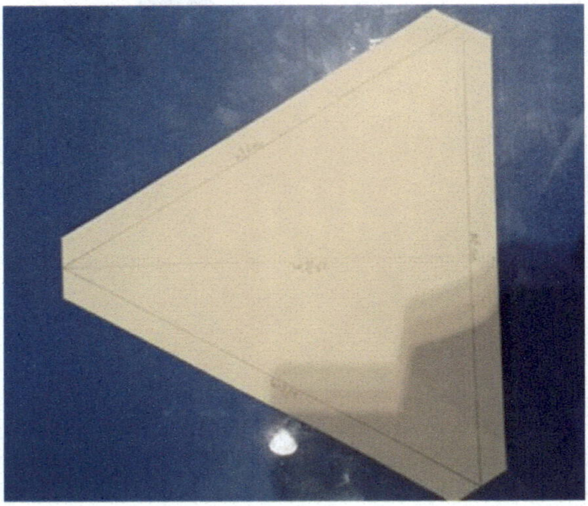

Fig. 10.12 Triangular electrode template

The conductive cloth connecting the same channel electrodes is sealed with insulating tape. Furthermore, to avoid tip discharge, the sharp corners of the triangular electrodes are removed.

To cut the triangular membrane electrode easily and effectively, cutting templates are designed (Fig. 10.12) according to the dimensions of the triangles of the support cable net and the insulation distance between adjacent electrodes. The triangular membrane electrodes are trimmed according to the template and pasted onto the front surface of the support cable net. The triangular electrodes linked to the same channel are directly overlapped and pasted together. The electrodes linked to different channels are spaced by a distance of 11 mm, and pasted with a wide polyimide tape. When all the triangular electrodes are pasted, the front and rear support cable nets are connected. The support cable net with membrane electrodes stretched to the truss is shown in Fig. 10.13.

The membrane electrodes fabricated with the above measures do not result in discharge at the conductive cloth, between electrodes or at the electrode tips.

The precision of the electrode surface is a key index of the membrane electrodes. Similar to the mesh reflector, surface precision adjustment for membrane electrodes is required when these electrodes are fabricated. The surface of the membrane electrodes is expected to be as close as possible to the desired membrane reflector,

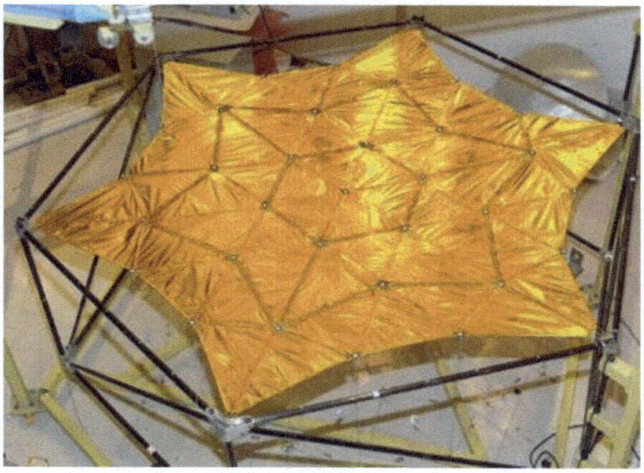

Fig. 10.13 Membrane electrodes on front support cable net

facilitating the control of the electrostatic force acting on the membrane reflector by altering the electrode voltage.

10.4.2 Fabrication of Membrane Reflector

Since a membrane with a large area and curvature is hard to produce, a feasible way to fabricate a curved membrane reflector is to cut and seam the flat membrane. The basic idea is to divide the whole reflector surface into a group of small surface patches and replace these patches with developable surface patches. The developable surface patches can be directly cut from flat membrane patches, and the patches are then seamed together to form the curved reflector. Since very high surface precision of the electrostatic forming membrane reflector is demanded, the cutting design plan shown in Fig. 10.14 is introduced to effectively control the error caused by cutting and seaming flat membrane patches.

In the fabrication of the membrane reflector, cutting the paraboloid reflector is the first step. The cutting line is arranged on the geometric configuration of the stress-free reflector obtained by finite element analysis, along which the spatial reflector is divided into several curved patches. For axisymmetric or almost axisymmetric reflectors, the best position of the cutting lines is on the generatrix. One end of the cutting line is the center point of the reflector, and the other end is located on the radial boundary. The cutting lines can be evenly arranged on rotationally symmetric structures, as shown in Fig. 10.15.

The error from the prestress of the membrane and flattening of undevelopable surfaces has an adverse effect on the reflector surface precision. This effect can be reduced by a reasonable arrangement of cutting lines and the adjustment of the electrostatic force in the design stage. The precision of the seamed membrane reflector needs to be numerically analyzed under the electrostatic force generated in the initial shape design of the membrane reflector. If the design requirements are met, the cutting

Fig. 10.14 Design process of membrane cutting

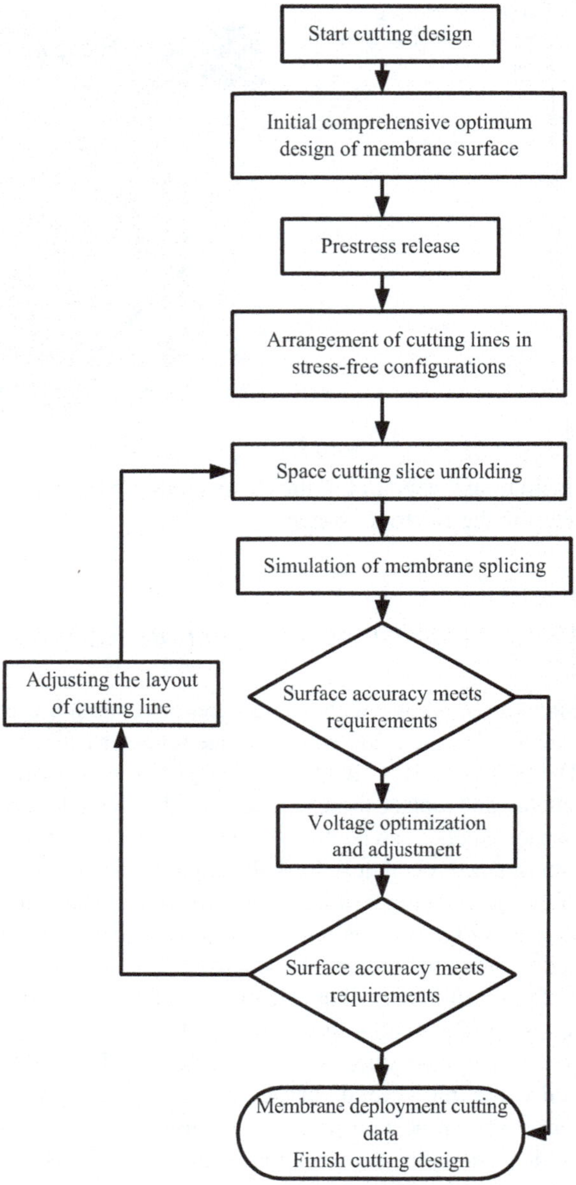

design is completed. Otherwise, more cutting lines and a corresponding arrangement are needed.

The reflector is divided along the cutting lines, and then the cutting patches are flattened. Cutting templates are made according to the shape of the flattened patches (Fig. 10.16) to assist in cutting the flat membrane. The sectored cutting patches are shown in Fig. 10.17. Smooth flat patches with fewer wrinkles are chosen from many

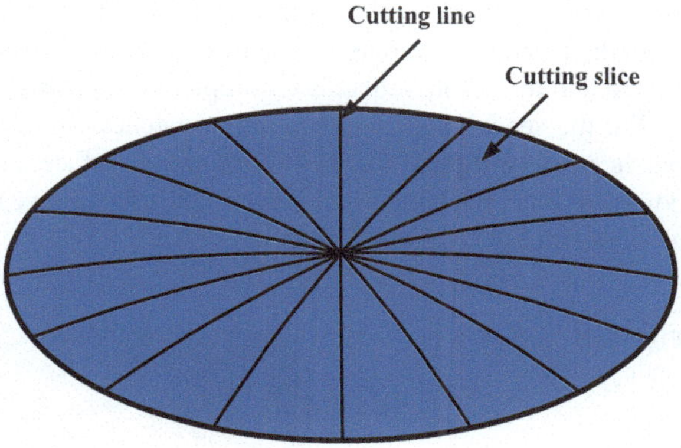

Fig. 10.15 Sketch map of cutting line layout

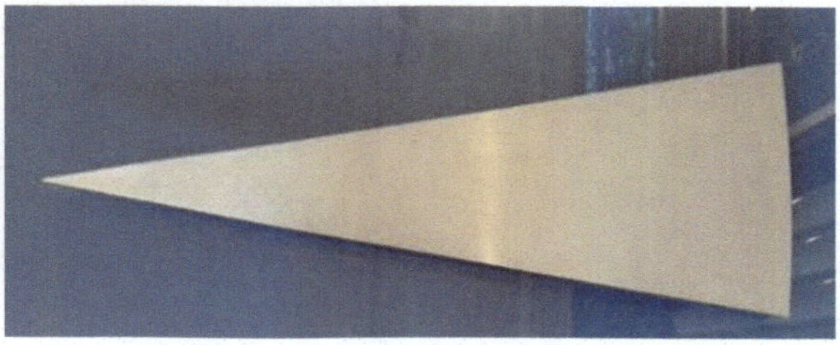

Fig. 10.16 Membrane cutting template

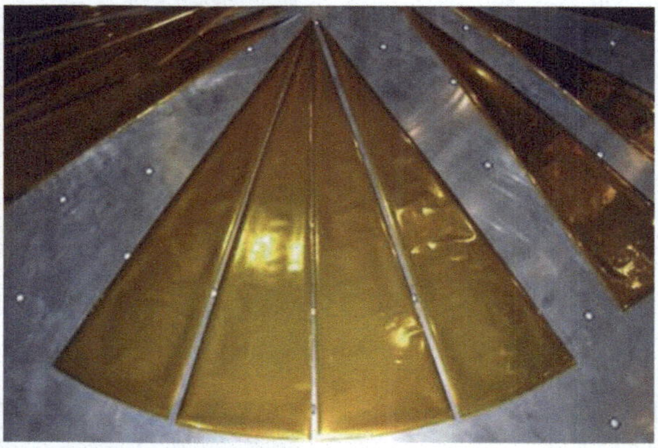

Fig. 10.17 Sectored cutting patches

cutting patches for seaming.

To ensure a good stitching effect, an aluminum alloy mold with a 2 m-diameter and a shape consistent with the desired paraboloid is made. All of the stitching is performed on the mold surface. The cutting plane can be spliced together through two schemes.

In scheme one, every two cutting patches are pasted together from the center to the edge so that the error can propagate to the outer edge area, and the surface accuracy can be improved by adjusting the boundary cables. In this way, the two cutting patches are pasted together, as shown in Fig. 10.18. According to the principle of pasting from the center to the edge, four cutting patches are pasted, eight cutting patches are pasted and the whole reflector is completed sequentially, as shown in Fig. 10.19.

For the second scheme, the cutting patches are arranged as a whole so that the patches can be tightly attached to the mold surface, as shown in Fig. 10.20. In the complete state, the patches are pasted with glue or tape to form the entire paraboloid, as shown in Fig. 10.21.

Fig. 10.18 Pasting of two cutting patches

Fig. 10.19 Reflector after pasting

Fig. 10.20 Overall layout of diaphragm

Fig. 10.21 Overall paste effect diagram

The membrane reflector needs to be installed on the foundation support structure with a certain boundary. Because the membrane is very thin, it cannot be directly connected to the support structure. For this reason, according to the force characteristics of the reflector, the outer edge of the reflector is machined into a skirt form, and a ring cable is added along the skirt edge. A radial cable is connected to the ring cable at the point where two adjacent skirt edges join, and the other end of the radial cable is connected with the support structure, thereby realizing the connection between the membrane reflector and the support structure.

The skirt edge needs to fit with all the cut membranes. The lowest point of the skirt edge is tangent to the outer edge of the cutting patch. Considering the fact that the skirt cable should be connected at the corner of the adjacent skirt, there should be a certain gap, so the corners of each skirt are cut into a circular arc with a radius of 5 mm. The shape of the skirt edge is shown in Fig. 10.22, and the corresponding skirt cutting template is manufactured as shown in Fig. 10.23. The skirt we made is shown in Fig. 10.24.

Based on a 2 m-diameter hexagonal ring truss deployable antenna, 3-ring electrodes were pasted on the front support cable net, 9 voltage channels were used to control the voltage of the electrodes, the membrane reflector was made by the pasting method, and the radial boundary cable was connected to the foundation

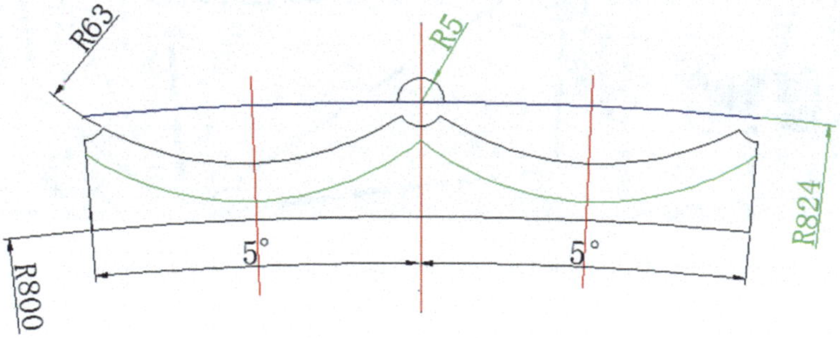

Fig. 10.22 Two-dimensional geometric design of skirt template

Fig. 10.23 Template for skirt cutting

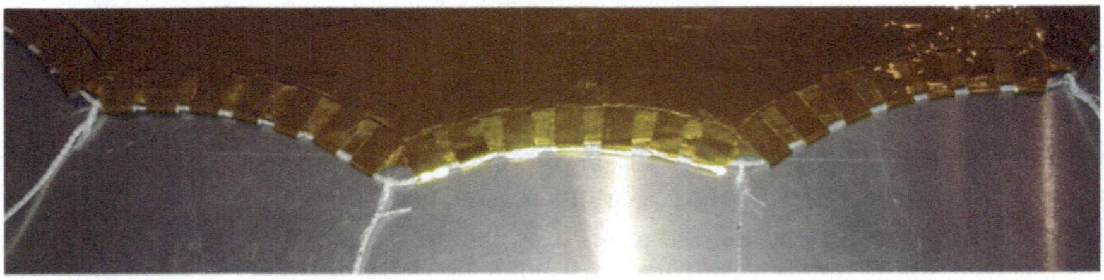

Fig. 10.24 Skirt of 2 m-diameter membrane reflector

support structure through a skirt. The developed 2 m-diameter prototype is shown in Fig. 10.25.

Fig. 10.25 2m-diameter electrostatic forming membrane reflector antenna prototype

10.4.3 Membrane Reflector Electrostatic Forming Experiment

To obtain the required precision, the position of the reflector surface must be measured during the electrostatic forming experiment. Since the flexibility of the membrane reflector is high, the shape precision can only be measured with high accuracy by the noncontact method. To ensure the accuracy of the image processing and coordinate solution, it is necessary to ensure that the light intensity is appropriate and camera angles vary when taking photos, and the shooting distance is almost unchanged to obtain high-quality photos. The photos are imported into the measurement system software, and the three-dimensional coordinates of the marker points and coding points are calculated to perform the three-dimensional reconstruction of the reflector, as shown in Fig. 10.26.

The operations in fabricating the membrane reflector by hand lead to an error that makes the product unacceptable in practice. The final reflector precision greatly depends on joint control of the boundary skirt cable tensions and electrostatic forces. Meeting the precision requirement relies on the perfectly joint adjustment of the cable tension and electrostatic force. When the 2 m diameter prototype is adjusted, the boundary cable is adjusted first, and then the electrode voltage is adjusted to improve the shape precision of the reflector.

Because of the installation error, the edge area of the membrane reflector is not on the same horizontal plane when the skirt cable is connected to the support structure. Therefore, the position of the target points on the membrane reflector in the initial state is obtained by photogrammetry, and the position error of the edge points relative to the ideal paraboloid is analyzed. According to the position error distribution, the installation height of the corresponding radial cable is adjusted so that the edges of the membrane reflector are as close to being on the same horizontal plane as possible. In the electrostatic forming process, because the electrostatic force is very small, it is necessary to adjust the boundary cable so that the actual position of the membrane reflector is close to the desired shape; thus, the surface accuracy can be further improved by adjusting the electrostatic force. Eight adjustments were made during the experiment, and the RMS shape precision change during the adjustment process is shown in Fig. 10.27. The shape precision in each adjustment is 0.54, 0.42, 0.36, 0.34, 0.29, 0.28, 0.25 and 0.26 mm. The shape precision is improved gradually to 0.25 mm.

10.5 Thermoplastic Forming of Membrane Reflector

Through many studies and experiments, it is found that to obtain a high precision membrane reflector, it is necessary to fabricate a membrane reflector that basically meets the requirement of the ratio of focus to diameter and then obtain a high precision surface under the combined electrostatic force and boundary cable force. The most

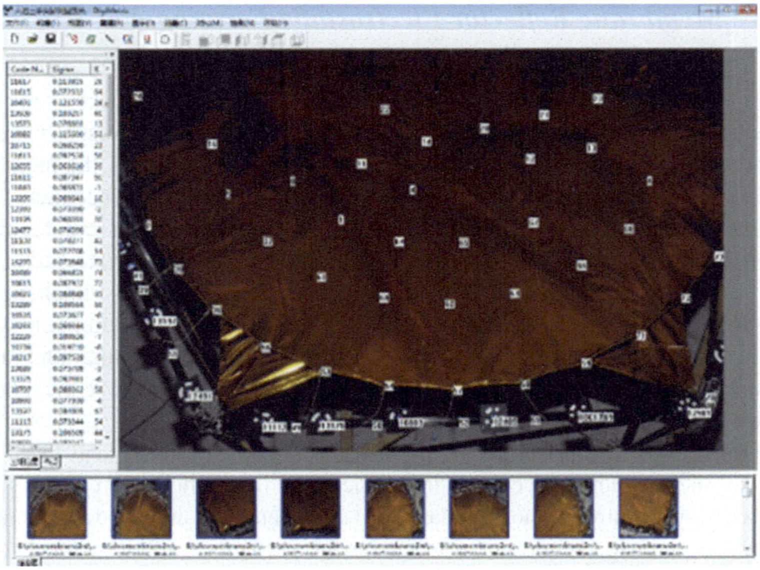

(a) Identification of marker points and coding points

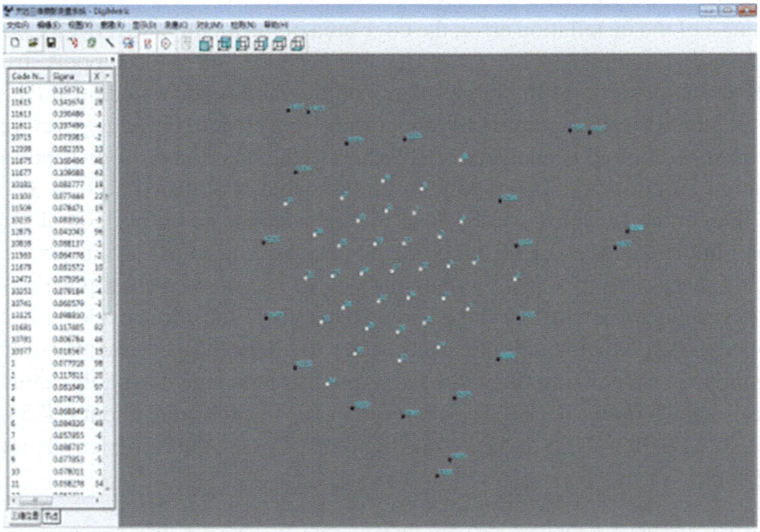

(b) Selection of marker points and coding points

Fig. 10.26 Photographic measurement of membrane reflector

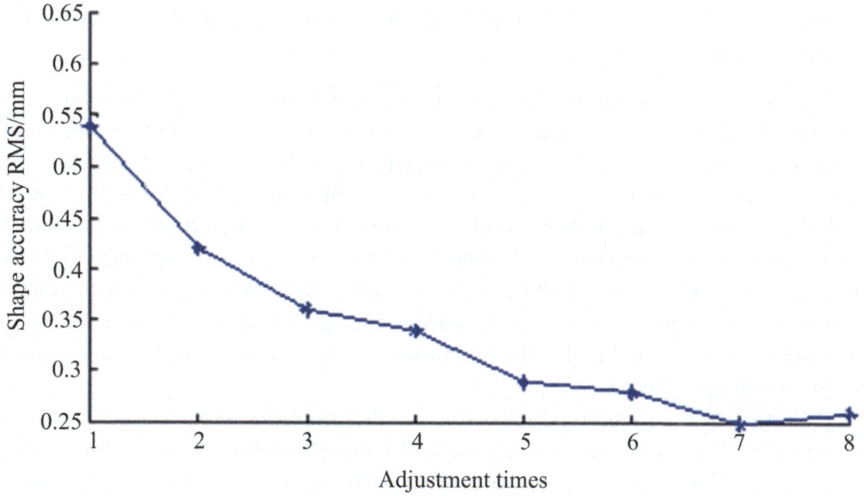

Fig. 10.27 RMS shape precision change in adjustment process

important problem is the fabrication and formation of a large membrane reflector. At present, there are three main ways to fabricate large membrane reflectors: cutting-and-pasting methods, integral casting methods and thermoplastic forming methods.

(1) The cutting-and-pasting methods splice flat membranes into a certain shape and form a large reflector with a specific ratio of focus to diameter. In this method, a developable surface is used to approximate an undevelopable surface, resulting in theoretical error. At the same time, due to the limitations of pasting technology. the membrane reflector when pasted together easily produces wrinkles, leading to poor surface precision. The theoretical error depends on the size of the membrane patch; the smaller the patch is, the smaller the theoretical error. However, at the same time, the number of patches increases, making the pasting error increase rapidly. Generally, the precision of membrane reflectors produced by the cutting-and-pasting method is low; the reflector in Sect. 10.4.2 is made by this method.
(2) The integral casting method is to spray polyimide resin on the surface of the mold and then solidify it at a high temperature. Although this method can eliminate theoretical error, it is difficult to control the membrane thickness in the manufacturing process, and it is difficult to remove the membrane after forming on the mold, resulting in wrinkles.
(3) Thermoplastic forming is a secondary kind of forming technology for thermoplastic membranes. The membrane is first stretched and pressed onto the surface of the mold and then heated to a specific temperature so that the membrane generates plastic deformation, forming the desired reflector shape. This method can also eliminate theoretical error, and the forming process is simpler than

integral casting, but a whole mold and high-temperature furnace are needed in the forming process.

The basic process of thermoplastic forming is as follows: First, the flat membrane is prepared and cut into a certain shape. If the required size exceeds the width of the membrane available, several membrane materials can be pasted together. Second, a special clamp is used to fix the membrane. Third, the clamped membrane is pressed onto the raised parabolic mold to ensure that the membrane is in contact with the mold surface and has the desired prestress. Fourth, the membrane is put into a high-temperature furnace with the mold, heated and maintained at a temperature according to the predetermined temperature curve, and then cooled to room temperature within the furnace. Finally, the membrane is removed from the fixture to obtain the thermoplastic reflector.

In addition to the membrane prestress and thermoplastic temperature control, a key step of the thermoplastic forming method is the fabrication of the forming mold. In the thermoplastic process, the membrane is formed on the mold surface with a certain prestress, so the forming precision depends directly on the precision of the mold surface. Generally, the mold surface precision is higher than the required membrane surface precision by at least one order of magnitude. Since the mold needs to be heated to a high temperature, to ensure uniform thermal deformation of the mold, it is preferable for it to be a whole mold.

The selected mold materials include aluminum alloys, ductile iron, glass-fiber-reinforced plastics (GFRPs), carbon fiber reinforced plastics (CFRPs), etc. The manufacturing cost of GFRP and CFRP molds are very expensive. Ductile iron molds usually require casting molds; the preparation cycle is long and manufacturing is difficult. In high-temperature environments, the surface coating easily falls off, the material easily rusts, and the surface finish deteriorates after repeated use. The aluminum alloy mold has a small density, light weight, good surface finish, and short preparation period, but the disadvantage is that the thermal expansion coefficient is large. After comprehensive consideration, an aluminum alloy is used to make the forming mold. Because the forming temperature reaches 360 °C, the 7075 aluminum alloy, with excellent high-temperature performance, is used.

To verify the theory of thermoplastic forming, a standard paraboloid mold with a diameter of 0.3 m and a focal length of 0.4 m is designed. A transitional fillet is designed at the intersection of the paraboloid and the side surface of the mold to prevent stress concentration in the forming process of the membrane. At the same time, the surface of the mold is required to be smooth and dense, and the paraboloid surface of the mold is in close contact with the membrane and is bonded during the forming process, so there is a high requirement for the mold surface profile, surface roughness and profile precision. Due to the small size of the mold, there is no need for hollowing and ribbing. A center pin hole for the positioning connection is designed at the center of the bottom of the mold.

One main process of thermoplastic forming is heating the polyimide membrane. Commonly used heating methods include heating rods, furnaces, infrared lamps, liquid baths and so on. Usually small parts are heated by heating rods and liquid

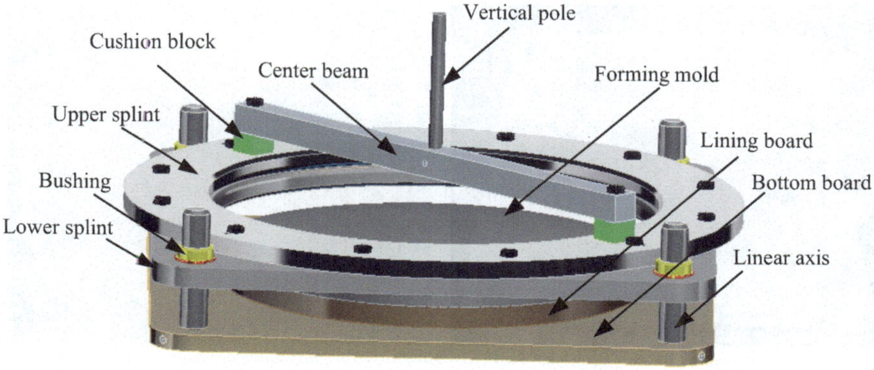

(a) Schematic diagram of thermoplastic forming mold

(b) 0.3m thermoplastic forming mold

Fig. 10.28 Composition of thermoplastic forming mold

baths. Large parts are heated by a furnace, and infrared heating equipment is used for some special cases. Therefore, furnace heating is used here. The furnace is heated by electricity, and the temperature uniformity is ensured by hot-air circulation. Inside the furnace, a plurality of tubular electric heaters are used as heating units to heat the air in the heating chamber, thus heating the whole membrane material through convection. The heating speed, temperature holding time and cooling speed are controlled by the temperature controller of the furnace.

The precision of the forming mold deteriorates when used several times. It needs to be disassembled for maintenance. For convenient disassembly, the mold cannot be placed and fixed directly on the bottom board. It is to be placed on the lining board first; then, the lining board is placed on the bottom board, and the lining board, bottom board and mold are fastened together with the center pin. The whole mold consists of the forming mold, lining and boards, straight shaft, lower and upper splints, bushing, cushion block, center beam and vertical pole, as shown in Fig. 10.28.

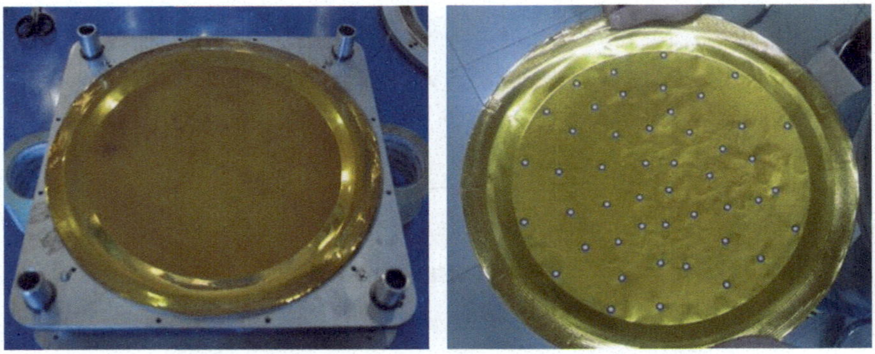

Fig. 10.29 Experiment results of thermoplastic forming of membrane reflector

In the process of thermoplastic forming, the polyimide membrane needs to be fixed on the clamping device. Therefore, a clamping device that can fix various sizes of membrane material is designed. Traditional mechanical connection methods, such as a screw connection or interference fit, will destroy the membrane material and affect the forming quality, and cannot guarantee the reliability of clamping. The frame clamping device we designed includes two frames, the upper and lower splints. The membrane material is sandwiched between the two splints, and there are 12 rings of meshing saw teeth between the upper and lower splints; this can meet the clamping requirements of the membrane with a thickness of 25–100 μm, and the clamping force is evenly distributed. The processes of manually opening the frame and clamping the membrane material are flexible and reliable.

According to the above design, a 0.3 m thermoplastic device was made for the thermoplastic experiment. The experimental results are shown in Fig. 10.29. The average precision of the membrane surface after forming is approximately 63 μm. A reflector made by this technology can achieve higher surface precision after adjustment with the electrostatic force.

References

1. Soh M, Lee JH, Youn SK (2004) An inflatable circular membrane mirror for space telescopes. Proc SPIE Int Soc Opt Eng 5638:262–271
2. Knize RJ (2002) Large-diameter holographically corrected membrane telescope. Proc SPIE Int Soc Opt Eng 41(7):1603–1607
3. Palisoc L (2000) Large telescope using a holographically-corrected membrane mirror-final report to the NASA institute for advanced concepts. L'Garde, Inc
4. Fang H, Quijano U, Bach V (2011) Experimental study of a membrane antenna surface adaptive control system. In: The 52nd AIAA/ASME/ASCE/AHS/ASC structures, structural dynamics and materials conference. Denver, USA, pp 1828–1837
5. Robison L, Wickersham M, Korde U, et al (2010) Experiments on a twelve mode membrane mirror with boundary located electrostatic actuators. In: The 51st

AIAA/ASME/ASCE/AHS/ASC structures, structural dynamics, and materials conference, Orlando, USA, pp 3073–3082
6. Mihora DJ (1981) Test progress on the electrostatic membrane mirror. NASA
7. Mihora DJ, Redmond PJ (1980) Electrostatically formed antennas. AIAA 17(5):465–473
8. Angel R, Burge J, Hege K et al (2000) Stretched membrane with electrostatic curvature (SMEC). Proc SPIE Int Soc Opt Eng 4013:699–706
9. Moore J D, Patrick B G, Chodiella S, et al (2005) Design and testing of a one meter membrane mirror with active boundary control. Proc SPIE Int Soc Opt Eng 5899: 58990Z-1–58990Z-7
10. Patrick B, Moore J, Chodimella S (2006) Final testing and evaluation of a meter-class actively controlled membrane mirror. In: The 47th AIAA/ASME/ASCE/AHS/ASC structures, structural dynamics, and materials conference, Newport, USA, p 1901
11. Chen X, Tang M, Shen W (2007) Membrane mirror deformed by electrostatic pressure. Proc SPIE 6721:67210 K-1–67210 K-8
12. Chen XH, Tang MX, Shen WM (2007) Electrostatic forming of thin membrane mirrors. Infrared and Laser Engineering 36(2):274–277
13. Qi YC, Jin G (2007) Surface design and optimization of space thin membrane mirrors. Opt Precision Engineering 15(6):818–823
14. An Y, Jin G, Qi YC (2009) Single-electrode controlled electrostatic forming of thin membrane mirrors. Optical Precision Engineering 17(8):1964–1970
15. Moore J (2004) Multi-function, adaptive, large aperture sensor. NASA Mirror Technology Days. in the Government Workshop, 2004
16. Wu ZH (2010) Basic electrostatics. Peking University Press, Beijing
17. Soedel W (2004) Vibrations of shells and plates. Marcel Dekker Inc, New York
18. Yang QS, Jiang YN (2004) Cable-membrane structure analysis and design. Science Press, Beijing
19. Li WJ (2004) Multi-domain coupling analysis and multi-discipline design optimization of micro-electro-mechanics. Doctoral Dissertation, Northwest Polytechnic University, Xi'an
20. Wang XP, Cao LM (2006) Application of genetic algorithms theory and software implementation. Xi'an Jiaotong University Press, Xi'an

The manufacturer's authorised representative in the EU is Springer
Nature Customer Service Centre GmbH, Europaplatz 3, 69115 Heidelberg,
Germany. If you have any concerns regarding our products, please
contact ProductSafety@springernature.com

Printed and bound by CPI Group (UK) Ltd, Croydon, CR0 4YY
05/10/2026
02228126-0005